Lecture Notes in Physics

Editorial Board

H. Araki
Research Institute for Mathematical Sciences
Kyoto University, Kitashirakawa
Sakyo-ku, Kyoto 606, Japan

E. Brézin
Ecole Normale Supérieure, Département de Physique
24, rue Lhomond, F-75231 Paris Cedex 05, France

J. Ehlers
Max-Planck-Institut für Physik und Astrophysik, Institut für Astrophysik
Karl-Schwarzschild-Strasse 1, D-85748 Garching, FRG

U. Frisch
Observatoire de Nice
B. P. 229, F-06304 Nice Cedex 4, France

K. Hepp
Institut für Theoretische Physik, ETH
Hönggerberg, CH-8093 Zürich, Switzerland

R. L. Jaffe
Massachusetts Institute of Technology, Department of Physics
Center for Theoretical Physics
Cambridge, MA 02139, USA

R. Kippenhahn
Rautenbreite 2, D-37077 Göttingen, FRG

H. A. Weidenmüller
Max-Planck-Institut für Kernphysik
Saupfercheckweg 1, D-69117 Heidelberg, FRG

J. Wess
Lehrstuhl für Theoretische Physik
Theresienstrasse 37, D-80333 München, FRG

J. Zittartz
Institut für Theoretische Physik, Universität Köln
Zülpicher Strasse 77, D-50937 Köln, FRG

Managing Editor

W. Beiglböck
Assisted by Mrs. Sabine Landgraf
c/o Springer-Verlag, Physics Editorial Department V
Tiergartenstrasse 17, D-69121 Heidelberg, FRG

W0050972

The Editorial Policy for Proceedings

The series Lecture Notes in Physics reports new developments in physical research and teaching – quickly, informally, and at a high level. The proceedings to be considered for publication in this series should be limited to only a few areas of research, and these should be closely related to each other. The contributions should be of a high standard and should avoid lengthy redraftings of papers already published or about to be published elsewhere. As a whole, the proceedings should aim for a balanced presentation of the theme of the conference including a description of the techniques used and enough motivation for a broad readership. It should not be assumed that the published proceedings must reflect the conference in its entirety. (A listing or abstracts of papers presented at the meeting but not included in the proceedings could be added as an appendix.)
When applying for publication in the series Lecture Notes in Physics the volume's editor(s) should submit sufficient material to enable the series editors and their referees to make a fairly accurate evaluation (e.g. a complete list of speakers and titles of papers to be presented and abstracts). If, based on this information, the proceedings are (tentatively) accepted, the volume's editor(s), whose name(s) will appear on the title pages, should select the papers suitable for publication and have them refereed (as for a journal) when appropriate. As a rule discussions will not be accepted. The series editors and Springer-Verlag will normally not interfere with the detailed editing except in fairly obvious cases or on technical matters.
Final acceptance is expressed by the series editor in charge, in consultation with Springer-Verlag only after receiving the complete manuscript. It might help to send a copy of the authors' manuscripts in advance to the editor in charge to discuss possible revisions with him. As a general rule, the series editor will confirm his tentative acceptance if the final manuscript corresponds to the original concept discussed, if the quality of the contribution meets the requirements of the series, and if the final size of the manuscript does not greatly exceed the number of pages originally agreed upon. The manuscript should be forwarded to Springer-Verlag shortly after the meeting. In cases of extreme delay (more than six months after the conference) the series editors will check once more the timeliness of the papers. Therefore, the volume's editor(s) should establish strict deadlines, or collect the articles during the conference and have them revised on the spot. If a delay is unavoidable, one should encourage the authors to update their contributions if appropriate. The editors of proceedings are strongly advised to inform contributors about these points at an early stage.
The final manuscript should contain a table of contents and an informative introduction accessible also to readers not particularly familiar with the topic of the conference. The contributions should be in English. The volume's editor(s) should check the contributions for the correct use of language. At Springer-Verlag only the prefaces will be checked by a copy-editor for language and style. Grave linguistic or technical shortcomings may lead to the rejection of contributions by the series editors. A conference report should not exceed a total of 500 pages. Keeping the size within this bound should be achieved by a stricter selection of articles and not by imposing an upper limit to the length of the individual papers. Editors receive jointly 30 complimentary copies of their book. They are entitled to purchase further copies of their book at a reduced rate. As a rule no reprints of individual contributions can be supplied. No royalty is paid on Lecture Notes in Physics volumes. Commitment to publish is made by letter of interest rather than by signing a formal contract. Springer-Verlag secures the copyright for each volume.

The Production Process

The books are hardbound, and the publisher will select quality paper appropriate to the needs of the author(s). Publication time is about ten weeks. More than twenty years of experience guarantee authors the best possible service. To reach the goal of rapid publication at a low price the technique of photographic reproduction from a camera-ready manuscript was chosen. This process shifts the main responsibility for the technical quality considerably from the publisher to the authors. We therefore urge all authors and editors of proceedings to observe very carefully the essentials for the preparation of camera-ready manuscripts, which we will supply on request. This applies especially to the quality of figures and halftones submitted for publication. In addition, it might be useful to look at some of the volumes already published. As a special service, we offer free of charge LATEX and TEX macro packages to format the text according to Springer-Verlag's quality requirements. We strongly recommend that you make use of this offer, since the result will be a book of considerably improved technical quality. To avoid mistakes and time-consuming correspondence during the production period the conference editors should request special instructions from the publisher well before the beginning of the conference. Manuscripts not meeting the technical standard of the series will have to be returned for improvement.

For further information please contact Springer-Verlag, Physics Editorial Department V, Tiergartenstrasse 17, D-69121 Heidelberg, FRG

G. F. Helminck (Ed.)

Geometric and Quantum Aspects of Integrable Systems

Proceedings of the Eighth Scheveningen Conference
Scheveningen, The Netherlands, August 16-21, 1992

Springer-Verlag Berlin Heidelberg GmbH

Editor

G. F. Helminck
Faculty of Applied Mathematics, University of Twente
P. O. Box 217, NL-7500 AE Enschede, The Netherlands

ISBN 978-3-662-13930-1 ISBN 978-3-540-48090-7 (eBook)
DOI 10.1007/978-3-540-48090-7

This work is subject to copyright. All rights are reserved, whether the whole or part of the material is concerned, specifically the rights of translation, reprinting, re-use of illustrations, recitation, broadcasting, reproduction on microfilms or in any other way, and storage in data banks. Duplication of this publication or parts thereof is permitted only under the provisions of the German Copyright Law of September 9, 1965, in its current version, and permission for use must always be obtained from Springer-Verlag Berlin Heidelberg GmbH. Violations are liable for prosecution under the German Copyright Law.

© Springer-Verlag Berlin Heidelberg 1993
Originally published by Springer-Verlag Berlin Heidelberg New York in 1993
Softcover reprint of the hardcover 1st edition 1993

58/3140-543210 - Printed on acid-free paper

Dedicated to the memory of J.W. de Roever,
a true scientist and enthusiastic participant,
who flew off too soon.

August 28, 1992.

Preface

The present volume contains the notes of lectures delivered at the eighth Scheveningen Conference held at Wassenaar, the Netherlands, August 16-21, 1992. The conference was organized by P.J.M. Bongaarts (University of Leiden), G.F. Helminck (University of Twente), E.M. de Jager (University of Amsterdam), H. Lemei (Delft University of Technology), R. Martini (University of Twente) and H.G.J. Pijls (University of Amsterdam). Financial support was obtained from the Royal Dutch Academy of Science (KNAW) and the Foundation for Fundamental Research on Matter (FOM).

This conference is the eighth in a series of meetings that started in 1973. All of these meetings have been centered around a main topic of research and have been partly instructional in character in order to stimulate further research. Like the last four, this Scheveningen conference was devoted to geometric ideas in mathematical physics.

Algebraic geometric aspects of integrable systems are dealt with in the contributions of J. Harnad and E. Previato. The paper of J. Harnad describes by means of several concrete examples a general framework in which many Hamiltonian systems with a "curve-background" can be explicitly integrated. The account of E. Previato furnishes the geometric insight into several integrability techniques for the modified KdV equation and gives the geometry behind transformations related to this equation. The equations that play a role in this work are part of whole systems of equations, so-called hierarchies. They reappear in the papers of R. Dijkgraaf and G.F. Helminck. The latter paper presents an analytical construction of solutions of the Toda lattice hierarchy, while the former discusses the relation between integrable hierarchies and string theory. This leads, in particular, to useful integral expressions for the τ-functions of these hierarchies. As is now well-known, these τ-functions are determinants of suitable operators. Probabilities for the distribution of eigenvalues are also of this form and they are the central topic in the paper by C.A. Tracy and H. Widom. In their work these authors give a new and transparent way to derive the JMMS-equations satisfied by these τ-functions.

Clearly, in the geometric description of interesting equations from mathematical physics, singularities might emerge. The paper by D-J. Smit and M.V. de Hoop, and that of J. de Graaf are beautiful illustrations of this phenomenon. In the contribution of D-J. Smit and M.V. de Hoop, a new expression for the fundamental solution of a specific hyperbolic system is given. It is described in terms of integrals over cycles corresponding to singularities of the hypersurface defined by the characteristic equation. This is used to analyze the asymptotic behaviour of the fundamental solution. The article by J. de Graaf gives the mathematical background to Hopper's equation, a description of its state space and a discussion of some finite dimensional subsystems.

In the world of integrable systems one is not only interested in differential equations but also in integrable difference equations. In their contribution F.W. Nijhof and H.W. Capel give a quantization procedure for integrable lattice models, such as the lattice analogue of MKdV. In addition, they discuss the properties of these systems and associate a quantum Yang-Baxter structure to them. Finally, the paper of S. Rodriguez-Romo and D.W. Ebner introduces q-deformations of Clifford algebras and discusses quantum symmetries related to Yang-Baxter bundles.

The organizing committee is indebted to all those who made this conference possible and in particular to the Royal Dutch Academy of Science (KNAW) and the Foundation for Fundamental Research on Matter (FOM) for their financial support. We would like to express our gratitude to all those who helped to make this conference a success, to the participants for creating a nice and friendly atmosphere, the invited speakers for accepting our invitation and the contributors for their very interesting and stimulating accounts. Finally, we wish to express our gratitude to Springer-Verlag for their assistance and the efficient production of these proceedings.

G.F. Helminck

Enschede, The Netherlands, June 1993

Contents

Isospectral Flow and Liouville-Arnold Integration in Loop Algebras[†]

J. Harnad[1]

Department of Mathematics and Statistics, Concordia University
7141 Sherbrooke W., Montréal, Canada H4B 1R6, and
Centre de recherches mathématiques, Université de Montréal
C. P. 6128-A, Montréal, Canada H3C 3J7

Abstract. Some standard examples of Hamiltonian systems that are integrable by classical means are cast within the framework of isospectral flows in loop algebras. These include: the Neumann oscillator, the cubically nonlinear Schrödinger systems and the sine-Gordon equation. Each system has an associated invariant spectral curve and may be integrated via the Liouville-Arnold technique. The linearizing map is the Abel map to the associated Jacobi variety, which is deduced through separation of variables in hyperellipsoidal coordinates. More generally, a family of moment maps is derived, embedding certain finite dimensional symplectic manifolds, which arise through Hamiltonian reduction of symplectic vector spaces, into rational coadjoint orbits of loop algebras $\tilde{\mathfrak{g}}^+ \subset \widetilde{\mathfrak{gl}}(r)^+$. Integrable Hamiltonians are obtained by restriction of elements of the ring of spectral invariants to the image of these moment maps; the isospectral property follows from the Adler-Kostant-Symes theorem. The structure of the generic spectral curves arising through the moment map construction is examined. *Spectral Darboux coordinates* are introduced on rational coadjoint orbits in $\widetilde{\mathfrak{gl}}(r)^{+*}$, and these are shown to generalize the hyperellipsoidal coordinates encountered in the previous examples. Their relation to the usual algebro-geometric data, consisting of linear flows of line bundles over the spectral curves, is given. Applying the Liouville-Arnold integration technique, the Liouville generating function is expressed in completely separated form as an abelian integral, implying the Abel map linearization in the general case.

Keywords. Integrable systems, Liouville-Arnold integration, loop algebras, isospectral flow, spectral Darboux coordinates, Abel map linearization.

[†] Research supported in part by the Natural Sciences and Engineering Research Council of Canada and the Fonds FCAR du Québec.
[1]e-mail address: harnad@alcor.concordia.ca *or* harnad@mathcn.umontreal.ca

1 Background Material and Examples

In this first section, we shall examine several examples of integrable Hamiltonian systems that may be represented by isospectral flows on coadjoint orbits of loop algebras. In each case, the flow may be linearized through the classical Liouville-Arnold integration technique. An explicit linearization of the flows can be made in terms of abelian integrals associated to an invariant spectral curve associated to the system. A key element in the integration is the fact that a complete separation of variables occurs within a suitably defined coordinate system - essentially, hyperellipsoidal coordinates, or some generalization thereof. A general theory will then be developed in subsequent sections, based essentially on moment map embeddings of finite dimensional symplectic vector spaces, or Hamiltonian quotients thereof, into the dual space of certain loop algebras, the image consisting of orbits whose elements are rational in the loop parameter.

The origins of this approach may be found in the works of Moser **[Mo]** on integrable systems on quadrics, Adler and van Moerbeke **[AvM]** on linearization of isospectral flows in loop algebras and the general algebro-geometric integration techniques of Dubrovin, Krichever and Novikov **[KN]**, **[Du]**. The theory of moment map embeddings in loop algebras is developed in **[AHP]**, **[AHH1]**. Its relation to algebro-geometric integration techniques is described in **[AHH2]**, and the use of "spectral Darboux coordinates" in the general Liouville-Arnold integration method in loop algebras is developed in **[AHH3]**. Some detailed examples and earlier overviews of this approach may be found in **[AHH4]**, **[AHH5]**. The proofs of the theorems cited here may be found in **[AHP]**, **[AHH1-AHH3]**.

1.1. The Neumann Oscillator

We begin with the Neumann oscillator system (**[N]**, **[Mo]**), which consists of a point particle confined to a sphere in $\mathbb{R}^n$, subject to harmonic oscillator forces. The phase space is identifiable either with the cotangent bundle or the tangent bundle (the equivalence being via the metric):

$$M = T^*S^{n-1} = \{(\mathbf{x},\mathbf{y}) \in \mathbb{R}^n \times \mathbb{R}^n \mid \mathbf{x}^T\mathbf{x} = 1,\ \mathbf{y}^T\mathbf{x} = 0\} \subset \mathbb{R}^n \times \mathbb{R}^n, \qquad (1.1)$$

where $\mathbf{x}$ represents position and $\mathbf{y}$ momentum. The Hamiltonian is

$$H(\mathbf{x},\mathbf{y}) = \frac{1}{2}[\mathbf{y}^T\mathbf{y} + \mathbf{x}^T A\mathbf{x}], \qquad (1.2)$$

where A is the diagonal $n \times n$ matrix

$$A = \mathrm{diag}(\alpha_1, \ldots \alpha_n) \in M^{n\times n}, \qquad (1.3)$$

with distinct eigenvalues $\{\alpha_i\}_{i=1,\dots n}$ determining the oscillator constants.

Equivalently, we may choose the Hamiltonian as:

$$\phi(\mathbf{x},\mathbf{y}) = \frac{1}{2}[(\mathbf{x}^T\mathbf{x})(\mathbf{y}^T\mathbf{y}) + \mathbf{x}^T A\mathbf{x} - (\mathbf{x}^T\mathbf{y})^2], \tag{1.4}$$

in view of the constraints

$$\mathbf{x}^T\mathbf{x} = 1, \quad \mathbf{y}^T\mathbf{x} = 0. \tag{1.5}$$

The unconstrained equations of motion for the Hamiltonian ϕ are:

$$\frac{d\mathbf{x}}{dt} = (\mathbf{x}^T\mathbf{x})\mathbf{y} - (\mathbf{x}^T\mathbf{y})\mathbf{x} \tag{1.6a}$$

$$\frac{d\mathbf{y}}{dt} = -(\mathbf{y}^T\mathbf{y})\mathbf{x} - A\mathbf{x} + (\mathbf{x}^T\mathbf{y})\mathbf{y}. \tag{1.6b}$$

Since

$$\{\phi, \mathbf{x}^T\mathbf{x}\} = 0, \tag{1.7}$$

it is convenient to interpret the relation

$$n(\mathbf{x}) := \mathbf{x}^T\mathbf{x} = 1, \tag{1.8}$$

alone as a first class constraint. The function $n(\mathbf{x})$ generates the flow

$$(\mathbf{x},\mathbf{y}) \longmapsto (\mathbf{x}, \mathbf{y} + t\mathbf{x}), \tag{1.9}$$

and is invariant under the ϕ–flow. We may then apply Marsden-Weinstein reduction, quotienting by the flow (1.9). The reduced manifold is identified with a section of the orbits under (1.9) defined by the other constraint:

$$\mathbf{y}^T\mathbf{x} = 0. \tag{1.10}$$

The integral curves for the constrained system are determined from those for the unconstrained system by othogonal projection:

$$\begin{aligned}(\mathbf{x}(t),\mathbf{y}(t))_{free} &\longmapsto (\widehat{\mathbf{x}}(t),\widehat{\mathbf{y}}(t))_{constr.} \\ &:= \left((\mathbf{x}(t),\mathbf{y}(t) - \left(\frac{\mathbf{x}^T(t)\mathbf{y}(t)}{\mathbf{x}^T(t)\mathbf{x}(t)}\right)\mathbf{x}(t)\right).\end{aligned} \tag{1.11}$$

This lifts the projected flow on the Marsden-Weinstein reduced space $n^{-1}(1)/\mathbb{R}$ to one that is tangential to the section defined by eq. (1.10).

Assuming the constants α_i determining the oscillator strengths are distinct, the integration proceeds (cf. [Mo]) by introducing the Devaney-Uhlenbeck commuting integrals

$$I_i := \sum_{j=1, j\neq i}^{n} \frac{(x_i y_j - y_i x_j)^2}{\alpha_i - \alpha_j} + x_i^2, \tag{1.12}$$

which satisfy

$$\sum_{i=1}^{n} I_i = \mathbf{x}^T \mathbf{x} \tag{1.13a}$$

$$\sum_{i=1}^{n} \alpha_i I_i = 2\phi. \tag{1.13b}$$

Define the degree $n-1$ polynomial $\mathcal{P}(\lambda)$ by

$$\frac{\mathcal{P}(\lambda)}{a(\lambda)} := -\frac{1}{4}\sum_{i=1}^{n} \frac{I_i}{\lambda - \alpha_i}, \tag{1.14}$$

where

$$a(\lambda) := \prod_{i=1}^{n} (\lambda - \alpha_i),$$

$$\mathcal{P}(\lambda) = P_{n-1}\lambda^{n-1} + P_{n-2}\lambda^{n-2} + \cdots + P_0. \tag{1.15}$$

Then on T^*S^{n-1},

$$P_{n-1} = -\frac{1}{4}\mathbf{x}^T\mathbf{x} = -\frac{1}{4},$$

$$P_{n-2} = \frac{1}{2}\phi. \tag{1.16}$$

An equivalent set of commuting integrals consists of the coefficients of the polynomial $\{P_0, \ldots, P_{n-2}\}$. The Liouville-Arnold tori $\mathbf{T}$ are the leaves of the Lagrangian foliation defined by the level sets:

$$P_i = C_i. \tag{1.17}$$

We now proceed to the linearization of the flows through the Liouville-Arnold method. First, introduce hyperellipsoidal coordinates $\{\lambda_\mu\}_{\mu=1,\ldots n-1}$ and their conjugate momenta, $\{\zeta_\mu\}_{\mu=1,\ldots n-1}$, which are defined by:

$$\sum_{i=1}^{n} \frac{x_i^2}{\lambda - \alpha_i} = \frac{\prod_{\mu=1}^{n-1}(\lambda - \lambda_\mu)}{a(\lambda)} \tag{1.18a}$$

$$\zeta_\mu = \frac{1}{2}\sum_{i=1}^{n} \frac{x_i y_i}{\lambda_\mu - \alpha_i} = \sqrt{\frac{\mathcal{P}(\lambda_\mu)}{a(\lambda_\mu)}}. \tag{1.18b}$$

In terms of these, the canonical 1–form is:

$$\theta = \sum_{i=1}^{n} y_i dx_i|_{T^\bullet S^{n-1}} = \sum_{\mu=1}^{n-1} \zeta_\mu d\lambda_\mu . \tag{1.19}$$

Restricting this to $\mathbf{T}$ determines the differential of the Liouville generating function S:

$$\sum_{\mu=1}^{n-1} \zeta_\mu d\lambda_\mu|_{P_i=cst.} = dS = \sum_{\mu=1}^{n-1} \sqrt{\frac{\mathcal{P}(\lambda_\mu)}{a(\lambda_\mu)}} d\lambda_\mu , \tag{1.20}$$

which, upon integration, gives

$$S = \sum_{\mu=1}^{n-1} \int_0^{\lambda_\mu} \sqrt{\frac{\mathcal{P}(\lambda)}{a(\lambda)}} d\lambda . \tag{1.21}$$

This is seen to be an abelian integral on the (generically) genus $g = n-1$ hyperelliptic curve $\mathcal{C}$ defined by:

$$z^2 + a(\lambda)\mathcal{P}(\lambda) = 0. \tag{1.22}$$

The linearizing coordinates conjugate to the invariants P_j are then:

$$Q_j := \frac{\partial S}{\partial P_j} = \frac{1}{2} \sum_{\mu=1}^{n-1} \int_0^{\lambda_\mu} \frac{\lambda^j d\lambda}{\sqrt{a(\lambda)\mathcal{P}(\lambda)}} = b_j t, \tag{1.23}$$

where, for $\phi = 2P_{n-2}$,

$$b_{n-2} = 2, \qquad b_j = 0, \quad j < n-2. \tag{1.24}$$

The map:

$$(\lambda_1, \ldots \lambda_{n-1}) \longrightarrow (Q_1, \ldots Q_{n-1}) \tag{1.25}$$

defined by eq. (1.23) is, up to normalization, the Abel map from the symmetric product $S^{n-1}\mathcal{C}$ to the Jacobi variety $\mathcal{J}(\mathcal{C})$ of $\mathcal{C}$:

$$\mathbf{A} : S^{n-1}\mathcal{C} \longmapsto \mathcal{J}(\mathcal{C}) \sim \mathbf{C}^{n-1}/\Gamma, \tag{1.26}$$

where $\Gamma =$ is the period lattice.

We now turn to the interpretation of such systems as isospectral flows in a loop algebra (cf. **[AHP]**, **[AHH3]**, **[AHH4]**). Let

$$\mathcal{N}(\lambda) = \lambda Y + \mathcal{N}_0(\lambda), \tag{1.27}$$

where

$$\mathcal{N}_0(\lambda) := \frac{\lambda}{2}\begin{pmatrix} -\sum_{i=1}^n \frac{x_i y_i}{\lambda-\alpha_i} & -\sum_{i=1}^n \frac{y_i^2}{\lambda-\alpha_i} \\ \sum_{i=1}^n \frac{x_i^2}{\lambda-\alpha_i} & \sum_{i=1}^n \frac{x_i y_i}{\lambda-\alpha_i} \end{pmatrix} \in \widetilde{\mathfrak{sl}}(2)^{+*} \tag{1.28a}$$

$$Y := \begin{pmatrix} 0 & -\frac{1}{2} \\ 0 & 0 \end{pmatrix}. \tag{1.28b}$$

We define a map $\widetilde{J}_A : \mathbb{R}^n \times \mathbb{R}^n \longrightarrow \widetilde{\mathfrak{sl}}(2)^{+*}$ to the dual space of the positive half of the loop algebra $\widetilde{\mathfrak{sl}}(2)$, relative to the standard splitting $\widetilde{\mathfrak{sl}}(2) = \widetilde{\mathfrak{sl}}(2)^{+*} + \widetilde{\mathfrak{sl}}(2)^{-*}$ into holomorhic parts inside and outside a circle S^1 in the complex λ-plane, containing the α_i's in its interior region:

$$\widetilde{J}_A : (\mathbf{x}, \mathbf{y}) \longmapsto \mathcal{N}_0(\lambda) \in \widetilde{\mathfrak{sl}}(2)^{+*}. \tag{1.29}$$

This is a Poisson map with respect to the Lie poisson structure on $\widetilde{\mathfrak{sl}}(2)^{+*}$. The Hamiltonian ϕ is then given by restriction of an elementary spectral invariant:

$$\phi(\mathbf{x}, \mathbf{y}) = -\mathrm{tr}(\mathcal{N}(\lambda)^2)_0 := -\frac{1}{2\pi i}\oint_{S^1} \mathrm{tr}(\mathcal{N}(\lambda)^2)\frac{d\lambda}{\lambda}, \tag{1.30}$$

and all the other invariants P_j may be similarly represented. The equations of motion are seen to be equivalent to the Lax equation:

$$\frac{d\mathcal{N}}{dt} = [\mathcal{B}, \mathcal{N}], \tag{1.31}$$

where

$$\mathcal{B} := d\phi(\mathcal{N})_+ = \begin{pmatrix} \mathbf{x}^T\mathbf{y} & \lambda + \mathbf{y}^T\mathbf{y} \\ -\mathbf{x}^T\mathbf{x} & -\mathbf{x}^T\mathbf{y} \end{pmatrix}, \tag{1.32}$$

and $(d\phi(\mathcal{N})_+$ signifies projection of the element $d\phi(\mathcal{N}) \in \widetilde{\mathfrak{sl}}(2)$ to $\widetilde{\mathfrak{sl}}(2)^+$. This is an example of the *Adler-Kostant-Symes* (AKS) *theorem* (to be explained more fully in Section 2). The spectral invariants (elements of the AKS ring) are generated by the residues of the rational function

$$\det\left(\frac{\mathcal{N}(\lambda)}{\lambda}\right) = \frac{P(\lambda)}{a(\lambda)} = -\frac{1}{4}\sum_{i=1}^n \frac{I_i}{\lambda - \alpha_i}. \tag{1.33}$$

To see the relation with the standard algebro-geometric linearization methods ([**Du**], [**KN**], [**AHH2-AHH3**]), we begin with the invariant spectral curve:

$$\det(\mathcal{L}(\lambda) - z\mathbb{I}_2) = z^2 + a(\lambda)P(\lambda) = 0, \tag{1.34}$$

where

$$\mathcal{L}(\lambda := \frac{a(\lambda)}{\lambda}\mathcal{N}(\lambda), \quad z := a(\lambda)\frac{\zeta}{\lambda}, \tag{1.35}$$

and let

$$\mathcal{M}(\lambda,\zeta) := \left(\frac{\mathcal{N}(\lambda)}{\lambda} - \zeta \mathbb{I}_2\right), \tag{1.36}$$

where $\mathbb{I}_2$ is the 2×2 unit matrix. Then the columns of the matrix $\widetilde{\mathcal{M}}(\lambda,\zeta)$ of cofactors of $\mathcal{M}(\lambda,\zeta)$ are the eigenvectors of $\mathcal{N}(\lambda)$ ($\zeta\lambda$ = eigenvalue) on $\mathcal{C}$:

$$\widetilde{\mathcal{M}}(\lambda,\zeta) = \begin{pmatrix} \frac{1}{2}\sum_{i=1}^n \frac{x_i y_i}{\lambda-\alpha_i} - \zeta & \frac{1}{2}\sum_{i=1}^n \frac{y_i^2}{\lambda-\alpha_i} + \frac{1}{2} \\ -\frac{1}{2}\sum_{i=1}^n \frac{x_i^2}{\lambda-\alpha_i} & -\frac{1}{2}\sum_{i=1}^n \frac{x_i y_i}{\lambda-\alpha_i} + \zeta \end{pmatrix}. \tag{1.37}$$

The hyperellipsoidal coordinates $\{\lambda_\mu, \zeta_\mu\}_{\mu=1,\dots,n-1}$ define the finite part of the zero-divisor:

$$\mathcal{D} = \sum_{\mu=1}^{n-1} p(\lambda_\mu, \zeta_\mu) + p(\infty_1), \tag{1.38}$$

i.e., the zeros of a section of the bundle $E \to \mathcal{C}$ dual to eigenvector line bundle. This bundle can be shown generically to have degree n, and thus to be an element of the Picard variety $E \in \mathrm{Pic}^n$. The Abel map then identifies the symmetric product $S^{n-1}\mathcal{C}$ with the Jacobi variety $\mathcal{J}(\mathcal{C}) \sim \mathrm{Pic}^0$. The linearity of the flow in Pic^n follows from noting that the Lax equation

$$\frac{d\mathcal{N}}{dt} = [d\phi(\mathcal{N})_+, \mathcal{N}], \tag{1.39}$$

implies a linear exponential form for the transition function

$$\tau(\lambda, z, t) = \exp(\phi_z(\lambda, z)t). \tag{1.40}$$

1.2 Nonlinear Schrödinger (NLS) Equation

We now apply a similar analysis to the quasi-periodic solutions of the cubically nonlinear Schrödinger equation (cf. **[P1]**, **[AHP]**, **[AHH4]**)

$$u_{xx} + \sqrt{-1}u_t = 2|u|^2 u. \tag{1.41}$$

Let

$$\mathcal{N}(\lambda) := \frac{\lambda}{2}\begin{pmatrix} i\sum_{j=1}^n \frac{|z_j|^2}{\lambda-\alpha_j} & -\sum_{j=1}^n \frac{\bar{z}_j^2}{\lambda-\alpha_j} \\ -\sum_{j=1}^n \frac{z_j^2}{\lambda-\alpha_j} & -i\sum_{j=1}^n \frac{|z_j|^2}{\lambda-\alpha_j} \end{pmatrix}, \tag{1.42}$$

and let ω denote the standard symplectic form on $\mathbb{C}^n$:

$$\omega = i\, d\bar{z}^T \wedge dz = i\sum_{j=1}^n d\bar{z}_j \wedge dz_j \tag{1.43}$$

We define the Poisson map:

$$\begin{aligned} \tilde{J} &: \mathbf{C}^n \longrightarrow \tilde{\mathfrak{su}}(1,1)^{+*} \\ \tilde{J} &: \mathbf{z} \longmapsto \mathcal{N}(\lambda), \end{aligned} \tag{1.44}$$

and let

$$\mathcal{L}(\lambda) = \frac{a(\lambda)}{\lambda}\mathcal{N}(\lambda) = L_0\lambda^{n-1} + L_1\lambda^{n-2} + \cdots + L_{n-1}. \tag{1.45}$$

The spectral curve is defined by the characteristic equation

$$\begin{aligned} \det(\mathcal{L}(\lambda) - zI) &= z^2 + a(\lambda)\mathcal{P}(\lambda) = 0, \\ \mathcal{P}(\lambda) &:= P_0 + P_1\lambda + \cdots + P_{n-2}\lambda^{n-2}, \end{aligned} \tag{1.46}$$

and has genus $g = n - 2$. Choosing the AKS Hamiltonians:

$$H_x = \frac{1}{2}\left[\frac{a(\lambda)}{\lambda^n}\lambda \operatorname{tr}(\mathcal{N}(\lambda)^2)\right]_0 = -P_{2,n-3} \tag{1.47a}$$

$$H_t = \frac{1}{2}\left[\frac{a(\lambda)}{\lambda^n}\lambda^2 \operatorname{tr}(\mathcal{N}(\lambda)^2)\right]_0 = -P_{2,n-4} \tag{1.47b}$$

gives the Lax equations

$$\frac{d}{dx}\mathcal{L}(\lambda) = [(dH_x)_+, \mathcal{L}(\lambda)] \tag{1.48b}$$

$$\frac{d}{dt}\mathcal{L}(\lambda) = [(dH_t)_+, \mathcal{L}(\lambda)], \tag{1.48b}$$

where

$$(dH_x)_+ = \lambda L_0 + L_1 \tag{1.49a}$$

$$(dH_t)_+ = \lambda^2 L_0 + \lambda L_1 + L_2. \tag{1.49b}$$

Choosing invariant constraints so that:

$$\begin{aligned} L_0 &= \frac{i}{2}\begin{pmatrix} 1 & 0 \\ 0 & -1 \end{pmatrix}, \quad L_1 = \begin{pmatrix} 0 & \overline{u} \\ u & 0 \end{pmatrix}, \\ L_2 &= i\begin{pmatrix} |u|^2 & -\overline{u_x} \\ u_x & -|u|^2 \end{pmatrix}, \end{aligned} \tag{1.50}$$

the compatibility conditions:

$$\frac{\partial (dH_x)_+}{\partial t} - \frac{\partial (dH_t)_+}{\partial x} + [(dH_x)_+, (dH_t)_+] = 0 \tag{1.51}$$

reduce to the NLS equation (1.41). To obtain the quasi-periodic solutions, we introduce the *spectral Darboux Coordinates* $\{q, P, \lambda_\mu, \zeta_\mu\}_{\mu=1,\dots n-2}$, analogous to the hyperellipsoidal coordinates above:

$$\sum_{i=1}^{n} \frac{z_i^2}{\lambda - \alpha_i} = -\frac{2u \prod_{\mu=1}^{n-2}(\lambda - \lambda_\mu)}{a(\lambda)} \tag{1.52a}$$

$$\zeta_\mu = -\frac{i}{2} \sum_{i=1}^{n} \frac{|z_i|^2}{\lambda_\mu - \alpha_i} = \sqrt{-\frac{\mathcal{P}(\lambda_\mu)}{a(\lambda_\mu)}} \tag{1.52b}$$

$$q := \ln(u), \qquad P := (L_0)_{22}, \tag{1.52c}$$

Then the symplectic form: may be expressed as

$$\omega = \sum_{\mu=1}^{n-2} d\lambda_\mu \wedge d\zeta_\mu + dq \wedge dP. \tag{1.53}$$

As above, the spectral curve is invariant, and the coefficients of the characteristic polynomial generate a complete set of commuting integrals, so we may apply the Liouville-Arnold integration method. The coordinates (λ_μ, ζ_μ) defined by (1.52a,b) again give the finite part of the divisor of zeros of the eigenvectors of $\mathcal{N}(\lambda)$, while the remaining pair (q, P) are determined by the additional spectral data at $\lambda = \infty$. The Liouville generating function in this case becomes

$$S = \sum_{\mu=1}^{n-1} \zeta_\mu d\lambda_\mu|_{P_i = cst.} + qP = \sum_{\mu=1}^{n-1} \int_0^{\lambda_\mu} \sqrt{-\frac{\mathcal{P}(\lambda)}{a(\lambda)}} d\lambda + P \ln u \tag{1.54}$$

and the linearizing coordinates conjugate to the P_i's are:

$$Q_i = \frac{\partial S}{\partial P_i} = \frac{1}{2} \sum_{\mu=1}^{n-2} \int_0^{\lambda_\mu} \frac{\lambda^i d\lambda}{\sqrt{-a(\lambda)\mathcal{P}_2(\lambda)}} = b_i x + c_i t, \qquad i = 0, \dots n-3, \tag{1.55}$$

which are abelian integrals of the first kind, and

$$Q_{2,n-2} = \frac{\partial S}{\partial P_i} = \frac{1}{2} \sum_{\mu=1}^{n-2} \int_0^{\lambda_\mu} \frac{\lambda^{n-2} d\lambda}{\sqrt{-a(\lambda)\mathcal{P}_2(\lambda)}} - \frac{\ln u}{2P_2} = b_{n-2} x + c_{n-2} t, \tag{1.56}$$

which is an abelian integral of the third kind, the integrand having simple poles at the two points (∞_1, ∞_2) over $\lambda = \infty$. For the Hamiltonian $H_x = -P_{n-3}$, we have $b_i = -\delta_{i,n-3}$, $c_i = 0$, while for $H_t = -P_{n-4}$, $b_i = 0$, $c_i = -\delta_{i,n-3}$. An explicit formula for the function $u(x, t)$ may be obtained in terms of the Riemann theta function θ associated to the spectral curve by applying the reciprocity theorem relating the two kinds of abelian integrals (cf. [**AHH4**]).

$$u(x,t) = \exp(q) = \tilde{K}\ \exp(bx+ct)\frac{\theta(\mathbf{A}(\infty_2,p)+t\mathbf{U}+x\mathbf{V}-\mathbf{K})}{\theta(\mathbf{A}(\infty_1,p)+t\mathbf{U}+x\mathbf{V}-\mathbf{K})}, \tag{1.57}$$

where $\mathbf{A} : S^{n-2}\mathcal{C} \longmapsto \mathcal{J}(\mathcal{C}) \sim \mathbb{C}^{n-2}/\Gamma$ is the Abel map, $\mathbf{U}, \mathbf{V} \in \mathbb{C}^{n-2}$, $b, c \in \mathbb{C}$ are obtained from the vectors with components (b_i, c_i) on the RHS of eq. (1.55), (1.56) by applying the linear transformation that normalizes the abelian differentials in (1.55), and $\mathbf{K}$ is the Riemann constant.

1.3 Sine-Gordon Equation

As a last example, consider the sine-Gordon equation (cf. [**HW**], [**P2**], [**AA**])

$$\frac{\partial^2 u}{\partial x^2} - \frac{\partial^2 u}{\partial t^2} = \sin(u). \tag{1.58}$$

Let

$$\mathcal{N}(\lambda) := \lambda Y + \mathcal{N}_0(\lambda), \tag{1.59}$$

where

$$Y = \begin{pmatrix} 0 & -1 \\ 1 & 0 \end{pmatrix}, \quad \mathcal{N}_0(\lambda) := 2\lambda \begin{pmatrix} b(\lambda) & c(\lambda) \\ -\bar{c}(\bar{\lambda}) & -b(\lambda) \end{pmatrix}, \tag{1.60}$$

with $b(\lambda)$, $c(\lambda)$ given by

$$b(\lambda) = \lambda \sum_{i=1}^{p} \left(\frac{-\varphi_i \bar{\gamma}_i}{\alpha_i^2 - \lambda^2} + \frac{\bar{\varphi}_i \gamma_i}{\bar{\alpha}_i^2 - \lambda^2} \right) \tag{1.61a}$$

$$c(\lambda) = \sum_{i=1}^{p} \left(\frac{\alpha_i \bar{\gamma}_i^2}{\alpha_i^2 - \lambda^2} + \frac{\bar{\alpha}_i \bar{\varphi}_i^2}{\bar{\alpha}_i^2 - \lambda^2} \right), \tag{1.61b}$$

and $\varphi, \gamma \in \mathbb{C}^p$ complex vectors with components $\{\varphi_i, \gamma_i\}_{i=1,\dots p}$. (Here α_i, $\overline{\alpha}_i$, $-\alpha_i$, $-\overline{\alpha}_i$ are assumed to be distinct.)

Define

$$a(\lambda) := \prod_{i=1}^{p} [(\lambda^2 - \alpha_i^2)(\lambda^2 - \bar{\alpha}_i^2)]. \tag{1.62}$$

The symplectic form on $\mathbb{C}^p \times \mathbb{C}^p$ is given by:

$$\omega = 4 \sum_{i=1}^{p} (d\gamma_i \wedge d\bar{\varphi}_i + d\bar{\gamma}_i \wedge d\varphi_i). \tag{1.63}$$

Again, define a Poisson map:

$$\tilde{J} : \mathbb{C}^p \times \mathbb{C}^p \longrightarrow \widehat{\mathfrak{su}}(2)^{+*}$$
$$\tilde{J} : (\varphi, \gamma) \longmapsto \mathcal{N}_0(\lambda), \tag{1.64}$$

where the twisted loop algebra:

$$\widehat{\mathfrak{su}}(2)^+ \subset \widetilde{\mathfrak{su}}(2)^+ \subset \widetilde{\mathfrak{u}}(2)^+ \subset \widetilde{\mathfrak{gl}}(2,\mathbb{C})^+ \tag{1.65}$$

is defined as the fixed point set in $\widetilde{\mathfrak{sl}}(2,\mathbb{C})^+$ under the involutions;

$$\sigma_1 : X(\lambda) \longmapsto X^\dagger(\bar{\lambda})$$
$$\sigma_2 : X(\lambda) \longmapsto \begin{pmatrix} 1 & 0 \\ 0 & -1 \end{pmatrix} X(-\lambda) \begin{pmatrix} 1 & 0 \\ 0 & -1 \end{pmatrix}. \tag{1.66}$$

Let $n = 2p$, and

$$\mathcal{L}(\lambda) := \frac{a(\lambda)}{\lambda} \mathcal{N}(\lambda) \tag{1.67a}$$
$$= a(\lambda)Y + L_0 \lambda^{2n-1} + L_1 \lambda^{2n-2} + \cdots + L_{2n-1}. \tag{1.67b}$$

The spectral curve is:

$$\det(\mathcal{L}(\lambda) - zI) = z^2 + a(\lambda)P(\lambda) = 0 \tag{1.68a}$$
$$P(\lambda) = P_0 + \lambda^2 P_1 + \cdots + \lambda^{2n-2} P_{n-1} + \lambda^{2n}. \tag{1.68b}$$

Choosing the AKS Hamiltonians:

$$H_\xi(X) := \frac{1}{2} \mathrm{tr} \left(\frac{a(\lambda)}{\lambda^2} (X(\lambda) + \lambda Y)^2 \right)_0 = -P_0$$
$$H_\eta(X) := -\frac{1}{2} \mathrm{tr} \left(\frac{a(\lambda)}{\lambda^{2N}} (X(\lambda) + \lambda Y)^2 \right)_0 = P_{n-1} \tag{1.69}$$

gives the Lax equations

$$\frac{d}{d\xi} \mathcal{L}(\lambda) = [A, \mathcal{L}(\lambda)] \tag{1.70a}$$
$$\frac{d}{d\eta} \mathcal{L}(\lambda) = [B, \mathcal{L}(\lambda)], \tag{1.70b}$$

where

$$-A = dH_\xi(\mathcal{N})_- = \frac{1}{\lambda}(L_{2n-1} + a(0)Y)$$
$$B = dH_\eta(\mathcal{N})_+ = L_0 + \lambda Y. \tag{1.71}$$

Choosing the level set:

$$P_0 = \frac{1}{16} \tag{1.72}$$

gives

$$L_{2n-1} + a(0)Y = \frac{1}{4}\begin{pmatrix} 0 & e^{iu} \\ -e^{-iu} & 0 \end{pmatrix}, \tag{1.73}$$

where

$$e^{iu} = a(0)(c(0) - 1), \tag{1.74}$$

with u real. Then the compatibility conditions:

$$\frac{\partial A}{\partial \eta} - \frac{\partial B}{\partial \xi} + [A, B] = 0 \tag{1.75}$$

reduce to the Sine-Gordon equation

$$u_{xx} - u_{tt} = \sin u, \tag{1.76}$$

where

$$\xi = x + t, \qquad \eta = x - t. \tag{1.77}$$

The quasi-periodic solutions are obtained as in the previous example. The unreduced spectral curve is defined by

$$z^2 + \tilde{a}(\lambda)\tilde{P}(\lambda) = 0, \tag{1.78}$$

and is again hyperelliptic, with genus $g = 2n - 1$. Quotienting by the involution

$$(z, \lambda) \longmapsto (z, -\lambda) \tag{1.79}$$

gives a reduced curve with genus $g = n - 1$ defined by

$$z^2 + \tilde{a}(E)\tilde{P}(E) = 0, \tag{1.80}$$

where

$$\lambda^2 := E, \quad \tilde{a}(E); = a(\lambda), \quad \tilde{P}(E) := P(\lambda). \tag{1.81}$$

We also introduce the augmented curve, defined by

$$\tilde{z}^2 + E\tilde{a}(E)\tilde{P}(E) = 0, \tag{1.82}$$

of genus $g = n$, where

$$\tilde{z} := z\lambda. \tag{1.83}$$

The spectral Darboux coordinates are defined by

$$\tilde{c}(E_\mu) - 1 = 0 \tag{1.84a}$$

$$\zeta_\mu \sqrt{E_\mu} = 2\tilde{b}(E_\mu), \qquad \mu = 1, \dots n, \tag{1.84b}$$

where

$$\tilde{b}(E) := b(\lambda), \quad \tilde{c}(E) := c(\lambda). \tag{1.85}$$

These again are interpreted as zeros of the sections of the dual to the eigenvector line bundle associated to $\mathcal{N}(\lambda)$. The symplectic form is then

$$\omega = \sum_{\mu=1}^{n} dE_\mu \wedge d\zeta_\mu = -d\theta. \tag{1.86}$$

The Liouville generating function is

$$S(P_0, \dots, P_{n-1}, E_1, \dots, E_n) = \sum_{\mu=1}^{n} \int_{E_0}^{E_\mu} \sqrt{-\frac{\tilde{P}(E)}{E\tilde{a}(E)}} dE, \tag{1.87}$$

giving rise to the Abel map linearization:

$$Q_i = \frac{\partial S}{\partial P_i} = -\frac{1}{2} \sum_{\mu=1}^{N} \int_{E_0}^{E_\mu} \frac{E^i}{\sqrt{-E\tilde{a}(E)\tilde{P}(E)}} dE \tag{1.88a}$$

$$= C_i + 2\delta_{i,0}\xi - 2\delta_{i,n-1}\eta, \tag{1.88b}$$

which only involves abelian integrals of the first kind on the augmented curve. In terms of theta functions, the solution may be expressed as

$$u = -i \left(\sum_{\mu=1}^{n} \ln(-E_\mu) - \pi \right) \tag{1.89a}$$

$$= -2i \ln \frac{\Theta(\mathbf{A}(p_0, 0) - \mathbf{U}\eta - \mathbf{V}\xi - \mathbf{K})}{\Theta(\mathbf{A}(p_0, \infty) - \mathbf{U}\eta - \mathbf{V}\xi - \mathbf{K})} + C, \tag{1.89b}$$

where $\mathbf{U}, \mathbf{V}$ are again obtained from the coefficients on the RHS of (1.88) by applying the normalizing linear transformation to the abelian differentials appearing in (1.88). Full details for this case may be found in [**HW**].

In the following two sections, a general approach to integrable systems is developed, yielding all the above results as particular cases, but allowing generalizations to more complex systems of higher rank.

2 Moment Map Embeddings in Loop Algebras

2.1 Phase Space and Loop Group Action

We begin by defining the *generalized Moser space* (cf. **[AHP]**) to be the symplectic vector space consisting of pairs (F, G) of rectangular $N \times r$ matrices:

$$M = \{(F, G) \in M^{N,r} \times M^{N,r}\} \tag{2.1}$$

with symplectic form:

$$\omega = \mathrm{tr}\, dF^T \wedge dG. \tag{2.2}$$

The loop algebra, denoted $\widetilde{\mathfrak{g}}$, consists of smooth maps from a circle S^1, centred at the origin of the complex λ–plane, into $\mathfrak{gl}(r)$, $\mathfrak{sl}(r)$, or some subalgebra thereof.

$$\begin{aligned}\widetilde{\mathfrak{g}} &= \widetilde{\mathfrak{gl}}(r) \quad (\text{or } \widetilde{\mathfrak{sl}}(r)) \\ &= \{X(\lambda) \in \mathfrak{gl}(r),\ \lambda \in S^1 \subset \mathbb{C} \cup \infty\}.\end{aligned} \tag{2.3}$$

There is a natural splitting of $\widetilde{\mathfrak{g}}$

$$\widetilde{\mathfrak{g}} = \widetilde{\mathfrak{g}}^+ + \widetilde{\mathfrak{g}}^-, \tag{2.4}$$

as a vector space direct sum of the subalgebra $\widetilde{\mathfrak{g}}^+$, consisting of elements $X(\lambda)$ admitting a holomorphic extension to the interior of S^1, and $\widetilde{\mathfrak{g}}^-$, consisting of elements admitting a holomorphic extension to the exterior, with normalization $X(\infty) = 0$. We identify $\widetilde{\mathfrak{g}}$ as a dense subspace of its dual space $\widetilde{\mathfrak{g}}^*$ through the pairing

$$\begin{aligned}&< \mu, X > := \frac{1}{2\pi i} \oint_{S^1} \mathrm{tr}\,(\mu(\lambda) X(\lambda)) \frac{d\lambda}{\lambda}, \\ &\mu \in \widetilde{\mathfrak{g}}_- ,\ X \in \widetilde{\mathfrak{g}}^+.\end{aligned} \tag{2.5}$$

Under this pairing, we have the identification

$$(\widetilde{\mathfrak{g}}^+)^* \sim \widetilde{\mathfrak{g}}_-, \quad (\widetilde{\mathfrak{g}}^-)^* \sim \widetilde{\mathfrak{g}}_+, \tag{2.6}$$

where

$$\widetilde{\mathfrak{g}}^* = \widetilde{\mathfrak{g}}_+ + \widetilde{\mathfrak{g}}_- \tag{2.7}$$

similarly represents a decomposition of $\widetilde{\mathfrak{g}}^*$ into subspaces consisting of elements holomorphic inside and outside S^1, but with the normalization such that elements $\mu \in \widetilde{\mathfrak{g}}_+$ satisfy $\mu(0) = 0$ (and hence the constant loops are included on $\widetilde{\mathfrak{g}}_-$).

The loop group $\widetilde{\mathfrak{G}}^+$ is similarly defined to consist of smooth maps $g : S^1 \to Gl(r)$ which admit holomorphic extensions to the interior of S^1: We define a Hamiltonian action:

$$\begin{aligned} \widetilde{\mathfrak{G}}^+ &: M \longrightarrow M \\ g(\lambda) &: (F, G) \longrightarrow (F_g, G_g), \end{aligned} \tag{2.8}$$

where (F_g, G_g) are determined by the decomposition

$$(A - \lambda I)^{-1} F g^{-1}(\lambda) = (A - \lambda I)^{-1} F_g + F_{\text{hol}} \tag{2.9a}$$

$$g(\lambda) G^T (A - \lambda I)^{-1} = G_g^T (A - \lambda I)^{-1} + G_{\text{hol}}. \tag{2.9b}$$

Here $A \in M^{N,N}$ is some fixed $N \times N$ matrix, with eigenvalues in the interior of S^1, and $(F_{\text{hol}}, G_{\text{hol}})$ denote the parts of the expressions on the left that are holomorphic in the interior of S^1. This Hamiltonian action is generated by the equivariant moment map:

$$\begin{aligned} \tilde{J}^A &: M \longrightarrow \tilde{\mathfrak{g}}^+ \sim \tilde{\mathfrak{g}}_- \\ \tilde{J}^A(F, G) &= \lambda G^T (A - \lambda I)^{-1} F, \end{aligned} \tag{2.10}$$

which is thus a Poisson map with respect to the Lie Poisson structure on $\tilde{\mathfrak{g}}^{+*}$. This map is not injective, its fibres being (generically) the orbits of the subgroup $G_A := \text{Stab}(A) \subset Gl(N)$ acting by conjugation on A, and by the natural symplectic action on M.

$$g : (F, G) \longrightarrow (gF, (g^T)^{-1}), \quad g \in G_A \subset Gl(N) \tag{2.11}$$

The relevant phase space is therefore the quotient

$$M/G_A \sim \tilde{\mathfrak{g}}^*_A, \tag{2.12}$$

which is identified with a finite dimensional Poisson subspace $\tilde{\mathfrak{g}}^*_A \subset \tilde{\mathfrak{g}}^{+*}$ consisting of elements that are rational in the loop parameter λ, with poles at the eigenvalues of A.

2.2 Simplest Case

We now consider the simplest case, where A is a diagonal matrix:

$$A = \text{diag}(\alpha_1, \ldots \alpha_1, \ldots \alpha_k, \ldots \alpha_k, \ldots \alpha_n, \ldots \alpha_n), \tag{2.13}$$

possibly with multiple eigenvalues $\{\alpha_i\}_{i=1,\dots n}$ of multiplicity $k_i \le r$, all in the interior of S^1. The matrices (F, G) are decomposed accordingly as

$$F = \begin{pmatrix} F_1 \\ \vdots \\ F_i \\ \vdots \\ F_n \end{pmatrix}, \qquad G = \begin{pmatrix} G_1 \\ \vdots \\ G_i \\ \vdots \\ G_n, \end{pmatrix}, \tag{2.14}$$

where (F_i, G_i) are the $k_i \times r$ dimensional blocks corresponding to the eigenvalues α_i. For this case, $\mathcal{N}_0(\lambda)$ has only simple poles:

$$\mathcal{N}_0(\lambda) = \tilde{J}^A(F;G) = -\lambda \sum_{i=1}^{n} \frac{G_i^T F_i}{\lambda - \alpha_i} := \lambda \sum_{i=1}^{n} \frac{N_i}{\lambda - \alpha_i}, \tag{2.15}$$

with residue matrices N_i generically of rank:

$$\mathrm{rk}(F_i) = \mathrm{rk}(G_i) = k_i. \tag{2.16}$$

The $Ad^*\tilde{\mathfrak{G}}^+$-action for this case becomes:

$$g(\lambda) : \mathcal{N}_0(\lambda) \longmapsto \lambda \sum_{i=1}^{n} \frac{g(\alpha_i) N_i g^{-1}(\alpha_i)}{\lambda - \alpha_i}, \tag{2.17}$$

which can be identified with the Ad^*-action of the direct product group $Gl(r) \times \cdots \times Gl(r)$ (n times) on $[\mathfrak{gl}(r)^*]^n$.

The fibres of the map $\tilde{J}^A$ coincide with the orbits of the block diagonal subgroup:

$$G_A = Stab(A) = Gl(k_1) \times Gl(k_2) \times \cdots \times Gl(k_n) \subset Gl(N), \tag{2.18}$$

under the action:

$$(h_1, \dots h_i, \dots h_n) : (F_i, G_i) \longmapsto (h_i F_i, (h_i^T)^{-1} G_i) \tag{2.19}$$

This is also a Hamiltonian action, generated by the "dual" moment map:

$$J_H(F, G) := (F_1 G_1^T, \dots, \dots F_n G_n^T) \in (\mathfrak{gl}(k_1) \times \cdots \times \mathfrak{gl}(k_n))^*. \tag{2.20}$$

The $Ad^*\tilde{\mathfrak{G}}^+$ orbits are then the level sets of the Casimir invariants:

$$\mathrm{tr}(F_i G_i^T)^l, \quad k = 1, \dots n \quad l = 1, \dots k_i. \tag{2.21}$$

More generally, the image of the moment map $\widetilde{J}^A$ is a Poisson submanifold of $\widetilde{\mathfrak{g}}_A \subset \widetilde{\mathfrak{g}}^{+*}$ consisting of elements of the form

$$\mathcal{N}_0(\lambda) = \lambda \sum_{i=1}^{n} \sum_{l_i=1}^{p_i} \frac{N_i^{l_i}}{(\lambda - \alpha_i)^{l_i}}, \tag{2.22}$$

where p_i is the dimension of the largest Jordan block of A corresponding to eigenvalue α_i.

2.3 Dynamics: Isopectral AKS Flows

The Hamiltonian flows to be considered are those generated by elements of the ring of Ad*–invariant functions $\mathcal{I}(\widetilde{\mathfrak{g}}^*)$, restricted to the translate $\lambda Y + \widetilde{\mathfrak{g}}_A$ of the subspace $\widetilde{\mathfrak{g}}_A$ by a fixed element $\lambda Y \in \widetilde{\mathfrak{g}}^{-*}$, where $Y \in \mathfrak{gl}(r)$. (The latter is an infinitesimal character for $\widetilde{\mathfrak{g}}^-$, since it annihilates the commutator of any pair of elements.) We denote the ring of elements so obtained by

$$\mathcal{I}^Y_{\text{AKS}} := \mathcal{I}(\widetilde{\mathfrak{g}}^*)|_{\lambda Y + \widetilde{\mathfrak{g}}_A}, \tag{2.23}$$

and refer to it as the AKS (Adler-Kostant-Symes) ring.

Let

$$\mathcal{N}(\lambda) = \lambda Y + \mathcal{N}_0(\lambda) \in \lambda Y + \widetilde{\mathfrak{g}}_A. \tag{2.24}$$

We then have the fundamental theorem that underlies the integrability of the resulting Hamiltonian systems, the Adler-Kostant-Symes theorem:

Theorem 2.1 (AKS).

1. *If $H \in \mathcal{I}^Y_{\text{AKS}}$, Hamilton's equations are:*

$$X_H(\mathcal{N}) = \frac{d\mathcal{N}}{dt} = [(dH)_+, \mathcal{N}] = -[(dH)_-, \mathcal{N}] \tag{2.25a}$$

2. *If $H_1, H_2 \in \mathcal{I}^Y_{\text{AKS}}$,*

$$\{H_1, H_2\} = 0. \tag{2.25b}$$

Thus, all the AKS flows commute, and are generated by isospectral deformations determined by Lax equations of the form (2.25a). In fact, it may be shown (**[RS]**, **[AHP]**, **[AHH2]**) that on generic coadjoint orbits of the form (2.22), these systems are completely integrable; i.e., the elements of the Poisson commutative ring $\mathcal{I}^Y_{\text{AKS}}$ generate a Lagrangian foliation. Since the map (2.10) with image consisting of elements of the form (2.22) is a Poisson map, and passes to the quotient Poisson space M/G_A to define an injective Poisson map, the same results may be applied to the pullback $\widetilde{J}^A \circ H$ of any Hamiltonian in the AKS ring $\mathcal{I}^Y_{\text{AKS}}$.

Corollary 2.2. *The results of Theorem 2.1 remain valid if the Hamiltonians H_1, H_2 are replaced by $\widetilde{J}^A \circ H_1, \widetilde{J}^A \circ H_2$ on the space $\lambda Y + \widetilde{\mathfrak{g}}_A$, identified with M/G_A through*

$$\mathcal{N}(\lambda) = \lambda Y + \mathcal{N}_0(\lambda) = \lambda Y + \tilde{J}_A(F, G). \tag{2.26}$$

2.4 Reductions

To obtain interesting examples, one usually must reduce the generic systems described above in a manner that is consistent with the structure of the dynamical equations. This generally consists of Hamiltonian symmetry reductions involving either continuous or discrete symmetry groups. (It may also involve symplectic, or more generally, Poisson constraints.) We briefly summarize the procedure for both types of symmetry reductions below. The discrete Hamiltonian reduction procedure is described in greater detail in [**HHM**]; the continuous, Marsden-Weinstein reduction is fairly standard [**AM**].

2.4.1 Discrete Reduction:

We consider discrete groups generated by elements σ either of finite order or generating compact orbits, which act on the space M by symplectic diffeomorphisms, and as automorphisms of the loop algebra $\widetilde{\mathfrak{g}}^+$. Let

$$\sigma : M \longrightarrow M \tag{2.27}$$

be such a symplectomorphism, and

$$\sigma_{\mathfrak{g}} : \widetilde{\mathfrak{g}}^+ \longrightarrow \widetilde{\mathfrak{g}}^+ \tag{2.28a}$$

the corresponding automorphism of $\widetilde{\mathfrak{g}}^+$, with dual Poisson map

$$\sigma_{\mathfrak{g}}^* : \widetilde{\mathfrak{g}}^{+*} \longrightarrow \widetilde{\mathfrak{g}}^{+*}. \tag{2.28b}$$

We assume that the moment map $\widetilde{J}^A$ intertwines these two actions, so that the following diagram commutes

$$\begin{array}{ccc} M & \xrightarrow{\ \sigma\ } & M \\ {\scriptstyle \widetilde{J}^A}\big\downarrow & & \big\downarrow{\scriptstyle \widetilde{J}^A} \\ \widetilde{\mathfrak{g}}^{+*} & \xrightarrow{\ \sigma_{\mathfrak{g}}^*\ } & \widetilde{\mathfrak{g}}^{+*} \end{array} \tag{2.29}$$

It follows that $\widetilde{J}^A$ may be restricted to the fixed point sets

$$M_\sigma \subset M, \qquad \widetilde{\mathfrak{k}}^+ := \widetilde{\mathfrak{g}}_\sigma^+ \subset \widetilde{\mathfrak{g}}^+, \tag{2.30}$$

and its restriction defines a moment map from the fixed point set M_σ to the dual space

$$\tilde{\mathfrak{k}}^{+*} := \tilde{\mathfrak{g}}^{+*}_\sigma \subset \tilde{\mathfrak{g}}^{+*} \tag{2.31a}$$

of the subalgebra

$$\tilde{\mathfrak{k}}^{+} := \tilde{\mathfrak{g}}^{+}_\sigma \subset \tilde{\mathfrak{g}}^{+} \tag{2.31b}$$

of fixed elements under $\sigma_{\mathfrak{g}}$. The results of Theorem 2.1 and Corollary 2.2 may then be applied on the reduced spaces, provided the Hamiltonians in the ring $\mathcal{I}^Y_{\mathrm{AKS}}$ are chosen to be invariant under the symmetry $\sigma^*_{\mathfrak{g}}$.

2.4.2 Continuous Hamiltonian Reduction

All the Hamiltonians in the ring $\mathcal{I}^Y_{\mathrm{AKS}}$ are invariant under the Hamiltonian group action given by conjugation of $\mathcal{N}(\lambda)$ by λ-independent elements in the stability subgroup of Y:

$$G_Y := \mathrm{Stab}(Y) \subset Gl(r), \qquad \mathfrak{g}_Y := \mathrm{stab}(Y) \subset \mathfrak{gl}(r). \tag{2.32}$$

This action is generated by a moment map J_Y, given by the leading term N_0 of $\mathcal{N}_0$, restricted to $\mathfrak{g}_Y$

$$J_Y := N_0|_{\mathfrak{g}_Y}, \tag{2.33}$$

where

$$\tilde{J}^A = \mathcal{N}_0(\lambda) = N_0 + N_1\lambda^{-1} + \dots. \tag{2.34}$$

Since the elements of $\mathcal{I}^Y_{\mathrm{AKS}}$ are G_Y invariant, J_Y is conserved under all the AKS flows. Fixing a level set

$$J_Y = \mu_0 \in \mathfrak{g}^*_Y, \tag{2.35}$$

which we assume to be a regular value of J_Y, and restricting to the coadjoint orbit $\mathcal{O}_{\mathcal{N}_0(\lambda)} \subset (\tilde{\mathfrak{g}}^+)^*$, the reduced space is

$$\mathcal{O}_{\mathrm{red}} := J_Y^{-1}(\mu_0)/G_0, \tag{2.36}$$

where $G_0 \subset G_Y$ denotes the stability subgroup of μ_0. The reduced Hamiltonians H_{red} on $\mathcal{O}_{\mathrm{red}}$ are then given by

$$H_{\mathrm{red}} \circ \pi = H|_{J_Y^{-1}(\mu_0)}, \quad H \in \mathcal{I}^Y_{\mathrm{AKS}}, \tag{2.37}$$

where

$$\pi : J_Y^{-1}(\mu_0) \longrightarrow J_Y^{-1}(\mu_0)/G_0 \tag{2.38}$$

denotes the projection map.

2.5 Examples

We now indicate how the loop algebra formulation of examples like those of Section 1 is obtained from the general scheme described above.

2.5.1 Neumann Oscillator (and similar examples) in $\widetilde{\mathfrak{sl}}(2,\mathbf{R})^{+*}$

Consider the case $r = 2, k_i = 1, n = N$. A discrete antilinear involution gives the reality conditions:

$$\alpha_i = \overline{\alpha}_i, \quad F = \overline{F}, \quad G = \overline{G} \tag{2.39}$$

reducing $\widetilde{\mathfrak{gl}}(2,\mathbf{C})^*$ to $\widetilde{\mathfrak{gl}}(2,\mathbf{R})^*$. The stabilizer of $A = \mathrm{diag}\,(\alpha_1,\ldots,\alpha_n)$ consists of the diagonal subgroup $G_A = \{\mathrm{diag}(d_1,\ldots,d_n) \subset Gl(n)\}$ acting as

$$G_A : M \longrightarrow M$$

$$(d_1,\ldots d_n) : \begin{pmatrix} \vdots \\ F_i \\ \vdots \end{pmatrix}, \begin{pmatrix} \vdots \\ G_i \\ \vdots \end{pmatrix} \longmapsto \begin{pmatrix} \vdots \\ d_i F_i \\ \vdots \end{pmatrix}, \begin{pmatrix} \vdots \\ d_i^{-1} G_i \\ \vdots \end{pmatrix}, \tag{2.40}$$

where F_i and G_i are just 2–component row vectors. The moment map generating this action is just

$$J_A(F,G) = (F_1 G_1^T, \ldots, F_n G_n^T) \in \mathbf{R}^n, \tag{2.41}$$

which coincides with the traces of the 2×2 residue matrices N_i in (2.15). Choosing the zero level set for these, Marsden-Weinstein reduction is equivalent to the subgroup reduction $\widetilde{\mathfrak{gl}}(2)^+ \supset \widetilde{\mathfrak{sl}}(2)^+$. Choosing an appropriate symplectic section gives the reduced parametrization:

$$F = \frac{1}{\sqrt{2}}(\mathbf{x},\mathbf{y}), \quad G = \frac{1}{\sqrt{2}}(\mathbf{y},-\mathbf{x}) \tag{2.42}$$
$$\mathbf{x},\mathbf{y} \in \mathbf{R}^n.$$

The reduced symplectic form becomes

$$\omega = d\mathbf{x}^T \wedge d\mathbf{y}. \tag{2.43}$$

The reduced moment map is

$$\mathcal{N}_0(\lambda) = \tilde{J}^A(F,G) = \frac{\lambda}{2}\begin{pmatrix} -\sum_{i=1}^n \frac{x_i y_i}{\lambda-\alpha_i} & -\sum_{i=1}^n \frac{y_i^2}{\lambda-\alpha_i} \\ \sum_{i=1}^n \frac{x_i^2}{\lambda-\alpha_i} & \sum_{i=1}^n \frac{x_i y_i}{\lambda-\alpha_i} \end{pmatrix}. \tag{2.44}$$

Viewing this as defined on the symplectic vector space $\mathbb{R}^n \times \mathbb{R}^n$, there is a residual fibration generated by the finite group $(\mathbb{Z}_2)^n$ of reflections in the coordinate hyperplanes. The $\widetilde{\mathfrak{sl}}(2)^-$ character λY may be expressed as:

$$\lambda Y = \lambda \begin{pmatrix} a & b \\ c & -a \end{pmatrix} \in \widetilde{\mathfrak{sl}}(2)^{-*}, \tag{2.45}$$

and the resulting AKS flows involve isospectral deformations of elements of the form:

$$\mathcal{N}(\lambda) = \begin{pmatrix} a & b \\ c & -a \end{pmatrix} + \mathcal{N}_0(\lambda). \tag{2.46}$$

The Hamiltonians are chosen, as usual, from the AKS ring $\mathcal{I}^Y_{\mathrm{AKS}}(\widetilde{\mathfrak{sl}}(2)^*)$. In addition to the symmetry reductions already implemented, it is possible to impose further symplectic constraints of the form

$$f(\mathbf{x},\mathbf{y}) = 0, \quad g(\mathbf{x},\mathbf{y}) = 0, \quad \{f,\ g\} \neq 0, \tag{2.47}$$

and apply the standard methods for constrained systems. (Provided one of these functions is in the Poisson commutative ring $\mathcal{I}^Y_{\mathrm{AKS}}(\widetilde{\mathfrak{sl}}(2)^*)$, the constrained Hamiltonians will still commute.) The particular case of the above with

$$a = 0, \quad b = -\frac{1}{2}, \quad c = 0, \tag{2.48a}$$

$$f := \mathbf{x}^T\mathbf{x} - 1 = 0, \quad g := \mathbf{y}^T\mathbf{x} = 0, \tag{2.48b}$$

and Hamiltonian (1.30), gives the Neumann oscillator system. The invariant spectral curve is of the form

$$\det(\mathcal{L}(\lambda) - z\mathbb{I}) = z^2 + a(\lambda)\mathcal{P}(\lambda) = 0, \tag{2.49}$$

where

$$\mathcal{L}(\lambda) := \frac{a(\lambda)}{\lambda}\mathcal{N}(\lambda), \tag{2.50}$$

and $\mathcal{P}(\lambda)$ is generally a polynomial of degree $n-1$ or n, depending on whether $a^2 + bc$ vanishes or not.

2.5.2 The NLS Equation: Reduction to $\widetilde{\mathfrak{su}}(1,1)^+$

Again, we choose $r = 2$, $k_i = 1$, $n = N$. Similarly to the previous example, the zero moment map reduction under

$$G_A = Stab(A) = \mathbb{C}^\times \times \mathbb{C}^\times \times \cdots \times \mathbb{C}^\times \tag{2.51}$$

is equivalent the subgroup reduction $\widetilde{\mathfrak{gl}}(2,\mathbb{C})^+ \supset \widetilde{\mathfrak{sl}}(2,\mathbb{C})^+$. Choosing a suitable symplectic section gives the parametrization

$$F = \frac{1}{\sqrt{2}}(\mathbf{z},\mathbf{w}), \quad G = \frac{1}{\sqrt{2}}(\mathbf{w},-\mathbf{z}), \qquad (2.52)$$
$$\mathbf{z},\ \mathbf{w} \in \mathbb{C}^n.$$

The further reality conditions

$$\alpha_i = \overline{\alpha}_i, \quad \mathbf{z} = i\overline{\mathbf{w}}$$
$$F = \frac{1}{\sqrt{2}}(\mathbf{z}, i\overline{\mathbf{z}}), \quad G = \frac{1}{\sqrt{2}}(-i\overline{\mathbf{z}},\mathbf{z}) \qquad (2.53)$$

give the discrete reduction $\widetilde{\mathfrak{sl}}(2,\mathbb{C})^{+*} \supset \widetilde{\mathfrak{su}}(1,1)^{+*}$ as the fixed point set under an antilinear involution. On this real subspace, the symplectic form becomes

$$\omega = i d\overline{\mathbf{z}}^T \wedge d\mathbf{z}, \qquad (2.54)$$

and the reduced moment map has the form.

$$\mathcal{N}_0(\lambda) = \tilde{J}^A(F,G) = \frac{\lambda}{2}\begin{pmatrix} i\sum_{j=1}^n \frac{|z_j|^2}{\lambda-\alpha_j} & -\sum_{j=1}^n \frac{\overline{z}_j^2}{\lambda-\alpha_j} \\ -\sum_{j=1}^n \frac{z_j^2}{\lambda-\alpha_j} & -i\sum_{j=1}^n \frac{|z_j|^2}{\lambda-\alpha_j} \end{pmatrix}. \qquad (2.55)$$

In this case, we choose the character λY to vanish, so $\mathcal{N}(\lambda)$ coincides with $\mathcal{N}_0(\lambda)$. The commuting flows are generated by the pair of commuting Hamiltonians:

$$H_x = \frac{1}{2}\left[\frac{a(\lambda)}{\lambda^n}\lambda\ \mathrm{tr}(\mathcal{N}(\lambda)^2)\right]_0 \in \mathcal{I}^0_{\mathrm{AKS}}(\widetilde{\mathfrak{su}}(1,1)^{+*}) \qquad (2.56a)$$
$$H_t = \frac{1}{2}\left[\frac{a(\lambda)}{\lambda^n}\lambda^2\ \mathrm{tr}(\mathcal{N}(\lambda)^2)\right]_0 \in \mathcal{I}^0_{\mathrm{AKS}}(\widetilde{\mathfrak{su}}(1,1)^{+*}). \qquad (2.56b)$$

Further invariant constraints are added to ensure that the leading terms of the polynomial matrix (1.45) have the form given in eq. (1.49). Defining $\mathcal{L}(\lambda)$ again as in eq. (2.50), the resulting invariant spectral curve again has the form

$$\det(\mathcal{L}(\lambda) - z\mathbb{I}) = z^2 + a(\lambda)\mathcal{P}(\lambda) = 0, \qquad (2.57a)$$
$$\mathcal{L}(\lambda) := \frac{a(\lambda)}{\lambda}\mathcal{N}(\lambda), \qquad (2.57b)$$

where $\mathcal{P}(\lambda)$ now is of degree $n-2$. The Lax form (1.48a,b) of Hamilton's equations then follows from the AKS theorem, and the compatibility condition (1.51) gives the NLS equation (1.41).

2.5.3 Higher Rank Case. Two Component Coupled NLS System: Reduction to $\widetilde{\mathfrak{su}}(1,2)^+$

As an illustration of a system described by an algebra of higher rank, we consider the case of the coupled two component cubically nonlinear Schrödinger equation (*viz.* **[AHP]**, **[AHH2]**, **[AHH3]**):

$$iu_t + u_{xx} = 2u(|u|^2 + |v|^2) \tag{2.58a}$$

$$iv_t + v_{xx} = 2v(|u|^2 + |v|^2). \tag{2.58b}$$

In this case, we take $r = 3$ and $k_i = 1$ for all i, so $n = N$. The process of discrete and continuous symmetry reduction is applied analogously to the preceding case, giving the sequence $\widetilde{\mathfrak{gl}}(3,\mathbb{C})^{+*} \supset \widetilde{\mathfrak{sl}}(3,\mathbb{C})^{+*} \supset \widetilde{\mathfrak{su}}(1,2)^{+*}$. The reduced form of the resulting pair of $n \times 3$ matrices (F, G) is

$$F = (\rho, \eta, \zeta), \quad G = (\rho, -\eta, -\zeta), \tag{2.59}$$

where η, $\zeta \in \mathbb{C}^n$ is a pair of complex n–component column vectors and $\rho \in \mathbb{R}^n$ is a real column vector with components

$$\rho_i = \sqrt{|\eta_i|^2 + |\zeta_i|^2}, \quad i = 1, \dots n. \tag{2.60}$$

The reduced symplectic form is

$$\omega = i(d\overline{\eta}^T \wedge d\eta + d\overline{\zeta}^T \wedge d\zeta), \tag{2.61}$$

so the components $(\eta_i, \zeta_i)_{i=1,\dots n}$ and their complex conjugates provide a canonical coordinate system on $\mathcal{O}_{\mathcal{N}_0}$. The reduced moment map has the form

$$\begin{aligned} \mathcal{N}_0(\lambda) &= \tilde{J}^A(F,G) \\ &= -i\lambda \sum_{j=1}^{n} \frac{1}{\lambda - \alpha_i} \begin{pmatrix} \rho_i^2 & \eta_i\rho_i & \zeta_i\rho_i \\ -\overline{\eta}_i\rho_i & -|\eta_i|^2 & -\overline{\eta}_i z_i \\ -\overline{\zeta}_i\rho_i & -\overline{\zeta}_i\eta_i & -|\zeta_i|^2 \end{pmatrix}, \end{aligned} \tag{2.62}$$

so the coadjoint orbit may be identified with $\mathbb{C}^n \times \mathbb{C}^n$. Again we choose the character λY to vanish, so $\mathcal{N}(\lambda)$ coincides with $\mathcal{N}_0(\lambda)$. As for the single component NLS equation, the commuting pair of Hamiltonians for the two component CNLS case is chosen to be

$$H_x = \frac{1}{2}\left[\frac{a(\lambda)}{\lambda^n}\lambda \operatorname{tr}(\mathcal{N}(\lambda)^2)\right]_0 \in \mathcal{I}(\widetilde{\mathfrak{su}}(1,2)^{+*}) \tag{2.63a}$$

$$H_t = \frac{1}{2}\left[\frac{a(\lambda)}{\lambda^n}\lambda^2 \operatorname{tr}(\mathcal{N}(\lambda)^2)\right]_0 \in \mathcal{I}(\widetilde{\mathfrak{su}}(1,2)^{+*}). \tag{2.63b}$$

Defining, as before,

$$\mathcal{L}(\lambda) := \frac{a(\lambda)}{\lambda}\mathcal{N}_0(\lambda) = L_0\lambda^{n-1} + L_1\lambda^{n-2} + \cdots + L_{n-1}, \tag{2.64}$$

further invariant constraints must also be imposed, implying that the leading terms are of the form (cf. [**AHP**]):

$$L_0 = \frac{i}{3}\begin{pmatrix} 2 & 0 & 0 \\ 0 & -1 & 0 \\ 0 & 0 & -1 \end{pmatrix}, \quad L_1 = \begin{pmatrix} 0 & \overline{u} & \overline{v} \\ u & 0 & 0 \\ v & 0 & 0 \end{pmatrix},$$
$$L_2 = i\begin{pmatrix} |u|^2 + |v|^2 & -\overline{u}_x & -\overline{v}_x \\ u_x & -|u|^2 & -\overline{v}u \\ v_x & -\overline{u}v & -|v|^2 \end{pmatrix}. \tag{2.65}$$

The Lax equations generated by the Hamiltonians (2.63a,b) have the same form as eqs. (1.48a,b), and the compatibility conditions (1.51) are equivalent to the CNLS system (2.58a,b). The invariant spectral curve in this case is a three sheeted branched covering of $\mathbb{P}^1$ determined by an equation of the form

$$\det(\mathcal{L}(\lambda) - z\mathbb{I}) = z^3 + a(\lambda)z\mathcal{P}(\lambda) + a(\lambda)^2\mathcal{Q}(\lambda) = 0, \tag{2.66}$$

where $\mathcal{P}(\lambda)$ and $\mathcal{Q}(\lambda)$ are polynomials of degrees $n-2$ and $n-3$, respectively.

3. Spectral Darboux Coordinates and Liouville-Arnold Integration

In this section, the general method of linearization of AKS flows in rational coadjoint orbits will be explained. For simplicity, the spectral properties of the matrix A will be chosen as in Section 2.2, but the method is equally valid in the more general case (see [**AHH3**]). It consists of two steps. First, a suitable generalization of the hyperellipsoidal coordinates encountered in the examples of Section 1 is introduced, the *spectral Darboux coordinates* (Theorem 3.2) associated to the invariant spectral curve $\mathcal{C}$. These consist of families of canonical coordinates on coadjoint orbits $\mathcal{O}_{\mathcal{N}_0}$ of the type discussed in the preceding section, which are naturally associated to the spectral data of the matrix $\mathcal{N}(\lambda)$. The second step consists of using a Liouville generating function S to compute the canonical transformation to coordinates conjugate to the spectral invariants, in which the flow becomes linear. It turns out that for all Hamiltonians in the AKS ring $\mathcal{I}^Y_{\text{AKS}}$ this generating function, defined with respect to the natural isospectral Lagrangian foliation of $\mathcal{O}_{\mathcal{N}_0}$, may be expressed within the spectral

Darboux coordinate system in completely separated form (Theorem 3.3). It follows from the construction that this transformation is given in terms of abelian integrals, showing that, in the general case, the Abel map yields a linearization of the flows on the Jacobi variety $\mathcal{J}(\mathcal{C})$ of the spectral curve. One thus arrives at the algebro-geometric linearization results (*viz.* **[Du]**, **[KN]**, **[AvM]**) entirely through classical Hamiltonian methods.

3.1 Phase Space and Group Actions

In the following, the phase space will initially be thought of as a coadjoint orbit $\mathcal{O}_{\mathcal{N}_0}$ within the image of a moment map of the type introduced in Section 2.

$$\tilde{J}^A : M \longrightarrow \tilde{\mathfrak{g}}^{+*} \tag{3.1a}$$

$$\tilde{J}^A : (F, G) \longmapsto \lambda G^T (A - \lambda \mathbb{I}_r)^{-1} F. \tag{3.1b}$$

The image defines a finite dimensional Poisson submanifold

$$\mathrm{Im}(\tilde{J}^A) := \tilde{\mathfrak{g}}_A \subset \tilde{\mathfrak{g}}^{+*} \tag{3.2}$$

which, in the simplest case, consists of elements of the form

$$\tilde{\mathfrak{g}}_A = \{\mathcal{N}_0(\lambda) = \lambda \sum_{i=1}^{n} \frac{N_i}{\lambda - \alpha_i}\}, \tag{3.3}$$

where the ranks $\{k_i\}_{i=1,\dots n}$ of the residue matrices N_i coincide with the multiplicities of the eigenvalues $\{\alpha_i\}_{i=1,\dots n}$ of the diagonal $N \times N$ matrix

$$A = \mathrm{diag}(\alpha_i, \dots, \alpha_i, \dots \alpha_n). \tag{3.4}$$

The coadjoint action of the loop group $\tilde{\mathfrak{G}}^+$ on $\tilde{\mathfrak{g}}_A$ in this case becomes equivalent to the coadjoint action of $(Gl(r))^n$ on $(\mathfrak{gl}(r)^*)^n$, obtained by evaluating the group element $g(\lambda) \in \tilde{\mathfrak{G}}^+$ at the poles $\lambda = \alpha_i$:

$$g : \{N_i\} \longmapsto \{g(\alpha_i) N_i g(\alpha_i)^{-1}\}. \tag{3.5}$$

It follows that the $\mathrm{Ad}^*_{\tilde{\mathfrak{G}}^+}$-orbits are determined as simultaneous level sets of the Casimir invariants of the separate residue matrices N_i under this action:

$$\mathcal{O}_{\mathcal{N}_0} = \{\lambda \sum_{i=1}^{n} \frac{N_i}{\lambda - \alpha_i} \mid \mathrm{tr} N_i^l = c_{il},\ l = 1, \dots k_i\}. \tag{3.6}$$

The equations of motion induced by any element of the AKS ring $\phi \in \mathcal{I}^Y_{\mathrm{AKS}}$ have the Lax form:

$$\frac{d\mathcal{N}(\lambda)}{dt} = [d\Phi(\mathcal{N})_+, \mathcal{N}], \tag{3.7}$$

where

$$\phi = \Phi|_{\lambda Y + \widetilde{\mathfrak{g}}_A} \tag{3.8}$$

is the restriction of the Ad^*-invariant element $\Phi \in \mathcal{I}(\widetilde{\mathfrak{g}}^*)$ to the subspace consisting of elements of the form

$$\mathcal{N}(\lambda) = \lambda Y + \mathcal{N}_0(\lambda), \quad \mathcal{N}_0 \in \widetilde{\mathfrak{g}}_A. \tag{3.9}$$

Define the $\mathfrak{gl}(r)$-valued polynomial

$$\mathcal{L}(\lambda) := \frac{a(\lambda)}{\lambda}\mathcal{N}(\lambda) = a(\lambda)Y + L_0\lambda^{n-1} + \cdots + L_{n-1}, \tag{3.10}$$

where

$$a(\lambda) := \prod_{i=1}^{n}(\lambda - \alpha_i) \tag{3.11}$$

is the minimal polynomial of A. This satisfies the equivalent Lax equation

$$\frac{d\mathcal{L}(\lambda)}{dt} = [d\Phi(\mathcal{N})_+, \mathcal{L}]. \tag{3.12}$$

The invariant spectral curve $\mathcal{C}_0$ is then determined by the characteristic equation

$$\det\left(\mathcal{L}(\lambda) - z\mathbb{I}_r\right) = 0 \tag{3.13}$$

which, after suitable compactification, is viewed as an r-fold branched cover of $\mathbb{P}^1$, possibly having singularities over the points $\lambda = \alpha_i$ due to the $r - k_i$ fold multiplicity of zero eigenvalues. Other singularities could, of course, also occur, but for simplicity we again place ourselves in a "generic" situation in order to be able to give the main results in as explicit form as possible, and therefore exclude this possibility. The essential results remain valid without such simplifying assumptions, but explicit formulae for the spectral polynomial, genus, dimensions of orbits, and form of the abelian differentials must be modified accordingly.

We assume henceforth, for simplicity, that the spectral curves $\mathcal{C}_0$ have no singularities other than those that arise over $\{\lambda = \alpha_i\}_{i=1,\dots n}$, if $k_i < r-1$, due to the multiple zero eigenvalues of the residue matrices N_i. This implies in particular that the N_i's, while not necessarily diagonalizable, must lie on orbits that have the same dimensions as the diagonalizable orbits whose nonzero eigenvalues are distinct; namely, $k_i(2r-k_i-1)$. We also assume that one of two conditions holds, which excludes further singularities over $\lambda=\infty$:

Case (a) $Y=0$ and L_0 lies on a $Gl(r)$ orbit of the same dimension $(r(r-1))$ as those with simple spectrum.
Case (b) $Y\neq 0$ and lies on a $Gl(r)$ orbit of the same dimension as those with simple spectrum.

Remark. An effect of this assumption is to eliminate from consideration the example 2.5.3, which has singularities over $\lambda=\infty$. However, this case may also be dealt with (*viz.* [**AHH3**]), by imposing a further set of symplectic constraints defining a "generic" deformation class of admissible spectral curves.

In some cases, it is not the orbit $\mathcal{O}_{\mathcal{N}_0}$ itself that is the relevant phase space, but its reduction under the Hamiltonian action consisting of conjugation by the stability subgroup $G_Y = \text{Stab}\,(Y)\subset Gl(r)$:

$$g:\mathcal{N}_0(\lambda)\longmapsto g\mathcal{N}_0(\lambda)g^{-1},\quad g\in G_Y. \tag{3.14}$$

The corresponding moment map is just the leading term in $\mathcal{N}_0(\lambda)$:

$$J(\mathcal{N}_0):=L_0=\sum_{i=1}^{n}N_i, \tag{3.15}$$

restricted to the subalgebra $\mathfrak{g}_Y:=\text{stab}(Y)\subset\mathfrak{gl}(r)$. Another case of interest, particularly when $Y=0$, consists of restricting to a symplectic submanifold $\mathcal{O}^S_{\mathcal{N}_0}\subset\mathcal{O}_{\mathcal{N}_0}$ determined by the zero level set of the components of L_0 within the annihilator of a Cartan subalgebra. (This is symplectic at regular elements L_0.) For future reference, we list the various subcases of interest.

Case (a) $Y=0$. In this case, all the elements $\phi\in\mathcal{I}|_{\tilde{\mathfrak{g}}^{+*}}$ in the AKS ring are invariant under the full $Gl(r)$ action (3.14), and all components of L_0 are conserved. We may therefore reduce by the full group $Gl(r)$ or any of its subgroups. The two subcases of greatest interest are:

Case (a.1) *Complete reduction at a regular point* $L_0=\mu_0\in\mathfrak{gl}(r)^*$. The reduced manifold is then

$$\mathcal{O}^{\text{red}}_{\mathcal{N}_0}=J^{-1}(\mu_0)/G_0, \tag{3.16}$$

where $G_0\subset Gl(r)$ is the stabilizer of μ_0. The dimension of the reduced orbit is:

$$\dim\,\mathcal{O}^{\text{red}}_{\mathcal{N}_0}=\dim\,\mathcal{O}_{\mathcal{N}_0}-(r-1)(r+2). \tag{3.17}$$

Case (a.2) *Symplectic invariant manifold.* We take the zero level set of all components of L_0 in the annihilator of a Cartan subalgebra, (e.g., we choose L_0 to be diagonal). Denote this submanifold, which is symplectic at all regular values of L_0, as:

$$\mathcal{O}^{S}_{\mathcal{N}_0} := \{\mathcal{N}_0 \in \mathcal{O}_{\mathcal{N}_0} | L_0 \in \mathcal{T} \quad \text{(Cartan subalgebra)}\}. \tag{3.18}$$

Its dimension is

$$\dim \mathcal{O}^{S}_{\mathcal{N}_0} = \dim \mathcal{O}_{\mathcal{N}_0} - r(r-1) = \dim \mathcal{O}^{\text{red}}_{\mathcal{N}_0} + 2(r-1). \tag{3.19}$$

Case (b) $Y \neq 0$. In this case, the elements $\phi \in \mathcal{I}|_{\tilde{\mathfrak{g}}^{+*}+\lambda Y}$ of the AKS ring are only invariant under the action of the stabilizer $G_Y = \mathrm{Stab}(Y) \subset Gl(r)$. Two cases of special interest arise:

Case (b.1) The full orbit $\mathcal{O}_{\mathcal{N}_0}$ (i.e., no reduction).

Case (b.2) The reduction of $\mathcal{O}_{\mathcal{N}_0}$ under the full stabilizer $G_Y \subset Gl(r)$ of a regular element $Y \in gl(r)$, taken at a value $L_0|_{\mathfrak{g}_Y} = \mu_0 \in \mathfrak{g}^*_Y$. The group G_Y is a maximal abelian subgroup with $r-1$ dimensional orbits and $G_0 = G_Y$. The reduced orbit is denoted

$$\mathcal{O}^{Y,\text{red}}_{\mathcal{N}_0} = J|^{-1}_{\mathfrak{g}_Y}(\mu_0)/G_Y, \tag{3.20}$$

and has dimension

$$\dim \mathcal{O}^{Y,\text{red}}_{\mathcal{N}_0} = \dim \mathcal{O}_{\mathcal{N}_0} - 2(r-1). \tag{3.21}$$

3.2 Structure of the Spectral Curve

The affine part of the spectral curve $\mathcal{C}_0$ is determined by the characteristic equation (3.13). Taking into account the ranks of the residue matrices $\{N_i\}_{i=1,\dots n}$ in (3.3), we see that the characteristic polynomial has the general form

$$\begin{aligned}\mathcal{P}(\lambda, z) &= \det\left(\mathcal{L}(\lambda) - z\mathbb{I}_r\right) \\ &= (-z)^r + z^{r-1}\mathcal{P}_1(\lambda) + \sum_{j=2}^{r} A_j(\lambda)\mathcal{P}_j(\lambda) z^{r-j},\end{aligned} \tag{3.22}$$

where

$$A_j(\lambda) := \prod_{i=1}^{n} (\lambda - \alpha_i)^{\max(0, j-k_i)}, \quad \text{rank } \mathcal{L}(\alpha_i) = k_i. \tag{3.23}$$

This shows that near $\lambda \sim \infty$, we have

$$z \sim \lambda^m, \quad m := \begin{cases} n & \text{if } Y = 0 \text{ (case (a))} \\ n-1 & \text{if } Y \neq 0 \text{ (case (b))}. \end{cases} \tag{3.24}$$

This suggests a compactification, not within $\mathbb{P}^2$, but rather in the total space of a line bundle over $\mathbb{P}^1 = U_0 \cup U_\infty$ (where U_0, U_∞ denote the open disks obtained by deleting $\lambda = \infty$ and $\lambda = 0$, respectively), with coordinate pairs (λ, z) over U_0 and $(\widetilde{\lambda}, \widetilde{z})$ over U_∞ related by

$$(\lambda, z) \longmapsto (\widetilde{\lambda} = \frac{1}{\lambda}, \ \widetilde{z} = \frac{z}{\lambda^m}) \quad (\text{over } U_0 \cap U_\infty). \tag{3.25}$$

This is just the total space $\mathcal{T}$ of the bundle $\mathcal{O}(m) \to \mathbb{P}^1$ whose sheaf of sections consists of homogeneous functions of degree m. The transformation (3.25) extends the affine curve $\mathcal{C}_0$ defined by (3.13) over $\lambda = \infty$, defining the compactification:

$$\mathcal{C}_0 \hookrightarrow \mathcal{C} \hookrightarrow \mathcal{T}. \tag{3.26}$$

The possible spectral curves so arising are branched r-sheeted covers of $\mathbb{P}^1$, which within any given orbit of type (3.3), have z-values over each $\lambda = \alpha_i$ that are fixed (being Casimir invariants of the coadjoint action (3.5)). Of these, there are k_i nonsingular points $(\lambda = \alpha_i, z = \zeta_{ia})_{a=1,\ldots k_i}$ corresponding to the nonzero eigenvalues of $\mathcal{L}(\alpha_i)$, and the point $(\lambda = \alpha_i, z = 0)$, which generically is an $r - k_i$-fold ordinary singular point corresponding to the $r - k_i$-fold zero eigenvalue. Figure 3.1 gives a visualization of the spectral curves $\mathcal{C}$, embedded in $\mathcal{T}$, as branched coverings of $\mathbb{P}^1$, constrained to pass through these points.

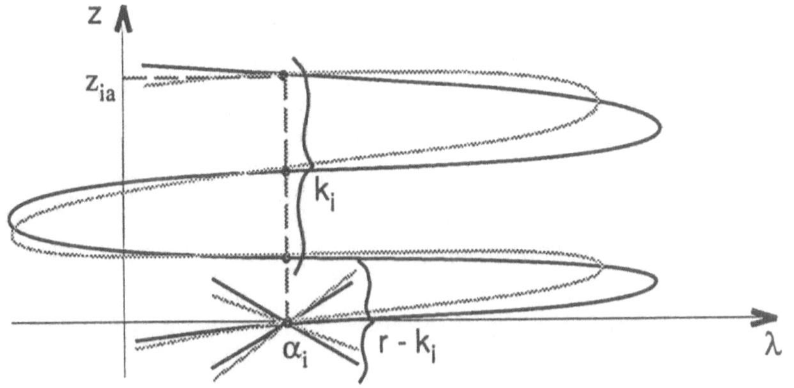

Figure 3.1

The detailed structure may be expressed more precisely by writing the form of the characteristic polynomial $\mathcal{P}(\lambda, z)$ for any spectral curve $\mathcal{C}$ in a neighborhood of a given curve $\mathcal{C}_R$ as a perturbation of the characteristic polynomial $\mathcal{P}_R(\lambda, z)$ defining $\mathcal{C}_R$.

Proposition 3.1 ([AHH3]). *In a neighborhood of the point $\mathcal{N}_R \in \mathcal{O}_{\mathcal{N}_0}$ with characteristic polynomial $\mathcal{P}_R(\lambda, z)$, the characteristic polynomial has the form:*

$$\mathcal{P}(\lambda, z) \equiv \mathcal{P}_R(\lambda, z) + a(\lambda) \sum_{j=2}^{r} a_j(\lambda) p_j(\lambda) z^{r-j}, \tag{3.27}$$

where

$$a_j(\lambda) = \prod_{i=1}^{n} (\lambda - \alpha_i)^{\max(0, j-k_i-1)} \tag{3.28}$$

$$p_j(\lambda) =: \sum_{a=0}^{\delta_j} P_{ja} \lambda^a, \tag{3.29}$$

and $\{p_j(\lambda)\}_{j=2,\dots r}$ are polynomials of degree:

$$\delta_j \equiv \deg p_j(\lambda) = \begin{cases} d_j - j & \text{if } Y = 0 \\ d_j & \text{if } Y \neq 0 \end{cases} \tag{3.30a}$$

$$d_j \equiv \sum_{i=1}^{n} \min(j-1, k_i). \tag{3.30b}$$

The number of independent spectral parameters $\{P_{ja}\}$, $(a = 0, \dots \delta_j + n - m - 1,\ j = 2, \dots r)$ is thus:

$$d = \tilde{g} + r - 1, \tag{3.31}$$

where

$$\tilde{g} = \frac{1}{2}(r-1)(mr-2) - \frac{1}{2} \sum_{i=1}^{n} (r - k_i)(r - k_i - 1) \tag{3.32}$$

is the genus of the (partially) desingularized spectral curve $\mathcal{C}$ obtained by separating branches at $\{\alpha_i, 0\}$. In a neighborhood of any generic point on $\mathcal{O}_{\mathcal{N}_0}$, these spectral invariants are all independent.

The complete integrability of the systems under consideration on the various coadjoint orbits $\mathcal{O}_{\mathcal{N}_0}$, and the reductions $\mathcal{O}_{\mathcal{N}_0}^{\text{red}}$, $\mathcal{O}_{\mathcal{N}_0}^{Y,\text{red}}$ and symplectic submanifolds $\mathcal{O}_{\mathcal{N}_0}^{S}$ thereof, may be seen from the following Table of generic dimensions for the various cases discussed above. (Note that the P_{ia}'s referred to do not include the leading terms of the polynomials $p_j(\lambda)$ in eq. (3.29).) Recall that the value of the genus $\tilde{g}$ of the desingularized curve $\mathcal{C}$ given in Proposition 3.1 depends on the value m, which is different in the cases (a) and (b):

$$\tilde{g} = \tilde{g}(m), \quad m = \begin{cases} n-1 & \text{for case (a)} \\ n & \text{for case (b).} \end{cases} \tag{3.33}$$

Table of Dimensions

Case	Dimension	$\#P_{ia}$'s	$\#P_i$'s
(a)	$\dim \mathcal{O}_{\mathcal{N}_0} = 2\widetilde{g} + (r-1)(r+2)$	$\widetilde{g}$	$r-1$
(a.1)	$\dim \mathcal{O}^{\mathrm{red}}_{\mathcal{N}_0} = 2\widetilde{g}$	$\widetilde{g}$	0
(a.2)	$\dim \mathcal{O}^{S}_{\mathcal{N}_0} = 2(\widetilde{g} + r - 1)$	$\widetilde{g}$	$r-1$
(b.1)	$\dim \mathcal{O}_{\mathcal{N}_0} = 2(\widetilde{g} + r - 1)$	$\widetilde{g}$	$r-1$
(b.2)	$\dim \mathcal{O}^{Y,\mathrm{red}}_{\mathcal{N}_0} = 2\widetilde{g}$	$\widetilde{g}$	0

Here, the notation $\{P_i\}_{i=2,\dots,r}$ is used to denote the components of L_0 evaluated on a basis of the relevent Cartan subalgebra (not including the trivial central element, which is a Casimir invariant, and hence constant on orbits). These are also elements of the ring $\mathcal{I}^Y_{\mathrm{AKS}}$, corresponding to the leading terms in the polynomials $p_j(\lambda)$ in eq. (3.29) for case (a) , and the next to leading terms for case (b) (the leading terms being constant in the latter case), but they are listed separately since, in cases (a.1) and (b.2), they are fixed through the Marsden-Weinstein reduction procedure, and hence do not contribute to the number of independent invariants on the reduced spaces. Moreover, these elements enter again in Section 3.4 when defining the *spectral Darboux coordinates* for these two cases. It follows from the dimensions in the Table and the independence of the commuting invariants that cases (a.1), (a.2), (b.1) and (b.2) all give completely integrable systems.

3.3 Spectral Lagrangian Foliation

The Lagrangian foliation given by fixing simultaneous level sets of the invariants in the ring $\mathcal{I}^Y_{\mathrm{AKS}}$ of spectral invariants (i.e., fixing the spectral curve $\mathcal{C}$) is depicted below in Figure 3.2 for the various cases discussed above.

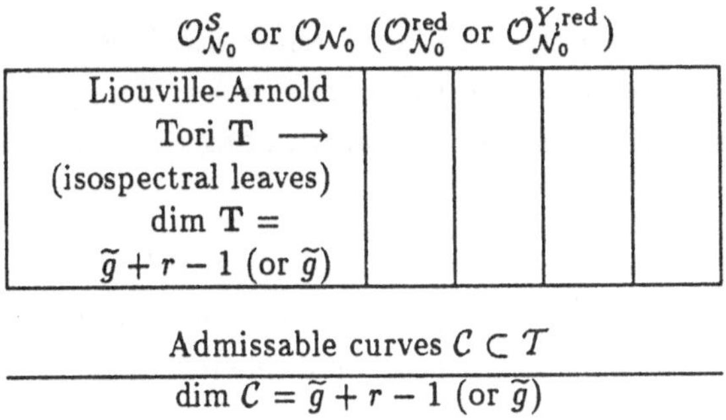

Figure 3.2

We may summarize the relevant spectral data associated to each element $\mathcal{N} \in \lambda Y + \widetilde{\mathfrak{g}}_A$ as follows:

• A *spectral curve* C (r–fold branched cover of $\mathbb{P}^1$) defined by the characteristic equation:

$$\mathcal{P}(\lambda, z) = \det(\mathcal{L}(\lambda) - z\mathbb{I}_r) = 0, \tag{3.34}$$

(after suitable compactification and desingularization). The $r-1$ points over $\lambda = \infty$ are determined by the leading terms of the polynomials $p_j(\lambda)$ of Proposition 3.1.

• An *eigenvector subspace*: $[V(\lambda, z)] \subset \mathbb{C}^r$ at each point in C which, by our genericity assumptions, is one dimensional.

Together, these determine an *eigenvector line bundle* $\check{E} \rightarrow C$, and its corresponding dual bundle $E \rightarrow C$. The latter may be shown (*viz.* **[AHH2]**, **[AHH3]**) to be generally of degree

$$\deg{(E)} = \widetilde{g} + r - 1, \tag{3.35}$$

and hence, by the Riemann-Roch theorem, to have an r–dimensional space of sections. Conversely, it turns out that this data is sufficient to reconstruct the matrix $\mathcal{L}(\lambda)$ (and hence $\mathcal{N}(\lambda)$) *up to conjugation by an element of* $Gl(r)$; that is, it is sufficient to determine the projected point in the reduced orbit $\mathcal{O}_{\mathcal{N}_0}^{Y,\mathrm{red}}$ or $\mathcal{O}_{\mathcal{N}_0}^{\mathrm{red}}$, but not the element $\mathcal{N}(\lambda)$ itself.

In order to reconstruct the element $\mathcal{N}(\lambda)$, it is necessary to add some further spectral data, consisting of a *framing* at $\lambda = \infty$; that is, a basis of sections $\{\sigma_i \in H^0(C, E)\}_{i=1,\dots r}$ of the bundle $E \rightarrow C$, chosen to vanish, e.g. at all but one of the r points $\{\infty_i\}_{i=1,\dots r}$ over $\lambda = \infty$ (for the case where the Cartan subalgebra in question consists of the diagonal matrices).

$$\sigma_i(\infty_j) = 0, \quad i \neq j. \tag{3.36}$$

This adds $r-1$ dimensions to the fibres, (since framings related by $\{\widetilde{\sigma}_i = \kappa\sigma_i\}$ are equivalent). Furthermore, the spectrum over $\lambda = \infty$ in the class of admissible spectral curves must be left undetermined, adding $r-1$ dimensions to the base space in Figure 3.2.

More generally, it is insufficient to just consider line bundles, since this excludes the possibility of degeneracy in the spectrum and further singular points. The appropriate generalization consists of a coherent sheaf defined by the exact sequence

$$0 \longrightarrow \mathcal{O}(-m)^{\oplus r} \xrightarrow{\mathcal{L}^T(\lambda) - z\mathbb{I}} \mathcal{O}^{\oplus r} \longrightarrow E \longrightarrow 0, \tag{3.37}$$

where $\mathcal{O}(-m)$ denotes the sheaf obtained by pulling back the corresponding sheaf over $\mathbb{P}^1$ to $\mathcal{T}$. In the case of bundles, the exact sequence (3.37) just means that the dual space to the space of eigenvectors over the spectral curve is given by the cokernel of the linear map defined by $\mathcal{L}^T(\lambda) - z\mathbb{I}_r$. A more complete discussion of the significance of this construction may be found in **[AHH2]**, **[AHH3]**.

3.4 Spectral Darboux Coordinates

In this section we give a method for constructing the appropriate Darboux coordinates naturally associated to the spectral data discussed above, in which the Hamiltonians in the spectral ring $\mathcal{I}^Y_{\text{AKS}}$ determine a Liouville generating function in completely separated form. First, we shall give a purely computational description of these coordinates in terms of simultaneous solutions of polynomial equations. The significance of this prescription in terms of the eigenvector line bundles of the preceding section will follow.

Let

$$M(\lambda,\zeta) := \frac{\mathcal{N}(\lambda)}{\lambda} - \zeta\mathbb{I}_r, \tag{3.38}$$

and denote by $\widetilde{M}(\lambda,\zeta)$ the transpose of the matrix of cofactors. Then, over the spectral curve defined by the characteristic equation (3.34), the columns of $\widetilde{M}(\lambda,\zeta)$ are the eigenvectors of $\mathcal{L}(\lambda)$ (or $\mathcal{N}(\lambda)$), and hence, generically, these are all proportional; i.e. $\widetilde{M}(\lambda,\zeta)$ has rank 1. Let $V_0 \in \mathbf{C}^r$ be a fixed vector, and denote the solutions to the system of polynomial equations

$$\widetilde{M}(\lambda,\zeta)V_0 = 0, \quad V_0 \in \mathbf{C}^r \tag{3.39}$$

as $\{\lambda_\mu, \zeta_\mu\}_{\mu=1,\dots}$. Note that, due to the rank condition, there really are only two independent equations here, the other $r-2$ following as linear consequences.

The significance of these equations in relation to the spectral data is quite simple; they are the conditions that a section of the dual eigenvector line bundle $E \to C$ should vanish. The solutions give the zeros of the components of the eigenvector determined by the vector V_0. As is well known in algebraic geometry, giving the divisor of zeros of any section of a line bundle amounts to giving the linear equivalence class of the bundle itself. It follows, since the bundle $E \to C$ is of degree $\widetilde{g}+r-1$, that there will in general be $\widetilde{g}+r-1$ zeros. However this is not necessarily the number of solutions to (3.39), since some of the zeros may be over $\lambda = \infty$. In fact, we may distinguish two cases of particular interest as follows. In order to characterize the spectrum over $\lambda = \infty$, define

$$\widetilde{\mathcal{L}}(\lambda) := \mathcal{L}(\lambda)/\lambda^m. \tag{3.40}$$

Case (i). V_0 is an eigenvector of $\tilde{\mathcal{L}}(\infty)$. In this case, $r-1$ of the zeros are over $\lambda=\infty$ (the only point omitted over ∞ being the one corresponding to the eigenvalue of V_0.) Hence, there are only $\tilde{g}$ finite solutions pairs $\{\lambda_\mu,\zeta_\mu\}_{\mu=1,\ldots\tilde{g}}$, and these are generically independent, when viewed as functions on the phase space $\mathcal{O}^{\mathcal{S}}_{\mathcal{N}_0}$ (case (a)) or $\mathcal{O}_{\mathcal{N}_0}$ (case (b)). Moreover, they are invariant under the action of the reduction group for both cases, since this leaves the space $[V_0]$ invariant, and hence they project to functions on the reduced space $\mathcal{O}^{\rm red}_{\mathcal{N}_0}$ (case (a)) or $\mathcal{O}^{Y,{\rm red}}_{\mathcal{N}_0}$ (case (b)). In view of the dimensions given in the Table of Dimensions, Section 3.2, the projected functions provide coordinate systems on the reduced spaces. On the prereduced spaces, we must supplement these with a further $r-1$ pairs of coordinate functions, which we define as follows. Choose a basis where L_0 (case (a)) or Y (case (b)) is diagonal, and $V_0=(1,0\ldots,0)^T$. Then let

$$q_i := \begin{cases} \ln(L_1)_{i1}+\frac{1}{2}\sum_{j=2,j\neq i}^{r}\ln(p_i-p_j) & \text{for case (a)} \\ \ln(L_0)_{i1} & \text{for case (b)} \end{cases} \tag{3.41a}$$

$$P_i := (L_0)_{ii}, \quad i=2,\ldots r. \tag{3.41b}$$

The pairs $\{q_i,P_i\}_{i=2,\ldots r}$ provide the remaining coordinates required.

Case (ii). V_0 is not an eigenvector of $\tilde{\mathcal{L}}(\infty)$ and, furthermore, $V_0\notin \operatorname{Im}(\tilde{\mathcal{L}}(\infty)-\tilde{z}_i\mathbb{I})$ for any eigenvalue $\tilde{z}_i(\infty)$ of $\tilde{\mathcal{L}}(\infty)$. In this case, none of the zeros are over $\lambda=\infty$, and there are generically $\tilde{g}+r-1$ independent solution pairs $\{\lambda_\mu,\zeta_\mu\}_{\mu=1,\ldots\tilde{g}+r-1}$ of eq. (3.39). These then provide a coordinate system on the prereduced space $\mathcal{O}^{\mathcal{S}}_{\mathcal{N}_0}$ (case (a)) or $\mathcal{O}_{\mathcal{N}_0}$ (case (b)).

We then have the following fundamental result.

Theorem 3.2. *1. If $V_0\notin Im(\tilde{\mathcal{L}}(\infty)-\tilde{z}_i\mathbb{I})$ for any eigenvalue $\tilde{z}_i(\infty)$ of $\tilde{\mathcal{L}}(\infty)$, the functions $\{\lambda_\mu,\zeta_\mu\}_{1,\ldots\tilde{g}+r-1}$ define a Darboux coordinate system on $\mathcal{O}^{\mathcal{S}}_{\mathcal{N}_0}$ (case (a) or $\mathcal{O}_{\mathcal{N}_0}$ (case (b)). The orbital symplectic form is therefore:*

$$\omega_{\rm orb}=\sum_{\mu=1}^{\tilde{g}+r-1} d\lambda_\mu\wedge d\zeta_\mu. \tag{3.42a}$$

2. If V_0 is an eigenvector of L_0 (case (a) or Y (case (b)), the functions $\{\lambda_\mu,\zeta_\mu\}_{1,\ldots\tilde{g}}$ project to Darboux coordinates on the reduced spaces $\mathcal{O}^{\rm red}_{\mathcal{N}_0}$ (case (a)) or $\mathcal{O}^{Y,{\rm red}}_{\mathcal{N}_0}$ (case (b)), so the reduced orbital symplectic form is:

$$\omega_{\rm red}=\sum_{\mu=1}^{\tilde{g}} d\lambda_\mu\wedge d\zeta_\mu. \tag{3.42b}$$

3. If V_0 is an eigenvector of L_0 (case (a)) or Y (case (b)), the functions $\{\lambda_\mu, \zeta_\mu, q_i, P_i\}_{\mu=1,\ldots\tilde{g}, i=2,\ldots r}$ define a Darboux coordinate system on $\mathcal{O}^S_{\mathcal{N}_0}$ (case (a) or $\mathcal{O}_{\mathcal{N}_0}$ (case (b)), so the orbital symplectic form is:

$$\omega_{\text{orb}} = \sum_{\mu=1}^{\tilde{g}} d\lambda_\mu \wedge d\zeta_\mu + \sum_{i=2}^{r} dq_i \wedge dP_i. \tag{3.42c}$$

In the following section, we consider a number of elementary examples of the above theorem. We shall see that the resulting *spectral Darboux coordinates* do, indeed, generalize the hyperellipsoidal coordinates that were encountered in the examples of Section 1.

3.5 Examples

We begin by considering the simplest possible case; namely, where $\mathcal{N}_0(\lambda)$ has only one pole, at $\lambda = \alpha_1$, and $r = 2$ or 3. This just corresponds to coadjoint orbits of the finite dimensional Lie algebras $\mathfrak{sl}(2)$ and $\mathfrak{sl}(3)$. Then we consider the case $\widetilde{\mathfrak{sl}}(2,\mathbf{R})^+$ for arbitrary n, with $\text{rank}(N_i) = k_i = 1$ for all $i = 1, \ldots n$. This reproduces the hyperellipsoidal coordinates for the finite dimensional examples of Secs. 1.1 and 2.5.1 (cf. [Mo], such as the Neumann oscillator. Finally, we consider the case $\widetilde{\mathfrak{su}}(1,1)^+$, which provides the appropriate complex coordinates for the nonlinear Schrödinger equation, as discussed in Secs. 1.2 and 2.5.2.

(a) *Single poles:* $n = 1$

(a.1) Take $\mathfrak{g} = \mathfrak{sl}(2,\mathbf{R})$, and (without loss of generality), $\alpha_1 = 0$. Then the dimension of a generic orbit is $\dim \mathcal{O}_{\mathcal{N}_0} = 2$. We parametrize $\mathcal{N}_0(\lambda)$ as follows:

$$\mathcal{N}_0(\lambda) = \frac{\lambda N_1}{\lambda - \alpha_1} = N_1 := \begin{pmatrix} -a & r \\ u & a \end{pmatrix}, \tag{3.43}$$

and choose

$$Y := \begin{pmatrix} 1 & 0 \\ 0 & -1 \end{pmatrix}, \quad V_0 = \begin{pmatrix} 1 \\ 0 \end{pmatrix}. \tag{3.44}$$

The characteristic equation is then

$$\det\left(\mathcal{L}(\lambda) - z\mathbb{I}_r\right) = z^2 - \lambda^2 - a^2 - ur = 0. \tag{3.45}$$

In this case, V_0 is an eigenvector of Y and the genus of the spectral curve is $\tilde{g} = 0$, so there are no $\{\lambda_\mu, \zeta_\mu\}$'s. The single pair of spectral Darboux coordinates is thus

$$q_2 = \ln\, u, \quad P_2 = a. \tag{3.46}$$

It is easily verified that, relative to the Lie Poisson structure, they satisfy

$$\{q_2, P_2\} = 1. \tag{3.47}$$

(a.2) We consider the same orbit as in (a.1), but choose

$$Y := \begin{pmatrix} 0 & 1 \\ 1 & 0 \end{pmatrix}. \tag{3.48}$$

In this case, V_0 is not an eigenvector of Y. The genus is still $\tilde{g} = 0$, but the equation (3.39) now has a finite solution, giving the Darboux coordinate pair

$$\lambda_1 = -u, \quad \zeta_1 = -\frac{a}{u}. \tag{3.49}$$

These are verified to also satisfy

$$\{\lambda_1, \zeta_1\} = 0. \tag{3.50}$$

(a.3) Take $\mathfrak{g} = \mathfrak{sl}(3, \mathbf{R})$, and again, $\alpha_1 = 0$. Then the dimension of a generic orbit with $n = 1$ is $\dim \mathcal{O}_{\mathcal{N}_0} = 6$. We parametrize $\mathcal{N}_0(\lambda)$ as:

$$\mathcal{N}_0(\lambda) = N_1 := \begin{pmatrix} -a-b & r & s \\ u & a & e \\ v & f & b \end{pmatrix}, \tag{3.51}$$

and choose

$$Y = \begin{pmatrix} 0 & 0 & 0 \\ 0 & 1 & 0 \\ 0 & 0 & -1 \end{pmatrix}, \quad V_0 = \begin{pmatrix} 1 \\ 0 \\ 0 \end{pmatrix}. \tag{3.52}$$

Again, V_0 is an eigenvector of Y, but the spectral curve has genus $\tilde{g} = 1$, and is realized as a 3–fold branched cover of $\mathbf{P}^1$. We therefore find one Darboux coordinate pair (λ_1, ζ_1), corresponding to a finite zero of the eigenvector components, plus two further pairs, (q_2, P_2, q_3, P_3), corresponding to zeros over $\lambda = \infty$:

$$\begin{gathered} \lambda_1 = \frac{1}{2}\left(b - a - \frac{ev}{u} + \frac{uf}{v}\right), \quad \zeta_1 = \frac{uva + uvb - ev^2 - fu^2}{-uva + uvb - ev^2 + fu^2} \\ q_2 = \ln u, \quad q_3 = \ln v, \quad P_2 = a, \quad P_3 = b. \end{gathered} \tag{3.53}$$

Again, it is easily verified directly that these form a Darboux system, with nonvanishing Lie Poisson brackets

$$\{\lambda_1, \zeta_1\} = 1, \quad \{q_2, P_2\} = 1, \quad \{q_3, P_3\} = 1. \tag{3.54}$$

(b) Now consider the case $\tilde{\mathfrak{g}}^+ = \widetilde{sl}(2,\mathbf{R})^+$, with arbitrary n, but ranks $k_i = 1$ for all i, and hence $\det(N_i) = 0$ for all the residue matrices N_i. For general Y, $\mathcal{N}(\lambda)$ then has the form

$$\mathcal{N}(\lambda) = \lambda \begin{pmatrix} a & b \\ c & -a \end{pmatrix} + \frac{\lambda}{2} \begin{pmatrix} -\sum_{i=1}^n \frac{x_i y_i}{\lambda-\alpha_i} & -\sum_{i=1}^n \frac{y_i^2}{\lambda-\alpha_i} \\ \sum_{i=1}^n \frac{x_i^2}{\lambda-\alpha_i} & \sum_{i=1}^n \frac{x_i y_i}{\lambda-\alpha_i} \end{pmatrix}, \tag{3.55}$$

where $\{x_i, y_i\}_{i=1,\dots n}$ form a Darboux system on the reduced Moser space, which is identified with $\mathbb{R}^{2n}/(\mathbb{Z}_2)^N$. In this case, the characteristic equation defining the invariant spectral curve $\mathcal{C}$ is

$$\det(\mathcal{L}(\lambda) - z\mathbb{I}_2) = z^2 + a(\lambda)P(\lambda) = 0, \tag{3.56}$$

where

$$P(\lambda) = -(a^2 + bc)\lambda^n + P_{n-1}\lambda^{n-1} + \dots . \tag{3.57}$$

In particular, this gives eq. (1.34) for the case $a = c = 0$, $b = -\frac{1}{2}$. Thus, $\mathcal{C}$ is hyperelliptic, a 2–sheeted branched cover of $\mathbb{P}^1$, with $2n - 1$ or $2n$ finite branch points, depending on whether or not $a^2 + bc$ vanishes. The genus is therefore generically $\tilde{g} = n - 1$. The dimension of the coadjoint orbit is $\dim \mathcal{O}_{\mathcal{N}_0} = 2n$. The matrix $\widetilde{\mathcal{M}}(\lambda, \zeta)$ is

$$\widetilde{M} = \begin{pmatrix} -a + \frac{1}{2}\sum_{i=1}^n \frac{x_i y_i}{\lambda-\alpha_i} - \zeta & -b + \frac{1}{2}\sum_{i=1}^n \frac{y_i^2}{\lambda-\alpha_i} \\ -c - \frac{1}{2}\sum_{i=1}^n \frac{x_i^2}{\lambda-\alpha_i} & a - \frac{1}{2}\sum_{i=1}^n \frac{x_i y_i}{\lambda-\alpha_i} - \zeta \end{pmatrix}. \tag{3.58}$$

Taking

$$V_0 = \begin{pmatrix} 1 \\ 0 \end{pmatrix}, \tag{3.59}$$

if $c \neq 0$, V_0 is not an eigenvector of Y, so the full set of n spectral Darboux coordinate pairs $\{\lambda_\mu, \zeta_\mu\}_{\mu=1,\dots n}$ are given by:

$$\sum_{i=1}^n \frac{x_i^2}{\lambda_\mu - \alpha_i} + 2c = 0, \tag{3.60a}$$

$$\zeta_\mu = -a + \frac{1}{2}\sum_{i=1}^n \frac{x_i y_i}{\lambda_\mu - \alpha_i}, \tag{3.60b}$$

$$\mu = 1, \dots n.$$

These are therefore hyperelliptic coodinates $\{\lambda_\mu\}$ and their conjugate momenta $\{\zeta_\mu\}$. In the case $c = 0$, V_0 is an eigenvector of Y, and eqs. (3.60a,b) are replaced by

$$\frac{1}{2}\sum_{i=1}^n \frac{x_i^2}{\lambda_\mu - \alpha_i} = 0, \tag{3.61a}$$

$$\zeta_\mu = -a + \frac{1}{2}\sum_{i=1}^n \frac{x_i y_i}{\lambda_\mu - \alpha_i}, \tag{3.61b}$$

$$\mu = 1, \dots n - 1,$$

yielding only $n-1$ pairs of Darboux coordinates $\{\lambda_\mu, \zeta_\mu\}_{\mu=1,\ldots n-1}$, since one of the zeros of the eigenvector components lies over $\lambda = \infty$. We must therefore complete the system by defining the additional pair

$$q := \ln\left(\frac{1}{2}\sum_{i=1}^{n} x_i^2\right), \quad P := \frac{1}{2}\sum_{i=1}^{n} x_i y_i. \tag{3.62}$$

It is easily verified directly that

$$\omega = -d\theta, \tag{3.63}$$

where

$$\theta := \sum_{i=1}^{n} y_i dx_i = \begin{cases} \sum_{\mu=1}^{n} \zeta_\mu d\lambda_\mu & \text{if } c \neq 0 \\ \sum_{\mu=1}^{n-1} \zeta_\mu d\lambda_\mu + Pdq & \text{if } c = 0. \end{cases} \tag{3.64}$$

(c) *NLS Equation:* $\widetilde{\mathfrak{su}}(1,1)^+$

Taking the orbit $\mathcal{O}_{\mathcal{N}_0}$ in $\widetilde{\mathfrak{su}}(1,1)^{+*}$ as parametrized in eqs. (2.54), (2.55), with $Y = 0$, the symplectic submanifold $\mathcal{O}^S_{\mathcal{N}_0} \subset \mathcal{O}_{\mathcal{N}_0}$ is defined by the constraint

$$\sum_{i=1}^{n} z_i^2 = 0. \tag{3.65}$$

The spectral Darboux coordinates $\{q, P, \lambda_\mu, \zeta_\mu\}_{\mu=1,\ldots n-1}$ are then given by eqs. (1.52a-c). It is easily verified in this case that the orbital symplectic form restricted to $\mathcal{O}^S_{\mathcal{N}_0}$ is

$$\omega_{\text{orb}} = -d\theta, \tag{3.66}$$

where

$$\theta|_{\mathcal{O}_{\mathcal{N}_0}} = -i\sum_{j=1}^{n} \bar{z}_j dz_j|_{\mathcal{O}_{\mathcal{N}_0}} = \sum_{\mu=1}^{n-2} \zeta_\mu d\lambda_\mu + Pdq, \tag{3.67}$$

so $\{q, P, \lambda_\mu, \zeta_\mu\}_{\mu=1,\ldots n-2}$ do, indeed, define a Darboux coordinate system.

A similar construction holds for the case of the sine-Gordon equation (Section 1.3), where the relevant algebra is the twisted loop algebra $\widetilde{\mathfrak{su}}(2)^+$, obtained by a suitable combination of discrete and continuous reductions. The orbits are parametrized by eqs. (1.60), (1.61a,b), and the relevant spectral Darboux coordinates determined by eqs. (1.84a,b). Details may be found in [**HW**].

In the last section we explain how, in the general case, these spectral Darboux coordinates lead directly to a linearization of the AKS flows through the Liouville-Arnold integration procedure. In each case the relevant linearizing map turns out to be the Abel map to the Jacobi variety of the spectral curve.

3.6 Liouville-Arnold Integration

Using the spectral Darboux coordianates, we may define the local equivalent of the "canonical" 1-form

$$\theta := \sum_{\mu=1}^{\tilde{g}} \zeta_\mu d\lambda_\mu + \sum_{i=2}^{r} P_i dq_i \tag{3.68a}$$

$$= \sum_{\mu=1}^{\tilde{g}} \frac{z_\mu}{a(\lambda_\mu)} d\lambda_\mu + \sum_{i=2}^{r} P_i dq_i, \tag{3.68b}$$

where

$$z_\mu := a(\lambda_\mu)\zeta_\mu. \tag{3.69}$$

(Note that for the examples given above, this actually *is* the canonical 1-form on $\mathbb{R}^{2n}/(\mathbb{Z}_2)^n$, or $\mathbb{C}^{2n}/(\mathbb{Z}_2)^n$, viewed as the cotangent bundle of $\mathbb{R}^n/(\mathbb{Z}_2)^n$ and $\mathbb{C}^n/(\mathbb{Z}_2)^n$, respectively.) On the Liouville-Arnold torus $\mathbf{T}$, defined by taking the simultaneous level sets

$$P_{ia} = C_{ia}, \quad P_i = C_i \tag{3.70}$$

of the spectral invariants, we have

$$\theta|_{\mathbf{T}} = dS(\lambda_1, \ldots \lambda_{\tilde{g}}, q_2, \ldots q_r, P_{ia}, P_i), \tag{3.71}$$

where $S(\lambda_1, \ldots, \lambda_{\tilde{g}}, q_2, \ldots, q_r, P_{ia}, P_i)$ is the Liouville generating function to the canonical coordinates conjugate to the invariants (P_{ia}, P_i). Eq. (3.71) can be integrated by viewing $z = z(\lambda, P_{ia}, P_i)$ as a meromorphic function on the Riemann surface of the spectral curve $\mathcal{C}$.

$$S(\lambda_\mu, q_i, P_{ia}, P_i) = \sum_{\mu=1}^{\tilde{g}} \int_{\lambda_\mu^0}^{\lambda_\mu} \frac{z(\lambda, P_{ia}, P_j)}{a(\lambda)} d\lambda + \sum_{i=2}^{r} q_i P_i, \tag{3.72}$$

where $z_\mu = z_\mu(\lambda_\mu, P_{ia}, P_i)$ is essentially determined implicitly by the spectral equation

$$\mathcal{P}(\lambda_\mu, z_\mu(\lambda_\mu, P_{ia}, P_i)) = 0. \tag{3.73}$$

The linearizing coordinates for AKS flows are then given, as usual, by differentiation of S with respect to the invariants:

$$Q_{ia} = \frac{\partial S}{\partial P_{ia}} = \sum_{\mu=1}^{\tilde{g}} \int_{\lambda_\mu^0}^{\lambda_\mu} \frac{1}{a(\lambda)} \frac{\partial z}{\partial P_{ia}} d\lambda = \frac{\partial h}{\partial P_{ia}} t + Q_{ia,0} \tag{3.74a}$$

$$Q_i = \frac{\partial S}{\partial P_i} = \sum_{\mu=1}^{\tilde{g}} \int_{\lambda_\mu^0}^{\lambda_\mu} \frac{1}{a(\lambda)} \frac{\partial z}{\partial P_i} d\lambda + q_i = \frac{\partial h}{\partial P_i} t + Q_{i,0}, \tag{3.74b}$$

where, from the explicit structure (3.27) of the characteristic polynomial given in Proposition 3.1, we obtain, by implicit differentiation, that the integrands of eqs. (3.74a,b) are of the form

$$\omega_{ia} := \frac{1}{a(\lambda)}\frac{\partial z}{\partial P_{ia}}d\lambda = \frac{a_i(\lambda)z^{r-i}\lambda^a}{\mathcal{P}_z(\lambda, z)}d\lambda \tag{3.75a}$$

$$\omega_i := \frac{1}{a(\lambda)}\frac{\partial z}{\partial P_i}d\lambda = -\sum_{j=2}^{r}\frac{R_{ij}a_j(\lambda)(-z)^{r-j}\lambda^{\delta_j-\epsilon}}{\mathcal{P}_z(\lambda, z)}d\lambda, \tag{3.75b}$$

where

$$R_{ij} := \begin{cases} (P_1 - P_i)\sum_{2\le i_1<i_2\cdots<i_{j-2}\neq i} P_{i_1}\dots P_{i_{j-2}} \\ \text{and} \quad \epsilon = 0 \qquad \text{for case (a)} \\ (Y_1 - Y_i)\sum_{2\le i_1<i_2\cdots<i_{j-2}\neq i} Y_{i_1}\dots Y_{i_{j-2}} \\ \text{and} \quad \epsilon = 1 \qquad \text{for case (b)}\ . \end{cases} \tag{3.76}$$

The point to note is that the differentials $\{\omega_{ia}\}$, $\{\omega_i\}$ appearing in eqs. (3.75a,b) are, respectively, abelian differentials of the first and third kinds on the Riemann surface defined by C, the latter having their poles at the points $(\infty_1, \dots, \infty_r)$ over $\lambda = \infty$.

Theorem 3.3 [AHH3]. *The $\tilde{g}$ differentials $\{\omega_{ia}\}_{i=1,\dots\tilde{g}}$ in eq. (3.75a) form a basis for the space $H^0(\tilde{S}, K_{\tilde{S}})$ of abelian differentials of the first kind (where $K_{\tilde{S}}$ denotes the canonical bundle). The linear flow equation (3.74a) may therefore be expressed as:*

$$\mathbf{A}(\mathcal{D}) = \mathbf{B} + \mathbf{U}t, \tag{3.77}$$

where $\mathbf{A} : S^{\tilde{g}}C \longrightarrow \mathbf{C}^{\tilde{g}}/\Gamma$ is the Abel map, and $\mathbf{B}, \mathbf{U} \in \mathbf{C}^{\tilde{g}}$ are obtained by applying the inverse of the $\tilde{g} \times \tilde{g}$ normalizing matrix $\mathbf{M}$, with elements

$$\mathbf{M}_{\mu,(ia)} := \oint_{a_\mu} \omega_{ia}, \tag{3.78}$$

to the vectors $\mathbf{C}, \mathbf{H} \in \mathbf{C}^{\tilde{g}}$ with components C_{ia} and $-\frac{\partial h}{\partial P_{ia}}$, respectively (the pair (ia) viewed as a single coordinate label in $\mathbf{C}^{\tilde{g}}$). The $r-1$ differentials $\{\omega_i\}_{i=2,\dots r}$ in eq. (3.75b) are abelian differentials of the third kind with simple poles at ∞_i and ∞_1, and residues $+1$ and -1, respectively.

Comparing this general formula with the specific cases (1.23), (1.55), (1.56), (1.88a,b), for the examples of Section 1, we see that this provides the generalization that was required, expressing all linearized AKS flows on rational coadjoint orbits of $\widetilde{\mathfrak{sl}}(r)^{+*}$ and its reductions through the Abel map.

It is possible, moreover, to invert the map expressing these flows, by expressing any symmetric function of the coordinates $\{\lambda_\mu, \zeta_\mu\}_{\mu=1,\ldots\tilde{g}}$ in terms of the Riemann theta function associated to the curve $\mathcal{C}$. For example, in view of eq. (3.74b), the coordinates $\{q_i\}_{i=2,\ldots r}$ themselves are expressed as such symmetric functions through abelian integrals of the third kind. Applying the reciprocity theorem relating the two kinds of abelian integrals (*viz.* [**AHH3**]), we obtain:

Corollary 3.4 [AHH3]. *For a suitable choice of constants $\{e_i, f_i\}_{i=2,\ldots r}$, the coordinate functions $\{q_i(t)\}$ satisfying eq.(3.74b) are given by:*

$$q_i(t) = ln\left[\frac{\theta(\mathbf{B}+t\mathbf{U}-\mathbf{A}(\infty_i)-\mathbf{K})}{\theta(\mathbf{B}+t\mathbf{U}-\mathbf{A}(\infty_1)-\mathbf{K})}\right] + e_i t + f_i, \tag{3.79}$$

where $\mathbf{K} \in \mathbf{C}^{\tilde{g}}$ is the Riemann constant.

This generalizes the theta function formula (1.57) giving the solution of the NLS equation. Similar formulae exist e.g., for the sine-Gordon equation ([**HW**]) and many other systems that can be cast in terms of commuting AKS flows in rational coadjoint orbits of loop algebras. Aside from technical complications resulting, e.g., from the reduction procedure or the imposition of further symplectic constraints, or from the presence of further singularities in the spectral curve, the procedure is largely algorithmic. It provides a very general setting for the explicit application of the Liouville-Arnold integration procedure to a wide class of known - and yet to be discovered - integrable Hamiltonian systems. Moreover, the moment map embedding method makes it possible to treat both the intrinsically finite dimensional systems, and those systems corresponding to finite dimensional sectors of integrable systems of PDE's (solitons, finite band solutions, etc.) on exactly the same footing.

Acknowledgements. Most of the results described here were obtained in collaboration with a number of colleagues and friends, whose important contributions to this work it is a pleasure to acknowledge. My thanks to M. Adams, J. Hurtubise, E. Previato and M.-A. Wisse for their valuable input and help over an extended period. I would also like to thank G. Helminck and the other organizers of the 8th Scheveningen conference for their very kind hospitality and for the cordial and stimulating environment they helped to create.

REFERENCES

[AA] Al'ber, S.J., Al'ber, M.S., "Hamiltonian Formalism for Nonlinear Schrödinger Equations and Sine-Gordon Equations", *J. London Math. Soc. (2)* **36**, 176–192 (1987).

[AM] Abraham, R., Marsden, J.E., *Foundations of Mechanics*, 2nd ed., Reading, MA, Benjamin Cummings, Ch. 4 (1978).

[AHH1] Adams, M.R., Harnad, J. and Hurtubise, J., "Dual Moment Maps to Loop Algebras", *Lett. Math. Phys.* **20**, 294–308 (1990).

[AHH2] Adams, M.R., Harnad, J. and Hurtubise, J., "Isospectral Hamiltonian Flows in Finite and Infinite Dimensions II. Integration of Flows", *Commun. Math. Phys.* **134**, 555–585 (1990).

[AHH3] Adams, M. R., Harnad, J. and Hurtubise, J., "Darboux Coordinates and Liouville-Arnold Integration in Loop Algebras," *Commun. Math. Phys.*/ (1993, in press).

[AHH4] Adams, M.R., Harnad, J. and Hurtubise, J., "Liouville Generating Function for Isospectral Hamiltonian Flow in Loop Algebras", in: *Integrable and Superintergrable Systems*, ed. B. Kuperschmidt, World Scientific, Singapore (1990).

[AHH5] Adams, M.R., Harnad, J. and Hurtubise, J., "Coadjoint Orbits, Spectral Curves and Darboux Coordinates", in: *The Geometry of Hamiltonian Systems*, ed. T. Ratiu, Publ. MSRI Springer-Verlag, New York (1991); "Integrable Hamiltonian Systems on Rational Coadjoint Orbits of Loop Algebras", in: *Hamiltonian Systems, Transformation Groups and Spectral Transform Methods*, ed. J. Harnad and J. Marsden, Publ. C.R.M., Montréal (1990).

[AHP] Adams, M.R., Harnad, J. and Previato, E., "Isospectral Hamiltonian Flows in Finite and Infinite Dimensions I. Generalised Moser Systems and Moment Maps into Loop Algebras", *Commun. Math. Phys.* **117**, 451–500 (1988).

[AvM] Adler, M. and van Moerbeke, P., "Completely Integrable Systems, Euclidean Lie Algebras, and Curves," *Adv. Math.* **38**, 267–317 (1980); "Linearization of Hamiltonian Systems, Jacobi Varieties and Representation Theory," *ibid.* **38**, 318–379 (1980).

[Du] Dubrovin, B.A., "Theta Functions and Nonlinear Equations", *Russ. Math. Surv.* **36**, 11–92 (1981).

[KN] Krichever, I.M.and Novikov, S.P., "Holomorphic Bundles over Algebraic Curves and Nonlinear Equations", *Russ. Math. Surveys* **32**, 53–79 (1980).

[HW] Harnad, J. and Wisse, M.-A. "Isospectral Flow in Loop Algebras and Quasiperiodic Solutions of the Sine-Gordon Equation", *J. Math. Phys.* (1993, in press).

[Mo] Moser, J., "Geometry of Quadrics and Spectral Theory", *The Chern Symposium, Berkeley, June 1979*, 147–188, Springer, New York, (1980).

[N] Neumann, C., "De problemate quodam mechanico, quod ad primam integralium ultraellipticorum classem revocatur", *J. Reine Angew. Math.* **56**, 46–63 (1859).

[P1] Previato, E., "Hyperelliptic quasi-periodic and soliton solutions of the nonlinear Schrödinger equation." *Duke Math. J.* **52**, 329–377 (1985).

[P2] Previato, E. "A particle-system model of the sine-Gordon hierarchy, Solitons and coherent structures", *Physica D* **18**, 312–314, (1986).

[RS] Reiman, A.G., and Semenov-Tian-Shansky, M.A., "Reduction of Hamiltonian systems, Affine Lie algebras and Lax Equations I, II", *Invent. Math.* **54**, 81–100 (1979); *ibid.* **63**, 423–432 (1981).

Geometry of the modified KdV equation

Emma Previato

Mathematics Department, Boston University, Boston, MA 02215

Introduction. This is a write-up of my three Scheveningen lectures, which were intended as an educational overview of various aspects of the mKdV (modified Korteweg-de Vries) equation, $v_t \mp 6v^2 v_x + v_{xxx} = 0$. The equation admits of the several integrability techniques which were progressively made famous by the KdV theory (see diagram below). However, I don't know of a place where all the recipes are spelled out for mKdV. Thus, there seemed to be two purposes to the exercise of writing these lectures: one was the precise identification of the appropriate geometric objects for solution; the other was spotlighting several concrete open problems in the KdV theory for which the mKdV construction gives a vantage point. The picture to keep in mind is that mKdV solutions fiber over KdV solutions the way a flag manifold fibres over a Grassmannian. This was pointed out in [DS], cf. [W2], but it was pursued in the language of Lie algebras rather than at the level of maps between curves, hamiltonian systems, and explicit families of special solutions. We precede the account of the lectures by a diagram which provides a Leitfaden and an abstract, and follow it by a list of open problems and suggested links, pertaining to the individual lectures.

Acknowledgements. I would like to thank Prof. R. Goldstein for making available his preprint [GP2] and very promptly clarifying a passage in his paper [GP1] which I had been unable to work out. I want to say that the NSA provided enormous incentive for me by financing the participation in the conference of my Ph.D. student A. Kasman of Boston University under grant MDA904-92-H-3032. Finally, I must thank the organizers, first and foremost for inviting me, and then for providing one of the most friendly, thoughtfully catered to, and homogeneous workshop experiences I ever had: as bright and crisp as a Rembrandt etching come to life.

Leitfaden

PDE	→	**Miura Transformation**	← →	**Drinfel'd-Sokolov**
↓				↑
Inverse Scattering Zakharov-Shabat Lax				**Segal-Wilson Grassmannian**
↓				↑
Scattering Data as Canonical Variables	→	**Finite-Dimensional Hamiltonian System**	→	**Algebraic Geometry Squared Eigenfunctions, Baker Function**

Lecture 1.

1.1 Set up for contrast:

KdV	KP
{ mKdV, NLS, sine-Gordon }	DS

1.2 AKNS formalism: Lax pairs and Lenard Hamiltonian structures.
1.3 Scattering data: Deift-Lund-Trubowitz Hamiltonian structures.
1.4 The finite-dimensional model: curve theory.
1.5 Coadjoint orbit interpretation and reductions for rank 2 perturbations.

Lecture 2.

2.1 The Segal-Wilson Grassmannian.
2.2 Adaptation to the multicomponent case.
2.3 The Drinfel'd-Sokolov theory.
2.4 Finite-dimensional Grassmannian.

Lecture 3.

3.1 Darboux transformations.
3.2 The Krichever-Novikov equations.

1. Why mKdV

1.1 Anecdotes. It's official! We no longer need to excuse our interest in KdV. I was able to bring to the school the 1992 poster created by the Joint Policy Board for Mathematics. It depicts an ocean wave towering over a ship and the caption says, among other things: "The solitary wave [was] found by Korteweg and de Vries in 1895 to be governed by the equation $\frac{\partial u}{\partial t} + \frac{\partial u}{\partial x} + u\frac{\partial u}{\partial x} + \frac{\partial^3 u}{\partial x^3} = 0$. Until recently, solution of this equation strained the resources of the most powerful computers, but mathematical advances have now made the solution of this equation routine. (...) Not only has the mathematical theory of water waves helped to understand and protect our environment, but its insights have also had significant impact on technological development. Although the solitary wave is now well understood, other water waves still have mysterious effects on our environment and remain objects of active mathematical research." As you can tell, the goal of this material is to sensitize the general public to the importance of mathematics; it was issued on the occasion of "Mathematics Awareness Week," an annual event established by a presidential proclamation in 1986. I thought it was pretty good of the KdV equation to make it to the poster! Less frivolously perhaps, KdV is the prototype of its genre, according to the Mathematics Subject Classification (1992 Revision) 35Q53: KdV-like equations. So, how to excuse my interest in mKdV? Until recently all I knew was the chance discovery reported in [AS, p. 6]: R. Miura decided to seek conserved quantities for an equation with one higher degree of nonlinearity than KdV:

$$v_t - 6v^2 v_x + v_{xxx} = 0 \tag{mKdV}$$

$$u_t + 6uu_x + u_{xxx} = 0 \tag{KdV}$$

and discovered what is known as the "Miura transformation":

$$u = -v^2 - v_x \tag{MT}$$

which takes a solution of mKdV to one of KdV: indeed, $u_t + 6uu_x + u_{xxx} = -(2v + \partial_x)(v_t - 6v^2 v_x + v_{xxx})$.

Moreover, the standard substitution $v = \frac{\psi_x}{\psi}$ for the Riccati equation $-v^2 - v_x - u = 0 = (\partial_x^2 + u)\psi$ suggested that the conserved quantities be related to the spectrum of the operator $L = \partial_x^2 + u$ (we write ∂_x or ∂ for $\frac{d}{dx}$) and gave rise to the Lax-pair representation, linearization and complete integrability of the problem! (see §2). In 1991 however, P. Nelson alerted me to recent work by R.E. Goldstein and D.M. Petrich [GP1,2] where mKdV appears naturally and KdV is derived from it. I report their theory below not only because it is so satisfying but also because it gives a geometric interpretation for the role of the Schwarzian derivative.

1.2 Curvature dynamics. It is shown in [GP1] that the physically important description of a plane region which moves in time with conserved area and perimeter can be given by imposing that the curvature satisfy the mKdV equation (up to rescaling): we reproduce this striking derivation. We study a family of closed plane curves C_t, whose points are 2-vectors $\underline{r}(t)$ depending on a (time) parameter t. The curve dynamics will be given by $\underline{r}_t = U\underline{N} + W\underline{T}$, where: U, W are functions of t; $\underline{T}, \underline{N}$ are the unit tangent and normal vector to the curve C_t at each point; and W is assumed to be periodic on each curve. On each curve the parametrization by arc length s obtains the Frénet-Serret equations $\underline{T} = \underline{r}_s$, $-\kappa\underline{N} = \underline{T}_s$, $\kappa\underline{T} = \underline{N}_s$ (κ is the curvature); we make the further assumption that the motion be purely local, that is U, W be functions of κ and its s-derivatives. Global geometric quantities are the length of the curve C_t and the enclosed area: $L(t) = \oint(\underline{r}_\sigma \cdot \underline{r}_\sigma)^{1/2}d\sigma$, $A(t) = \frac{1}{2}\oint \underline{r} \times \underline{r}_\sigma d\sigma$, where σ is any given parametrization of C_t; we regard σ, t as independent variables, thus we get variational equations:

$$L_t = \oint(\underline{r}_\sigma \cdot \underline{r}_\sigma)^{-1/2}(\underline{r}_{t\sigma} \cdot \underline{r}_\sigma)d\sigma = \oint(U + \kappa W)ds \text{ (by substitution: } d\sigma = \frac{d\sigma}{ds}ds,$$

$$\underline{r}_\sigma = \frac{ds}{d\sigma}\underline{r}_s, \text{ etc.); } A_t = \oint \underline{r} \times \underline{r}_{t\sigma}d\sigma = \oint \underline{r}_t \times \underline{r}_\sigma d\sigma \text{ (by parts)} = \oint U ds.$$

If we now impose the stronger condition that arc length be preserved locally, the first equation says $U + \kappa W \equiv 0$ and this is (not surprisingly) equivalent to requiring that s and t be independent variables, indeed: $[\partial_s, \partial_t]\underline{r} = (\kappa U + W_s)\underline{r}_s$. Under this assumption, the curvature evolution is found as: $\kappa = (\underline{T}_s \cdot \underline{T}_s)^{1/2} \Rightarrow \kappa_t = -\Omega U$ where $\Omega = \partial_{ss} + \kappa^2 + \kappa_s\partial^{-1}\kappa$, where a choice of integration is unique up to an arbitrary function of time $c(t)$: $W = -\int^s ds'\kappa U + c = -\partial^{-1}\kappa U$. The next major remark is that area-invariance is satisfied for $U = \partial_s f$ for which $\kappa U(= -W_s) = \partial_s g$; the simplest example is $U^{(1)} = 0$, $W^{(1)} = c$ (arbitrary constant), which means that C_t is simply a reparametrization of C_0. The next $U^{(2)} = \kappa_s$, $W^{(2)} = -\frac{\kappa^2}{2}$ gives mKdV $\kappa_t = -\kappa_{sss} - \frac{3}{2}\kappa^2\kappa_s$, as advertised. Moreover, the recursion: $U^{(n)} = \Omega U^{(n-1)} = \Omega^{(n-2)}\kappa_s$ $(n \geq 2)$, $W^{(n)} = -\partial^{-1}\kappa U^{(n)}$ gives the mKdV hierarchy, $\kappa_t = -\Omega U^{(n)}$. Now for the last feature, which we call the Schwarzian Derivative (SD): I would like to try and explain it in three separate ways:

(A) We know that the (rescaled) MT, $u = -\frac{1}{2}\kappa^2 - \mathrm{i}\kappa_s$ achieves the KdV equation; in fact, the MT of the entire hierarchy $\kappa_t = -\Omega U^{(n)}$ goes to the (independently defined) KdV hierarchy $u_t + \partial_s K^{(n)} = (\partial_s - \mathrm{i}\kappa)(\kappa_t + \Omega U^{(n)}) = 0$. [GP1] gives the curve-dynamics interpretation: if we identify the $\mathbf{R}^2$-plane with the $\mathbf{C}$-plane through $z(s,t) = x(s,t) + \mathrm{i}y(s,t)$ and change variables through $z(s,t) = \int^s ds' e^{\mathrm{i}\theta(s',st)}$ so that $\theta = -\mathrm{i}\log(z_s)$ and $\kappa = -\mathrm{i}\frac{z_{ss}}{z_s} = -\mathrm{i}\partial_s \log z_s$ (notice that z_s and $-\mathrm{i}z_s$ are the unit tangent and normal, resp.) then $u = -\mathrm{SD}(z) = -\left[\left(\frac{z_{ss}}{z_s}\right)_s - \frac{1}{2}\left(\frac{z_{ss}}{z_s}\right)^2\right]$. Thus, the area-preserving dynamics for $z(s,t)$ becomes KdV for the SD, which is the basic invariant under (conformal) Möbius maps $z \mapsto \frac{az+b}{cz+d}$.

(B) The equation $\kappa = -\mathrm{i}\partial_s \log z_s$ in (A) actually occurs if you implement the "QR algorithm" version of the MT: factor $L = (\partial_s + \kappa)(\partial_s - \kappa)$ (this gives $u = -\kappa^2 - \kappa_s$ and can be viewed as an operator analog of the QR factorization for matrices) and notice that for any basis ψ_1, ψ_2 of solutions of $L\psi = 0$ compatible with the factorization, namely such that $(\partial_s - \kappa)\psi_1 = 0$, $z = \psi_1/\psi_2$ satisfies the "singular Krichever-Novikov equation": $z_t = z_{sss} - \frac{3}{2}z_{ss}^2 z_s^{-1}$ ([Do], we follow [W3]). The group SL(2, C) is the differential Galois group of the extension $\mathbf{C} < u > \subset \mathbf{C} < \psi_1, \psi_2 >$: here $\mathbf{C} < u >$ denotes the differential field generated by u, which in turn means the field of rational functions in the indeterminates $u, u^{(j)}$, $j \geq 0$, with the derivation acting as $\partial_s u^{(j)} = u^{(j+1)}$, and differential automorphisms are those which preserve derivatives. The extension $\mathbf{C} < u > \subset \mathbf{C} < z >$ has Galois group $G = \mathrm{PSL}(2, \mathbf{C})$ and the intermediate extension $\mathbf{C} < v > \subset \mathbf{C} < z >$ is the fixed field of the subgroup $B \subset G$ of upper-triangular matrices (modulo $\pm I$).

(C) We follow [Se]: A basis of solutions of the equation $L_u y = (\partial_s^2 + u)y = 0$ gives a projective connection (cf. [D, I.5]) on $\{s\}$, by: $s \mapsto [\psi_1(s), \psi_2(s)]$; conversely, a parametrization of $\mathbf{P}^1$ can be so written and by multiplying ψ_1 and ψ_2 by a function of s the Wronskian $\psi_1'\psi_2 - \psi_2'\psi_1$ may be made to equal 1; then $\psi_1''/\psi_1 = \psi_2''/\psi_2$ so that ψ_1 and ψ_2 are solutions of an equation L_u. To render this independent of parametrization, we let the group PSL(2) of linear fractional transformations act on $\mathbf{P}^1$; indeed, a smooth map $\phi: s \mapsto \phi(s)$ induces $\phi^* L_u = L_{\tilde{u}}$, where $\tilde{u} = u(\phi(s))\phi'(s)^2 + \frac{1}{2}\mathrm{SD}(\phi)$; the SD is the only PSL(2)-invariant. [Se] goes on to show that KdV is reduction of "free motion" on the dual $\hat{\mathcal{V}}^*$ of the lie algebra $\hat{\mathcal{V}} = \mathbf{R} \oplus \mathcal{V}$, a central extension of $\mathcal{V} = \mathrm{Diff}^+(S^1)$. Moreover, the space $X(S^1) \subset \hat{\mathcal{V}}^*$ of periodic KdV operators is then identified with the symplectic quotient under a $G = \mathrm{PSL}(2)$ action on the space of connections in a $\mathbf{P}^1$-bundle over S^1; this seems to be the z-space, which under reduction for the action of the subgroup B defined in (B) will yield the v-space and the mKdV equation: thus, the perimeter/area preserving curve dynamics falls out of a variational principle.

2. Squared eigenfunctions

These objects, like the SD of §1, have also surfaced quite often, and in apparent isolation, in the theory of integrable equations: I would like to offer a thread through the maze:

(A) Burchnall and Chaundy [BC]: really the fundamental idea. Introduce the formal adjoint $L^\dagger$ of an ODO L to give a dictionary between algebraic and transcendental solutions to the BC problem (cf. 2.3.2). Via "trace formulae", one gets solutions to the PDE's; theta functions are then available in two ways, one *a priori* via the Krichever map, and one *a posteriori* by a knowledge of the divisors: Mumford uses this to characterize hyperelliptic period matrices! [Mu].

(B) Magnus and Winkler [MW, Ch.II]: the ODE satisfied by the "squared eigenfunctions" of Hill's operator gives the recursion operator (or the Lenard pair of Hamiltonian structures) because of the equation $\psi_\lambda^+(x)\psi_\lambda^-(x) = R_\lambda(x, x)$, the kernel of the resolvent [cf. McK].

(C) The Hamiltonian model in which ψ^+, ψ^- are viewed as canonical variables, and reduction of free motion yields classical problems (Neumann's constrained harmonic motion; Jacobi's geodesic on the ellipsoid, cf. [EF]) as related to KdV; mysteriously, this is capable of a "smearing" procedure on the line which allowed [DLT] to interpret the KdV (and other second-order) evolutions for rapidly vanishing potentials as hamiltonian flows for a symplectic structure on coordinates parametrized by the continuous spectrum.

(D) The generalization of (A) to a dual tau/Baker function on the Sato-Segal-Wilson Grassmannian; at the level of divisors, Cherednik's result: $\mathcal{O}(D) \mapsto \mathcal{O}^*(D) \otimes \mathcal{O}(K)$ where K is the canonical divisor [C].

(E) The generalization of (B),(C) to any order (called the n-th generalized KdV equation in [SW]) was found by R. Schilling [S] in the context of classical ODE and in [AHP] by symplectic quotients in loop algebras.

By and by, it became clear how the same objects could be given the various interpretations (A)-(E), but since a theoretic explanation of that would exceed the length allotted, we will just give concrete illustrations for the mKdV case.

2.1 Eigenvalue problem. To get started, one needs to pull the Lax pair out of the blue: recall that for KdV this was: $L_t = [B, L]$ where $L = \partial^2 + u$, $B = 4\partial^3 + 6u\partial + 3u_x$. More generally, the KP hierarchy is defined as the set of evolutions:

$$\partial_{t_k} \mathcal{L} = [(\mathcal{L}^k)_+, \mathcal{L}] \tag{KP}$$

on the coefficients $u_j(\underline{t})$ of a formal pseudodifferential operator $\mathcal{L} = \sum_{-\infty}^{1} u_j \partial^j$, with normalizations $u_1 \equiv 1$, $u_0 \equiv 0$ (which are achieved by the two automorphisms of the algebra $\mathcal{D}$ of differential operators, change of variables and conjugation by a function.) For the definition and basic properties of operations in $\mathcal{D}$ we refer the reader to [SW,§4]. The generalized nth KdV hierarchy is satisfied by those KP solutions $\mathcal{L}$ for which $(\mathcal{L}^n)_+ = \mathcal{L}^n$, equivalently t_{kn} is a stationary evolution for all integers $k \geq 1$.

The AKNS method (cf. [AS]§1.2), a spectral problem for 2×2 matrix operators, gives both KdV and mKdV as follows: compatibility of $\underline{y}_x = X\underline{y}$ and $\underline{y}_t = T\underline{y}$, where $\underline{y} = [y_1\ y_2]^T$, means $X_t - T_x = [T, X]$. The matrices X, T are assumed to have particular forms, for example

$$\begin{cases} y_{1x} = -i\zeta y_1 + q y_2 \\ y_{2x} = i\zeta y_2 + r y_1 \end{cases} \quad \text{and} \quad \begin{cases} y_{1t} = A y_1 + B y_2 \\ y_{2t} = C y_1 + D y_2 \end{cases}$$

with A, B, C, D scalar functions independent of t; it turns out to be useful to expand them in powers of the spectral parameter ζ: if

$$\begin{cases} A = 4i\zeta^3 - 2iqr\zeta - (qr_x - q_x r) = -D \\ B = 4q\zeta^2 + 2iq_x\zeta + 2q^2 r - q_{xx} \\ C = 4r\zeta^2 - 2ir_x\zeta + 2qr^2 - r_{xx} \end{cases}$$

then the "constraint" $r = -1$ gives $q_t + 6qq_x + q_{xxx} = 0$ (KdV) and $r = \mp q$ gives $q_t \pm 6q^2 q_x + q_{xxx} = 0$ (mKdV). Now let us rework and explain them (after

[W1]) in terms of differential rings. We can rewrite the AKNS compatibility in the form of a 2×2 matrix Lax pair:

$$L = \begin{bmatrix} \partial & -q \\ r & -\partial \end{bmatrix}, \; B = 4\begin{bmatrix} \partial^3 & 0 \\ 0 & \partial^3 \end{bmatrix} - 6\begin{bmatrix} qr & q_x \\ r_x & qr \end{bmatrix}\partial - 3\begin{bmatrix} (qr)_x & q_{xx} \\ r_{xx} & (qr)_x \end{bmatrix}.$$

The form of B is explained by the fact that a sequence of commuting flows in $\mathcal{M}$ (the algebra of formal expressions $\sum_{-\infty}^{N} u_j \partial^j$ where u_j are 2×2 matrices of analytic functions) is given by equations $\partial_j L = [(P_j)_+, L]$ where P_j belongs to the centralizer of L $\mathcal{C}_{\mathcal{M}}(L)$, and there is a unique such element homogeneous of any (positive) degree j (for a precise statement, see [W1,2.19]).

The next item is the implementation of algebraic geometry to get exact solutions; there are two approaches and we describe first the Krichever method, then the one inspired by Inverse Scattering.

2.2 Krichever method. As opposed to the KP (scalar) case, in which the solution was provided by a Baker function, we need to construct a Baker vector which have 2 components; geometrically, this is achieved by picking 2 (smooth) points on a curve C rather than 1 point P_∞; the osculating vectors to the curve inside Jac C at those points give rise to 2 sequences $x_j^{(\alpha)}$ ($\alpha = 1, 2$; $j \geq 1$) of "multicomponent" KP flows, cf. [DJKM§4]. The amazing thing is that you can impose constraints on the curve and the point so as to obtain the very equations of applied mathematics. The amusing thing is that the algebro-geometric description of the orbits changes, however slightly. The prototype situation (generalized) KdV, is the simplest: the fact that t_{kn} be stationary flows for all positive integers k means that there exists a function with pole of order n at P_∞ and regular elsewhere: indeed, if the nth root of that function is taken to be the local parameter (inverse) z corresponding to the operator $\mathcal{L}$ (as in [SW]§4-6), then $(\mathcal{L}^n)_+ = \mathcal{L}^n$. For the 2-point case, we set up the general situation first. Let P_1, P_2 be smooth points on a curve C, which too will be smooth in most of our considerations for the sole purpose of making the exposition simpler. The geometric ingredients for solution will be: (a) a choice of local parameter (inverse) κ_i at P_i ($i = 1, 2$); (b) an element of the $\mathbf{C}^*$ extension G of Jac C which is the generalized Jacobian of the curve obtained by identifying P_1 and P_2; in practice, a line bundle associated to a sheaf $\mathcal{E}$ on C, plus an identification of the fibres $\mathcal{E}_{P_1}$ and $\mathcal{E}_{P_2}$; with the additional condition that the divisor associated to $\mathcal{E}$ be $Q_1 + \ldots + Q_{g+1}$, where g =genus C, and $Q_1 + \ldots + Q_{g+1} - P_1 - P_2$ have no sections; (c) a deformation of $\mathcal{E}$, given by patching functions ψ_i ($i = 1, 2$) with the following properties: ψ_i is meromorphic on $C \backslash \{P_1, P_2\}$ with poles bounded by $\sum_1^{g+1} Q_j$; and ψ_i has an essential singularity at ∞_j with local expansion:

$$\psi_i(x, t_k, \kappa_j) = \exp(x\kappa_j + t_k R_j(\kappa_j))\left(\sum_{s=0}^{\infty} \xi_s^{ij}(x, t_k)\kappa_j^{-s}\right),$$

the R_j being suitable polynomials. Corresponding to these flows, we obtain the differential operators $(P_j)_+$ referred to in 2.1. For reasons which will become clear below, we compute two examples: after normalizing the Baker vector so that $\xi_0^{ij} = \delta_{ij}$, we can find "unique" (up to some normalizations) operators

$$L = \begin{bmatrix} \partial & 0 \\ 0 & -\partial \end{bmatrix} + \begin{bmatrix} 0 & -q \\ r & 0 \end{bmatrix}, \quad B_2 = \begin{bmatrix} \partial^2 & 0 \\ 0 & -\partial^2 \end{bmatrix} + \begin{bmatrix} 0 & -q \\ r & 0 \end{bmatrix} \partial + \begin{bmatrix} \alpha & \beta \\ \gamma & \delta \end{bmatrix},$$

$$B_3 = \begin{bmatrix} \partial^3 & 0 \\ 0 & \partial^3 \end{bmatrix} - \frac{3}{2} \begin{bmatrix} qr & q_x \\ r_x & qr \end{bmatrix} \partial + \frac{3}{4} \begin{bmatrix} q_x r + q r_x & q_{xx} \\ r_{xx} & q_x r + q r_x \end{bmatrix}$$

such that

$$L[\psi\Big|_{P_1} \ \psi\Big|_{P_2}] = \partial_y[\psi\Big|_{P_1} \ \psi\Big|_{P_2}], \quad B_2[\psi\Big|_{P_1} \ \psi\Big|_{P_2}] = \partial_{t_2}[\psi\Big|_{P_1} \ \psi\Big|_{P_2}],$$

and $B_3[\psi\Big|_{P_1} \ \psi\Big|_{P_2}] = \partial_{t_3}[\psi\Big|_{P_1} \ \psi\Big|_{P_2}]$, where $\psi = [\psi_1, \psi_2]^T$ and the "time" deformations of the line bundle are imposed by a suitable choice of the polynomials R_j:

$$[y\kappa_1, -y\kappa_2]; \ [t_2\kappa_1^2, -t_2\kappa_2^2]; \ [t_3\kappa_1^3, t_3\kappa_2^3].$$

2.2.1 Remark. The choices are not so obvious from an algebraic point of view, see also Remark 2.2.2. Why not seek (nontrivial) Lax pairs for $L = \begin{bmatrix} \partial & 0 \\ 0 & \partial \end{bmatrix} + \begin{bmatrix} 0 & u \\ v & 0 \end{bmatrix}$? This follows from the (only) assumption made in [W1] on the form of the matrices, namely: the leading coefficient is an invertible diagonal matrix $u_n = \mathrm{diag}(c_1, \ldots, c_l)$ and the next coefficient u_{n-1} is a matrix of functions with $u_{n-1,ij} = 0$ if $c_i = c_j$. To quote [W1], "The purpose of this assumption is to ensure that we can conjugate L into its leading term by a suitable 'integral operator.' The assumption is true for all the operators L that have arisen in applications, though this may be partly because the examples were found using the construction [given below]." However, the geometric meaning is clear: we ask that the line bundle be deformed linearly (thus the transition $\exp(\Sigma R_j(\kappa_j)t_j)$). Let's call ∇_i the (translation invariant) flow on G corresponding to the tangent vector ∂_{κ_i} to C at P_i; the various Lax operators B_j are simply the matricial representation of these deformations on a normalized set of transition functions ψ_1, ψ_2: thus $L\psi = \partial_x \psi \Rightarrow L = \begin{bmatrix} \partial & 0 \\ 0 & \partial \end{bmatrix}$; $L\psi = \partial_y \psi$ is the deformation $\nabla_1 - \nabla_2$, etc. As per Krichever's construction, a flow is trivial $\Leftrightarrow$ the corresponding deformation comes from a global function regular on $C \backslash \{P_1, P_2\}$. Thus, if C is hyperelliptic, $P_1 + P_2$ is the hyperelliptic divisor, and f is a function whose only (simple) poles are P_1 and P_2, then the choice $\kappa_1 = f = -\kappa_2$ will produce a y-independent hierarchy, the spectral equation $L\psi = \partial_y \psi$ will become $L\psi = f\psi = \begin{bmatrix} \kappa_1 & 0 \\ 0 & -\kappa_2 \end{bmatrix} [\psi\Big|_{P_1} \ \psi\Big|_{P_2}]$ and the Lax pair $[L_1, B - \partial_{t_2}] = 0$ will give the coupled NLS (=nonlinear Schrödinger)(the "reality conditions" $\overline{q} = \pm r$ give the usual NLS) $\begin{cases} q_t = \dfrac{q_{xx}}{2} - q^2 r \\ r_t = \dfrac{r_{xx}}{2} + r^2 q \end{cases}$. Notice that if this is the case the flow $[\tilde{t}_2\kappa_1^2, \ \tilde{t}_2\kappa_2^2]$ is trivial as well as $[\tilde{t}_3\kappa_1^3, -\tilde{t}_3\kappa_3^3]$, which is why we chose B_2, B_3 the way we did. In fact, under the same assumptions, $[L, -4B_3 - \partial_t]$ yields mKdV: $q_t + 6q_x qr + q_{xxx} = 0 = q_t \pm 6q_x q^2 + q_{xxx}$ for $q = \pm r$. However, if we don't impose restrictions on the curve, we obtain the Davey-Stewartson equation[1]! Indeed, the Lax pair $[L - \partial_y, B_2 - \partial_t]$ yields

$$\begin{cases} q_t = \dfrac{q_{xx}}{2} + \dfrac{q_{yy}}{2} + q(\alpha - \delta) \\ r_t = \dfrac{r_{xx}}{2} - \dfrac{r_{yy}}{2} - r(\alpha - \delta) \end{cases}$$

together with

$$\begin{cases} \alpha_x - \alpha_y = -\dfrac{(qr)_x}{2} - \dfrac{(qr)_y}{2} \\ \delta_x + \delta_y = \dfrac{(qr)_x}{2} - \dfrac{(qr)_y}{2} \end{cases}$$

so that letting $\alpha - \delta = Q - qr$ one obtains $q_t = \frac{q_{xx}}{2} + \frac{q_{yy}}{2} - q^2 r + qQ$, where $Q_{xx} + Q_{yy} = 2(qr)_{yy}$, which becomes exactly Davey-Stewartson [AS (2.1.59)] under the reality conditions $\overline{q} = \pm r$. As a last remark, we offer an algorithm for writing (in the general-curve case) the (formal) pseudodifferential (matrix) operator $\mathcal{L}$ such that $\mathcal{L}\psi = \mathrm{diag}(\kappa_1, -\kappa_2)\psi$ hence such that $(\mathcal{L})_+ = L$; it is nothing but the Euclidean algorithm adapted to noncommutative rings, cf. [P3].

2.2.2 Remark. There is of course a dictionary which takes scalar to matrix operators, though not in general the other way around, so that (generalized) KdV has been presented as a matrix problem. I would like to explain a point of potential confusion (for me at least), which is also pertinent to the mKdV situation. There are 2 essentially different 2×2 matrix models for KdV: [DS] set it up for $\tilde{L} = \begin{bmatrix} \partial & 0 \\ 0 & \partial \end{bmatrix} + \begin{bmatrix} 0 & q \\ 0 & 0 \end{bmatrix}$ and [AS] for $L = \begin{bmatrix} \partial & 0 \\ 0 & -\partial \end{bmatrix} - \begin{bmatrix} 0 & q \\ 1 & 0 \end{bmatrix}$. You may recall that the $\tilde{L}$-format was excluded in 2.2.1, unless $q \equiv 0$; however, the corresponding spectral problem is $\tilde{L}\underline{y} = \Lambda\underline{y}$ where Λ is the matrix $\begin{bmatrix} 0 & 1 \\ \lambda & 0 \end{bmatrix}$; thus, this is nothing but the scalar $\leftrightarrow$ matrix isometry defined in [SW] and explained thereby as a correspondence between $Gr^{(2)} \subset Gr$ and a homogeneous space for the loop group L(SL(2)). In particular, the corresponding scalar equation is $(\partial^2 + q)y_1 = \lambda y_1$, so that λ is still the square of a local parameter κ at P_∞ on C. On the contrary, the [AS] format of L and $B = -4\begin{bmatrix} \partial^3 & 0 \\ 0 & \partial^3 \end{bmatrix} - 6\begin{bmatrix} q & -q_x \\ 0 & q \end{bmatrix}\partial - 3\begin{bmatrix} q_x & -q_{xx} \\ 0 & q_x \end{bmatrix}$ does correspond to the 2-point geometry described above. Now this is puzzling: we saw that the line bundle flows on a generalized Jacobian, whereas if we present KdV as a scalar Lax pair the line bundle flows on a

[1] I am grateful to J. Harnad for telling me about this "2+1" (i.e. (x, y) and t) version of NLS a few years ago.

compact Jacobian. The explanation is the following: the curve $\tilde{C}$ where the 2-point construction is implemented is a 2:1 cover of a 1-point curve C (P_∞ is a branch point for the hyperelliptic involution); the constraint $r \equiv -1$ for the solution is achieved on a subvariety of G which can be identified with Jac C. For example, in genus 1 (i.e., $[L, B] = 0$ with a third-order B of the kind given above) the curve $\tilde{C}$ is calculated as the resultant of $L - \lambda, B - \mu$ given by the determinant of a 6×6 BC matrix as in [P3]; the equation is $\mu^2 = \lambda^3 + a\lambda^2 + b$, obviously a 2:1 cover of a KdV curve. More instances of this phenomenon will be illustrated in 2.3.3 (sine-Gordon, mKdV).

Before leaving this subsection we should say that Krichever's construction gives explicit solutions in terms of theta functions. In view of that, we record what the mKdV/Davey Stewartson functions are in terms of the coefficients of the Baker vectors: $q = 2\xi_1^{12}$, $r = 2\xi_1^{21}$, $\xi_{1x}^{11} + \xi_{1y}^{11} = \xi_{1x}^{22} + \xi_{1y}^{22} = \frac{qr}{2}$. Using formulas of J. Fay's it is possible to check directly that the corresponding theta-function expressions, though transcendental, satisfy the given PDE (cf. [Mu,P1]). However, imposing reality conditions and constraints becomes easier if we use algebraic functions on C, as explained in 2.3.

2.3 Classical integrable Hamiltonian systems. I would like to present three manifestations of these. The first to be discussed, because the most mysterious for me, is a continuum of harmonic oscillators; the second, the residue theorem for a given differential on C; the third, a moment-map construction which produces a matrix Lax-pair polynomial in the curve parameter, as opposed to the differential-operator Lax pair of 2.2.

2.3.1 [DLT] used Jost eigenfunctions labeled by $k \in \mathbf{R}$ as symplectic coordinates to write the evolution of several second-order problems as constrained harmonic oscillators via the IST (Inverse Scattering Transform). Their formulas are extremely (and mysteriously) inspirational for the finite-dimensional model, which they often yield by a simple renormalization and by letting $k = 1, \ldots, n$. Here we just want to exhibit the squared eigenfunctions. It so happens that the analytic formalism (Poisson brackets and trace formulas/constraints) hasn't been worked out for mKdV. It may be worth pursuing this Hamiltonian interpretation of the scattering data, in view of the result of [AKS] to the effect that rapidly vanishing KdV potentials produced by MT from rapidly vanishing mKdV potentials are sparse. The scattering problem for $L = \begin{bmatrix} \partial & -q \\ r & -\partial \end{bmatrix}$ is the following: $\phi = [\phi_1 \phi_2]^T$ and $\psi = [\psi_1 \psi_2]^T$ are 2-vectors of functions x and $\mathrm{i}\kappa$, solutions of $Ly = \mathrm{i}\kappa y$, satisfying the asymptotic conditions: $\phi_+ \sim \left[\begin{smallmatrix} 1 \\ 0 \end{smallmatrix}\right] e^{-\mathrm{i}\kappa x}$, $\phi_- \sim \left[\begin{smallmatrix} 0 \\ -1 \end{smallmatrix}\right] e^{\mathrm{i}\kappa x}$ as $x \to -\infty$ and $\psi_+ \sim \left[\begin{smallmatrix} 0 \\ 1 \end{smallmatrix}\right] e^{\mathrm{i}\kappa x}$, $\psi_- \sim \left[\begin{smallmatrix} 1 \\ 0 \end{smallmatrix}\right] e^{-\mathrm{i}\kappa x}$ as $x \to \infty$, and are therefore linked by an x-independent matrix: $[\phi_+ \phi_-] = [\psi_+ \psi_-] \begin{bmatrix} b_+ & b_- \\ a_+ & a_- \end{bmatrix}$. The entries of the scattering matrix are "Wronskians" $W[uv] = u_1 v_2 - u_2 v_1$, indeed: $\phi_+ = a_+ \psi_- + b_+ \psi_+ \to \begin{bmatrix} a_+ e^{-\mathrm{i}\kappa x} \\ b_+ e^{\mathrm{i}\kappa x} \end{bmatrix}$ as $x \to \infty$, $\phi_- = b_- \psi_- - a_- \psi_+ \to$

$\begin{bmatrix} b_- e^{-i\kappa x} \\ -a_- e^{i\kappa x} \end{bmatrix}$ as $x \to \infty$ and $\{L$ traceless $\Rightarrow$ constant Wronskian$\}$, imply $a_+ = W(\phi_+, \psi_+)$, $a_- = W(\phi_-, \psi_-)$, $b_+ = -W(\phi_+, \psi_-)$, $b_- = W(\phi_-, \phi_+)$; and $W(\phi_+, \phi_-) = -1 \Rightarrow a_+ a_- + b_+ b_- = 1$. Assuming that a_+ and a_- are root free (analogous to requiring absence of bound stakes for a Schrödinger potential), and letting $r_+ = b_+/a_+$, $r_- = b_-/a_-$, x-translation becomes constrained harmonic motion for the canonical variables: $x_{+\kappa} = \sqrt{\frac{r_+(\kappa)}{\pi\kappa}}\psi_{+1}(0,\kappa)$, $y_{+\kappa} = \sqrt{\frac{r_+(\kappa)}{\pi\kappa}}\psi_{+2}(0,\kappa)$, etc.; all higher times of the hierarchy are also given by commuting hamiltonians.

2.3.2 We shall use the scalar (KdV/Neumann) case for illustration; we would have worked out the (philosophically similar) matrix case but for lack of time. The idea goes back to [BC] and is one of their major breakthroughs–the other is the introduction of the "transference" operation on a commutative algebra of ODO's, the identification of transference with the abelian sum and the proof that g transferences produce the general solution associated to a fixed spectral curve of (arithmetic) genus g. We start with a commutative pair of ODO's, L and B, of coprime orders n and m, normalized as usual: $L = \partial^n +$ lower-order terms. Without detracting from the main ideas, we keep the presentation simple by assuming that $\mathcal{C_D}(L)$ defines the affine part of a smooth curve C, given by an equation $f(\lambda, \mu) = 0$, and the functions $\lambda \leftrightarrow L$, $\mu \leftrightarrow B$ generate this ring. The genus of C is $\frac{1}{2}(m-1)(n-1)$. Note that this is the typical KdV situation ($n = 2$, $m = 2g+1$). Now let $\psi(x, P)(= \psi(x, \kappa)$ near $P_\infty)$ be the Baker function and ϕ be the dual Baker function, associated to the ring of formal adjoints, $\mathcal{C_D}(L^\dagger)$ where $L^\dagger = (-\partial)^n + (\partial)^{n-2}u_{n-2}(x) + \ldots + u_0(x) = \pm[\partial^n + u^*_{n-2}\partial^{n-2} + \ldots + u^*_0]$. What [BC] realized was that the line bundle (whose linear flow on Jac C gives the solution!) can be given in its algebraic manifestation (i.e. as a set of points satisfying $f(\lambda, \mu) = 0$) by quadratic functions in the ψ, ϕ and their x-derivatives. We quote two results and then switch to the $n = 2$ case.

(A) Let (λ, μ) be a point on the curve and $f(L, \mu) = T(L - \lambda)(B - \mu) + (a\partial^{n-2} + a_1\partial^{n-3} + \ldots + a_{n-1})(B - \mu)$, with $(b, b_1, \ldots, b_{n-1})$ playing the role of the a's for the adjoint equation. Then $a = b = \psi\phi$ generically cuts the curve C in $2g$ points, half of which lie on a_1 and the other half on b_1. Using the differential equations $L\psi = \lambda\psi$, etc., the motion of the divisor $Q_1 + \ldots + Q_g$ common to a and a_1, say, is computed in terms of the Abel coordinates $\Sigma \int_{P_0}^{Q_i} \omega$, where $\omega = (\lambda^a \mu^b \frac{d\mu}{f_\lambda})_{nb+ma<2g-2}$, and is shown to be linear.

(B) [P3] The $n \times n$ "BC matrix" Λ which represents the action of $B - \mu$ on the eigenspace of $L - \lambda$ has (generically) the following property: the divisor $Q_1 + \ldots + Q_g + P_\infty$ defined by the first row of adj Λ (namely the points which satisfy the n simultaneous equations) is linearly equivalent to the divisor $S_1 + \ldots + S_{g+1}$ defined by the second row (and so forth), which gives a geometric construction of the abelian sum.

Finally, we follow [EF] for the case $n = 2$. The dual Baker function is most explicit precisely because the curve is hyperelliptic: if we write the equation of the curve as $\mu^2 = f(\lambda) = \prod_1^{2g+1}(\lambda - e_i)$ and if ψ satisfies $\mathcal{L}\psi = \kappa\psi$ with $\kappa = \lambda^{1/2}$, then $\phi = \psi(\iota P)$, where ι is the hyperelliptic involution; ψ, ϕ are independent solutions of $Ly = \lambda y$ except at the branchpoints, and $\psi\phi$ is a meromorphic function which comes from the base (i.e., a function of $\lambda \in \mathbf{P}^1$ only): $\psi\phi = \prod^g(\lambda - \lambda(Q_i(x)))/(\lambda - \lambda(Q_i))$. We can push the poles to $\lambda = \infty$ by defining the differential $\Omega = \frac{1}{2}\prod^g(\lambda - \lambda(Q_i))\frac{d\lambda}{\mu}$; then $U = \psi\phi\Omega/d\lambda$, $w = \psi_x\phi_x\Omega/d\lambda$, $V = \frac{i}{2}(\psi\phi_x + \psi_x\phi)\Omega/d\lambda$ are polynomials in λ and $UW + V^2 = -Wr^2(\psi, \phi) = f(\lambda)$ independent of x (notice that it equals zero precisely when λ is a branch point). The first g coefficients of f are constants of motion and by the Liouville method one gets action-angle variables for a completely integrable system [FN]. However, there is quite a different way to construct a symplectic manifold using these objects which is formally analogous to the continuous model described in 2.3.1: choose $g+1$ among the branch points, say $E_i = (e_i, 0)$, $i = 1, \ldots, g+1$ and let $h = \mu/\prod_i^{g+1}(\lambda - e_i)$. Define $\rho_i = \operatorname{res}_{(e_i,0)}h\Omega$ and $x_i = \sqrt{\rho_i}\psi(x, E_i)$. Then by the residue theorem $\sum_1^{g+1} x_i^2 \equiv 1$ is x-independent; $U = \sum^{g+1}\frac{x_i^2}{\lambda - e_i}$; $y_i = \dot{x}_i$ and $u = \sum^{g+1}(e_i x_i^2 + y_i^2)$ yield the Neumann system of harmonic oscillators constrained to the unit sphere: $\ddot{x}_i + ux_i = e_i x_i$ $(i = 1, \ldots, g+1)$. The finite-dimensional analog of the trace formula of [DLT] gives the algebraic solution to the inverse problem (as opposed to the transcendental of 2.2): the equation for the squared eigenfunctions $(\partial^3 + 4u\partial + 2u')\psi\phi = 4\lambda(\psi\phi)'$, when expanded in λ, gives $u(x) = \sum_1^{2g+1} e_i - 2\sum_1^g \lambda(Q_i(x))$.

2.3.3 Lastly, we use the moment map of [AHP] to present the hamiltonian flows of 2.3.2 as a reduction of Adler-Kostant-Symes flows for an $sl(2)$ loop algebra; here we do the mKdV case. We refer to [AHP §6D] for all definitions and here present only the result, and its geometric interpretation which had not appeared in [AHP]. On a symplectic space M with canonical coordinates $x_1, \ldots, x_n$; $y_1, \ldots, y_n$ one defines a moment map whose image is the matrix

$$J = \frac{\lambda}{2}\begin{bmatrix} \sum_1^n \frac{\lambda x_i y_i}{\lambda^2 - a_i^2} & \sum_1^n \frac{a_i y_i^2}{\lambda^2 - a_i^2} \\ -\sum_1^n \frac{a_i x_i^2}{\lambda^2 - a_i^2} & -\sum_1^n \frac{\lambda x_i y_i}{\lambda^2 - a_i^2} \end{bmatrix}.$$

If we let $a(\lambda) = \prod_1^n(\lambda^2 - a_i^2)$, the matrix $\mathcal{L}(\lambda) = \frac{a(\lambda)}{\lambda^{2n}}J = \mathcal{L}_0 + \frac{1}{\lambda}\mathcal{L}_1 + \ldots + \frac{1}{\lambda^{2n-1}}$ $\mathcal{L}_{2n-1}$ defines the Lax-pair flows

$$\partial_{t_k}\mathcal{L}(\lambda) = [(\lambda^{2k}\mathcal{L}(\lambda))_+, \mathcal{L}(\lambda)]$$

for which the curve $\tilde{C}$: $\mu^2 = \det(\frac{a(\lambda)}{\lambda}J)$ (compactified by adding the points P_1 and P_2 over $\lambda = \infty$) is isospectral. If the following (time-invariant) functions are fixed: $\Sigma x_i y_i = 0$, $\Sigma a_i x_i^2 = \Sigma a_i y_i^2 = 1$, $v^2 + \frac{1}{2}(q - p) = 0$ where: $\mathcal{L}_2 = \begin{bmatrix} v & 0 \\ 0 & -v \end{bmatrix}$, $\mathcal{L}_3 = \begin{bmatrix} 0 & p \\ q & 0 \end{bmatrix}$, then v satisfies mKdV for $x = t_1$ and $t = t_2$. Note: the curve is invariant under the transformation: $v \mapsto -v$.

We articulate the geometric interpretation into a series of remarks, and precede them by asking the natural questions which arise from bringing together the flow descriptions of 2.1 and 2.3.2. The questions are: does the mKdV hierarchy linearize on the *generalized* Jacobian of the curve $\tilde{C}$ which we found, as does the 2-point Baker vector of 2.2, or does it stay away from the $\mathbf{C}^*$ extension? What is the dimension of the orbit of the integrable Hamiltonian system, analogous to the Neumann system of 2.3.2, and how are the "action" variables described? The potentially confusing feature is the following: KdV (Neumann) evolves on a hyperelliptic Jacobian, with a corresponding 1-point Baker function, as we saw. Geometry, unlike analysis which has to reckon with special classes of solutions [AKS], *can* find an mKdV for any KdV, so that $u = MT(v) = -v^2 - v_x$; in fact, it finds a 1-dimensional fibre of such v, as we will see in §3; and they correspond to a generalization of the transferences of 2.3.2: indeed, they correspond to a choice of solution y of $Ly = 0$ $(L = \partial^2 + u)$, or a factorization $L = (\partial + v)(\partial - v)$, $v = \frac{y'}{y}$; this is a $\mathbf{P}^1$ (since we must assume $y(0) \neq 0$) because y and cy for a nonzero constant c give the same v. Ehlers-Knörrer [EK] interpret this geometrically: if C: $\mu^2 = \prod^{2g+1}(\lambda - e_i)$ was the KdV curve, then the various v correspond to a $\mathbf{C}^*$ extension of Jac C, obtained by imposing the singularity $\mu^2 = \lambda^2 \prod^{2g+1}(\lambda - e_i)$ (more details in §3). It is tempting to think that by resolving the node we get the curve of the 2-point model, and the mKdV hierarchy possesses the 2-point flow of the matrix model (2.2) along $\mathbf{C}^*$, thus producing a Neumann-like system with one extra "angle" variable which projects to a conventional Neumann system. This is not quite so, as I shall show by a sequence of examples and remarks; the conclusion is summarized in (v) below.

(i) The single wave. The 1-soliton KdV solution was found classically, and 1-wave solutions to several integrable PDE's can also be found quite effortlessly. The method is not powerful enough to suggest higher-genus solutions, but it may point quite clearly to its geometric interpretation, as in the case of NLS. We follow [SCMcL]. KdV is solved by integrating twice the equation which results from the ODE $u''' - cu' + 6uu' = 0$ for $u(x - ct)$; indeed, $\frac{u'^2}{2} = -u^3 - \frac{c}{2}u^2 + au + b$ $(a, b$ are the integration constants) can be solved by the Weierstrass $\wp$-function and its singular (elementary) limits. NLS requires two velocities (envelope and carrier): $q = \phi(x - ct)e^{i\theta(x-dt)}$, ϕ and θ real; the real and imaginary parts of $q_{xx} + iq_t + k|q|^2 q = 0$ give

$$\begin{cases} \phi'' - \phi\theta'^2 + d\phi\theta' + k\phi^3 = 0 \\ \phi\theta'' + 2\phi'\theta' - c\phi' = 0 \end{cases}$$

The second equation integrates to $\phi^2(2\theta' - c) = \text{const}$, and the choice $\theta' = c/2$ integrates the first: $\phi'^2 = -\frac{k}{2}\phi^4 + \frac{1}{4}(c^2 - 2cd)\phi^2 + a$. This is again an elliptic curve (or a singular limit) but has 2 points at infinity and the form of the solution q is indeed its generalized theta function! Now for mKdV (not in [SCMcL]) the ansatz $v = e^{x-dt}w(x - ct)$ is not integrable in general, whereas $v(x - ct)$ yields $\frac{v'^2}{2} = \frac{1}{2}v^4 + c\frac{v^2}{2} - av - b$. The choice $c = a = b = 0$ gives $v = -\frac{1}{x}$, which corresponds to the rational KdV, $MT(v) = -\frac{2}{x^2}$, obtained from the cuspidal cubic $\mu^2 = \lambda^3$. The choice $a = b = 0$ gives a nodal cubic and, indeed, a 1-soliton $v = -\sqrt{2c}$ sech $\sqrt{2c}(x - ct + \text{const})$. **Exercise 1:** Given v such that $\frac{v'^2}{2} = \frac{1}{2}v^4 + c\frac{v^2}{2} - av - b$, show that the curve of $u = MT(v)$ is $\frac{u'^2}{2} = -u^3 + \frac{c}{2}u^2 - 2bu + bc + \frac{a^2}{2}$; explain why replacing v by $-v$ changes the mKdV curve ($a \mapsto -a$) but not the KdV curve (this will be explained in §3). **Exercise 2:** Show that all v's corresponding to $u = -\frac{2}{x^2}$ are of the form $-\frac{1}{x-\alpha}$ (this will be explained in §3).

(ii) Parameter count. The Hamiltonian systems which give evolutions for second-order Lax pairs were unified by J. Moser under the model of "rank 2 perturbations." In a way very reminiscent of mKdV, to produce the sine-Gordon flows we need to introduce a symmetry in the rank 2 perturbation, which has the effect of linearizing the flows on the Jacobian of a curve different from the spectral curve of the system. We showed in [P2] that on certain orbits of a rank 2 perturbation of a matrix diag $(a_1, \dots, a_n, -a_1, \dots, -a_n)$ the flows preserve an "even" curve $\tilde{C}$: $\mu^2 = \phi(\lambda^2)$ and evolve on Jac C where the covering $\tilde{C} \to C$ is given by $(\lambda, \mu) \mapsto (\eta = \lambda^2,\ \nu = \mu/\lambda)$ and C has equation $\nu^2 = \eta\phi(\eta)$. $\tilde{C}$ has genus $2n - 3$ and the flows span its generalized Jacobian (the $\mathbf{C}^*$ extension corresponds to it 2 points over $\lambda = \infty$), but the $n - 1$ "even" flows span the Jacobian of C. E. Date writes the sine-Gordon solution using generalized theta functions on $\tilde{C}$, while D. Mumford uses theta functions on C; geometrically, $\tilde{C}$ is the fibre product:

$$
\begin{array}{ccccc}
 & & \tilde{C} & & \\
 & \swarrow & \downarrow & \searrow \lambda & \\
C & & H & & \mathbf{P}^1 \\
 & \searrow & \downarrow & \swarrow & \\
 & & \mathbf{P}^1 & &
\end{array}
$$

for the maps $\eta: C \to \mathbf{P}^1$, branched at $a_1^2, \dots, a_n^2$, $b_1^2, \dots, b_{n-2}^2$ and 0; and $\lambda^2: \mathbf{P}^1 \to \mathbf{P}^1$. For the same rank 2 perturbation, the symmetry we impose to get mKdV flows is the following: we start with canonical coordinates $(w_1, \dots, w_n; w_{n+1}, \dots, w_{2n})$, $(z_1, \dots, z_n; z_{n+1}, \dots, z_{2n})$ and form the polynomials $U(\lambda) = \sum_{i=1}^n \left(\frac{w_i^2}{\lambda - a_i} + \frac{w_{n+i}^2}{\lambda + a_i}\right) a(\lambda)$, where $a(\lambda) = \prod_1^n (\lambda^2 - a_i^2)$ similarly, $V(\lambda)$ by substituting $w_i^2 \mapsto z_i^2$ and $Z(\lambda)$ by $w_i^2 \mapsto w_i z_i$, so that the coefficients of $\phi(\lambda) = UV - Z^2$ are invariants; notice that $\phi(\lambda)$ is divisible by $a(\lambda)$, and the constraints $\sum_{j=1}^{2n} w_j z_j = \sum^{2n} w_j^2 = \sum^{2n} z_j^2 = \sum^n a_j(w_j^2 - w_{n+j}^2) = \sum^n a_j(z_j^2 - z_{n+j}^2) = 0$, $\sum^n a_j(w_j z_j - w_{n+j} z_{n+j}) = \mathrm{i}$, ϕ is monic of degree $4n-4$. We get an mKdV solution as $u = \mp v$, where $u = u_{2n-3} = \sum^n a_j(w_j^2 - w_{n+j}^2)$ is the leading coefficient of U, $v = v_{2n-3}$ of V, and alternate this condition down the coefficients, $u_{2n-4} = \pm v_{2n-4}$, etc. This symmetry is achieved on the manifold $w_{n+j} = \mathrm{i} z_j$ and $z_{n+j} = -\mathrm{i} w_j$ so that

$$U=\sum_{1}^{n}\prod_{i\neq j}(\lambda^2-a_i^2)\ (\lambda(w_j^2-z_j^2)+a_j(w_j^2+z_j^2))$$

$$V=\sum_{1}^{n}\prod_{i\neq j}(\lambda^2-a_i^2)(\lambda(z_j^2-w_j^2)+a_j(w_j^2+z_j^2))$$

$$Z=\mathrm{i}\sum_{1}^{n}\prod_{i\neq j}(\lambda^2-a_i^2)2a_jw_jz_j.$$

Indeed, the transformations $x_i=w_i-z_i$, $y_i=w_i+z_i$, and $T=\frac{U-V}{2}$, $R=-Z+\frac{U+V}{2}$, $S=-Z+\frac{U+V}{2}$ bring the curve $UV-Z^2$ to

$$RS-T^2=(a(\lambda))^2\det\begin{bmatrix}\Sigma\frac{\lambda x_iy_i}{\lambda^2-a_i^2} & \Sigma\frac{a_iy_i^2}{\lambda^2-a_i^2}\\ -\Sigma\frac{a_ix_i^2}{\lambda^2-a_i^2} & -\Sigma\frac{\lambda x_iy_i}{\lambda^2-a_i^2}\end{bmatrix}$$

Remark. The base change $\begin{bmatrix}y_i\\x_i\end{bmatrix}=\begin{bmatrix}1&1\\-1&1\end{bmatrix}\begin{bmatrix}z_i\\w_i\end{bmatrix}$ conjugates the Lax operator $\begin{bmatrix}\partial & v\\-v&-\partial\end{bmatrix}$ into $\begin{bmatrix}0&\partial-v\\\partial+v&0\end{bmatrix}$, with the effect of interchanging the role of two Cartan subalgebras of $sl(2)$, the diagonal matrices and the circulants.

(iii) Example: genus 1. A parameter count shows that in general the flows of the Hamiltonian system span the generalized Jacobian of the curve $\mu^2=UV-Z^2=\phi(\lambda^2)$ (which has dimension $2n-2$), but the "even" flows span the Jacobian of $\mu^2=\eta\phi(\eta)$. For example, for $n=2$ the constraints we imposed leave one degree of freedom, which is a direction on the generalized Jacobian of the curve $\tilde{C}$: $(\lambda^2+p)(\lambda^2+q)-\lambda^2v^2=a(\lambda)\mu^2$, or the Jacobian of C: $\eta(\eta+p)(\eta+q)-\eta^2v^2=\eta(\eta-a_1^2)(\eta-a_2^2)=\nu^2$. Calculating the x-flow gives: $+v_x-v^2=-p$, $-v_x-v^2=-q$. Conclusion: C hosts both the KdV and the mKdV flows; $\tilde{C}$ gives a different way to write solutions in terms of theta functions; unlike the mKdV curve found in (i), it is insensitive to the transformation $v\mapsto -v$. We give the theta-function recipe as the next item because it holds for any genus:

(iv) Theta formulas. The solution for the 2-point hierarchy, as per Krichever's construction, is worked out in [P1] and involves a periodic function $e^{z_0}\vartheta(z+P_2-P_1)/\vartheta(z)$, with an extra "Abel coordinate" z_0 which corresponds to a logarithmic differential $\omega_{P_2-P_1}$. But, as in the sine-Gordon case, the solution can be written using only theta functions of the curve C, which is to be expected as a consequence of (ii), in that the flows are linearized on Jac C. The representation-theoretic construction which will be briefly sketched in §3 provides the formula $v=\partial_x\log\frac{\tau_1}{\tau_0}$ for two different tau functions: but that appears as simply a translation in the algebro-geometric set-up, since the KdV solution is $u=2\partial_x^2\log\vartheta(z)$+const., with the constant depending only on the curve; the "transference" operation shows $\tilde{u}=2\partial_x^2\log\vartheta(z+P)$; and finally $v_x=\frac{1}{2}(\tilde{u}-u)=\partial_x^2\log(\vartheta(z+P)/\vartheta(z))$ and $v=\partial_x\log(\vartheta_P/\vartheta)$ (the mKdV equation constrains the integration constant). It is satisfying to compare these formulas directly in the genus 1 case: start with a commuting pair (L, B) as in 2.1, with $q = -v = r$. Write the 4×4 Burchnall–Chaundy matrix as per recipe in [P3] and compute its determinant; this is the curve, which you find to have the form: $(\mu - \frac{c}{2})^2 = \lambda^4 + c\lambda^2 + a + \frac{c^2}{4}$ $(L = \lambda, B = \mu)$. Now use the abelian coordinates of the flow computed in [P1, Th. 1.7] and you will see that the coordinate in the $\mathbf{C}^*$ direction is zero because of the particular value of the coefficients of the curve. Lastly, use Jacobi's functions [DV] for $\wp^{1/2}$ ($\wp$ of the curve $\nu^2 = \eta(\eta^2 + c\eta + a + \frac{c^2}{4})$, $\eta = \lambda^2, \nu = \mu\lambda$) to compare the 2-point formula for the theta function of the half-period lattice $\vartheta[\epsilon](2z, 2\tau)$ with the 1-point for $\vartheta(z, \tau)$.

(v) Conclusion. To complete the picture of linearization of mKdV flows we return to the relationship between the 3 curves we encountered: X: $\frac{v'^2}{2} = \frac{1}{2}v^4 + c\frac{v^2}{2} - av - b$ in (i), and $\tilde{C} \to C$ in (iii) (all statements go over unchanged for higher genus). The issue to be dealt with was this: the Lax-pair mKdV problem 2.1 seemed to indicate a 2-point situation, and so did the Moser-system solution 2.3. However, the flows do not require the generalized Jacobian of $\tilde{C}$ but linearize on the Jacobian of the KdV curve C.

3. The flag as a space of Darboux transformations

In this section we simply put together the language of [EK] with that of [W2] which shows how to view the $\mathbf{C}^*$ of Darboux transforms of a given KdV operator as the fibre of a projection from a flag variety to a Grassmannian. Originally we had intended to translate the classical flag-analog of the Plücker equations into the mKdV equations using finite-dimensional flags, in analogy to what is done in [E], but for lack of time we state that as an exercise in the next section.

3.1 We recall that a Darboux transformation is the conjugation of an operator L by $D = \partial - \frac{\psi'}{\psi}$ where $\psi \in \operatorname{Ker} L$ and $\psi(0) \neq 0$, so that L is divisible by D on the right and the new $\tilde{L} = DLD^{-1}$ is still regular at $x = 0$ (a reference point). If $\mathcal{L}$ and ψ evolve according to the KP hierarchy, then $\tilde{\mathcal{L}} = D\mathcal{L}D^{-1}$ does too and in the PDE context this transformation is named after Bäcklund. [EK] described the effect of such a Bäcklund transformation on the geometric data of the KdV equation. There are three possible cases, according to the properties of the commutative algebra $A = \mathcal{C}_{\mathcal{D}}(L)$. (1) If DAD^{-1} is not maximal commutative, then the curve C of A was singular and the new KdV potential is associated to a curve where the singularity is of order one less. (2) If DAD^{-1} contains some pseudodifferential operators, then the curve corresponding to $\mathcal{D} \cap DAD^{-1}$ is singular. (3) Finally, if DAD^{-1} is again maximal commutative, then the new KdV potential corresponds to the translation on Jac C by a point on the curve (if we start with the operator $L - c$, the point has λ-coordinate c). Now recall that the space of Darboux transformations is in 1:1 correspondence with the $\mathbf{P}^1$ of solutions y to $Ly = 0$, with $y(0) \neq 0$ and regarded up to nonzero multiplicative constant. [EK] also identify these as the $\mathbf{C}^*$ direction of the generalized Jacobian of the singular curve (the solutions corresponding to the two points $P, \iota P$ with λ-coordinate

$= c$ are deleted from $\mathbf{P}^1$ to give the $\mathbf{C}^*$). Lastly, the space of eigenfunctions of L is in 1:1 correspondence with the factorizations $L = (\partial + v)(\partial - v)$, $v = \frac{\psi'}{\psi}$, namely with the mKdV solutions whose $\mathrm{MT}(v) = u$; the Darboux transform of u is $\mathrm{MT}(-v)$. Thus, the extra (fibre) direction of mKdV potentials which give the same KdV lies indeed in the span of the KdV hierarchy, but in fact of the hierarchy which belongs to the singular curve of a Darbouxed potential (cf. comments preliminary to 2.2.3(i)).

Example: $u = 0$, $\tilde{u} = -\frac{2}{x^2}$ for $\psi = x$; $MT(v = \frac{1}{x}) = 0$, $MT(v = -\frac{1}{x}) = -\frac{2}{x^2}$; the other v which project to $u = 0$ are $\frac{1}{x-a}$; in fact (as a singular limit of the transference result): if $(\frac{1}{a^2}, \frac{1}{a^3})$ is a point on $\mu^2 = \lambda^3$, then $L - \frac{1}{a^2}$ Darboux transforms into $\partial^2 - \frac{2}{(x-a)^2} - \frac{1}{a^2}$. This clarifies the remarks in 2.2.3: $u = 0$ "flows" on Jac C_0, $\tilde{u} = -\frac{2}{x^2}$ on Jac C, $\pm v = \pm\frac{1}{x}$ "flows" on Jac C (by $\frac{1}{x} \to \frac{1}{x-a}$), thereby demonstrating that the $\mathbf{C}^*$ direction corresponding to a fixed $u = 0$ is a KdV direction for any other Darboux transform $\tilde{u}$, except the transferenced ones which are a translate of $u = 0$ on the same curve (in this case the choice $\lambda = 0$ corresponds to a branch point).

3.2 [DS] interpreted the mKdV solutions corresponding to a fixed KdV solution as a set of upper triangular matrices of the form

$$L = \begin{bmatrix} \partial & 0 \\ 0 & \partial \end{bmatrix} = \begin{bmatrix} 0 & \lambda \\ 1 & 0 \end{bmatrix} + \begin{bmatrix} w & z \\ 0 & t \end{bmatrix}$$

modulo the action of the group of gauge transformations, NLN^{-1}, for N an upper-triangular matrix of functions with 1 on the diagonal. The KdV solution appears for $w = t = 0, z = u$ while the diagonal case $z = 0$ has the mKdV diagonal terms $w = v$, $t = -v$ [DS, 3.23]. With one more switch in conventions (to follow [W2]) we replace $\begin{bmatrix} 0 & \lambda \\ 1 & 0 \end{bmatrix}$ by $\begin{bmatrix} 0 & \lambda^{1/2} \\ \lambda^{1/2} & 0 \end{bmatrix}$, which amounts to replacing the standard with the principal realization of the group G of smooth maps $S^1 \to \mathrm{SL}(2, \mathbf{C})$, via the isomorphism:

$$\begin{bmatrix} a(z) & b(z) \\ c(z) & d(z) \end{bmatrix} \leftrightarrow \begin{bmatrix} a(z^2) & zb(z^2) \\ z^{-1}c(z^2) & d(z^2) \end{bmatrix}.$$

We define several subgroups of G: first of all we let G act on the (2-) vector Grassmannian by the rule: $g \circ \underline{\psi} = [\psi_0\psi_1]g^{-1}$; we then interpret the action on the scalar Grassmannian by the usual isometry $\psi(z) = \psi_0(z^2) + z\psi_1(z^2)$. As usual A is the subgroup of diagonal constant matrices and $U_\pm$ the elements whose constant term is the identity and which extend holomorphically to the interior/exterior of the disc $D_0 = \{z\colon |z| < 1\}$. If Γ is the centralizer of $\begin{bmatrix} 0 & z \\ z & 0 \end{bmatrix}$ in G, we let $\Gamma_\pm = \Gamma \cap U_\pm$. If H_+ is the usual reference space of boundary values of functions holomorphic in D_0, we define P_0 and P_1 as the stabilizers of H_+, zH_+, resp., and B as the subgroup AU_+. In other words, B is the set of elements of G that extend holomorphically to D_0 and P_0 of those that extend in the standard realization. The difference is that for a Fourier expansion $f(z) = \sum_0^\infty a_i z^i$, an $f \in P_0$ is in B iff a_0 is upper triangular; moreover, $B = P_0 \cap P_1$. Modulo technical exceptions, the Grassmannian $Gr^{(2)}$ is the orbit of H_+ under G, namely the homogeneous space G/P_0; the projection to the flag manifold G/B (=the G orbit of $H_+^{(2)}$) has fibre $\mathbf{P}^1$. The usual construction of KdV solutions associated to points of $Gr^{(2)}$ generalizes as follows:

3.3.1 [W2, Prop. 2.5] There is an injective map $\Gamma_- \backslash G/B \to$ {mKdV solutions}, under which the mKdV flows are given by the natural action of $\Gamma_+ = \{\exp \sum_{odd} t_i \begin{bmatrix} 0 & z^i \\ z^i & 0 \end{bmatrix}\}$. Moreover, the following diagram is commutative:

$$\begin{array}{ccc} G/B & \to & \{\text{mKdV solutions}\} \\ \downarrow & & \downarrow \text{MT} \\ G/P_0 & \to & \{\text{KdV solutions}\} \end{array}$$

Lastly, since $B = P_0 \cap P_1$, the elements of G/B can be identified with flags $F = (W_0, W_1)$ of pairs of elements $W_i \in Gr^{(2)}$ such that $z^2 W_0 \subset z W_1 \subset W_0$. The KdV solutions associated to W_0, W_1 resp. are MT(v), MT($-v$); the Baker function of W_0 is obtained by adding the two entries of the Baker vector corresponding to F. At the differential operator level, this corresponds to: $L\psi = z^2\psi$ for $L_0 = \partial^2 + u = (\partial + v)(\partial - v)$, the KdV operator of W_0; $L_1 = (\partial - v)(\partial + v)$ corresponds to W_1; and

$$\begin{cases} (\partial - v)\psi_0 = z\psi_1 \\ (\partial + v)\psi_1 = z\psi_0 \end{cases}.$$

Writing this as a matrix $N\underline{\psi} = z\underline{\psi}$, $N = \begin{bmatrix} 0 & \partial + v \\ \partial - v & 0 \end{bmatrix} =$ $\begin{bmatrix} 1 & 1 \\ -1 & 1 \end{bmatrix}^{-1} M \begin{bmatrix} 1 & 1 \\ -1 & 1 \end{bmatrix}$ where $M = \begin{bmatrix} \partial & v \\ -v & -\partial \end{bmatrix}$. When setting up an analog of the Burchnall-Chaundy theory, we obtain a curve (see 2.3.3 (iv)) with affine ring given by the centralizer of N in the ring $\mathcal{M}$ of matrix differential operators, as defined in 2.1.

4. Questions

In the context of a summer school, we feel justified in stating some questions which are readily workable, together with open-ended ones: this is indicated at the end of each question. We take them up in the same order as the exposition.

4.1 In 1.2 we stated the conditions $U = \partial_s f, -W_s = \kappa U = \partial_s g$ for certain conservative curve motion. [GP1] remark that there are other obvious solutions besides the one which gives mKdV (e.g. $U = \kappa^n \kappa_s$ and $W = -\kappa^{n+2}/(n+2)$) and ask whether these lead to integrable hierarchies. (I would label this question as open-ended)

4.2 The Schwarzian derivative representation of a KdV solution in 1.2(B) has been generalized in [DN] for KP/modified KP; it may be interesting to have a curvature-dynamics interpretation for that. But a more far-reaching question is an understanding of the singular Krichever-Novikov (KN) equation and its generalized version in [DN] within the theory of rational KP solutions. The non-singular KN equations have never been solved explicitly; they appear (quite mysteriously, so far) in the construction of higher-rank KP solutions (the geometric data of a line bundle over a curve is replaced by a rank r bundle) over curves of genus 1 [KN]. In [LP] a (Darboux) transformation is given, which reduces them to the KdV equation exactly when the curve becomes singular. The most difficult and important question would be a *nonsingular* generalization of the KN equation for the KP case; a step in the direction of generalized KdV has been taken in [M], where the 2×2 Boussinesq system is obtained. (Very open, concrete)

4.3 In 1.2(C) we indicated the appearance of the Schwarzian derivative in the context of a projective connection (in the x-variable). Actually, a projective connection in the spectral parameter also arises in the KdV/KP theory, and I believe that finding a link between the two would be quite worthwhile, especially in view of the two appearances of Virasoro algebras in KP theory (one in x, cf. [Se], the other in the spectral parameter of τ, cf. [SW]). The projective connection in the spectral parameter is to be found in [Du]: if a KP solution $u = 2\partial^2 \log \vartheta(\zeta) + c$ is viewed as function of the choice of local parameter at ∞, then

$$3c(P) = \frac{\Sigma\omega_i''(P)\vartheta_i(\zeta)}{\Sigma\omega_i(P)\vartheta_i(\zeta)} - \frac{3}{2}\left(\frac{\Sigma\omega_i'(P)\vartheta_i(\zeta)}{\Sigma\omega_i(P)\vartheta_i(\zeta)}\right)^2 + \frac{3}{2}\left(\frac{\vartheta_{xx}(\zeta)}{\vartheta_x(\zeta)}\right)^2 - 2\frac{\vartheta_{xxx}(\zeta)}{\vartheta_x(\zeta)}.$$

This projective connection also appears in [Fa], p.19 formula 27. (Open; less focused than 4.2)

4.4 For the Davey-Stewartson solutions constructed from curves as in 2.2 to be of physical interest, a theory of reality conditions needs to be developed. Some special cases would probably come for free (much as was the case for NLS in [P1]) from an identification of the algebraic description of the flows, i.e. by putting together 2.2 and 2.3. Some properties would be a lot harder to detect, as demonstrated by KP versus KdV. (Workable)

4.5 In 2.3.1 we indicated a mysterious analogy between Poisson brackets of Jost solutions and finite dimensional Hamiltonian systems. The formal reasons for this are the equations satisfied by squared eigenfunctions. A closer analytic investigation of the scattering matrix as a limit of monodromy matrices for the periodic problem is a very commonplace thing to suggest. The specifics of the mKdV example, which brings together AKNS models and Hill's equation would be a nice starting point; one gain would be that the finite dimensional model of 2.3.3 would yield the continuous model (rather than the other way around) and perhaps enhance some spectral properties investigated in [AKS]. (Open/Workable)

4.6 In 2.3.2(B) we identified the Burchnall-Chaundy (BC) matrices as objects of rank-2 perturbations; through the BC theory every abelian flow on a Jacobian can be written as a Lax pair; but thanks to the Jacobi-Mumford model for hyperelliptic Jacobians, in that case the Lax pair has an especially simple universal expression with applications to integrable systems [F]. The generalized BC matrices of [P3] could be used for analogous purposes in the nonhyperelliptic case. (Readily workable)

4.7 In 2.3.2 we indicated the relationship found by Flaschka between KdV eigenfunctions and the Neumann system in n variables, namely n transferences by branch points. On the other hand, the "duality" between the two representations of the system (the 2×2 BC matrix and the $n\times n$ Baker-vector problem) has been given a group-theoretic interpretation in [AHH]. The picture is especially worth generalizing since many other integrable systems, such as Toda, occur by implementing KP flows on a finite set of suitably transferenced Baker functions. (Recommended work!)

4.8 The representation-theoretic viewpoint of §3 is full of profound open problems, but one small interesting question would be the quest for alternative representations of solutions of mKP given by flag manifolds (cf. [HH]) by using one τ function rather than a sequence $\tau_1, \ldots, \tau_r$. [W2] explains why this is not possible in general, but as we saw in the specific case of mKdV, it can be done by using the ϑ solution for a 2-point problem. (Workable)

4.9 The Hirota equations were shown, again via representation theory, to be an (infinite-dimensional) analog of the Plücker relations; however, for the specific case of a grassmannian point which belongs to a finite dimensional $Gr(m, 2m) \subset Gr_0$, the *classical* Plücker relations are the Hirota equations, as worked out by [E]. It would be pleasant to do the same by using the projective equations of flag manifolds (the corresponding Hirota equation [AS 3.4.45a] depends on an additional parameter!) (Readily workable)

References

I must make a disclaimer, as I did on the school's abstract: since I surveyed several constructions, the number of original sources would easily run into the hundreds. I restricted myself to references essential to the exposition, whether or not they contain the first instance of a construction; for attribution, the reader may consult [AS].

[AKS] M.J. Ablowitz, M.D. Kruskal and H. Segur, A note on Miura's transformation, *J. Math. Phys.* **20** (1979), 999-1003.

[AS] M.J. Ablowitz and H. Segur, Solitons and the inverse scattering transform, *SIAM Stud. Appl. Math.*, Philadelphia 1981.

[AHH] M.R. Adams, J. Harnad and J. Hurtubise, Dual moment maps into loop algebras, *Lett. Math. Phys.* **20** (1990), 299-308.

[AHP] M.R. Adams, J. Harnad and E. Previato, Isospectral Hamiltonian flows in finite and infinite dimensions I. Generalized Moser systems and moment maps into loop algebras, *Comm. Math. Phys.* **117** (1988), 451-500.

[BC] J.L. Burchnall and T.W. Chaundy, Commutative ordinary differential operators, *Proc. London Math. Soc.* **21** (1923), 420-440; –, *Proc. Roy. Soc. London Ser. A* **118** (1928), 557-583; – II. The identity $P^n = Q^m$, *Proc. Roy. Soc. London Ser. A* **134** (1932), 471-485.

[C] I.V. Cherednik, Differential equations for the Baker-Akhiezer functions of algebraic curves, *Functional Anal. Appl.* **12** (1978), 195-203.

DJKM] E. Date, M. Jimbo, M. Kashiwara and T. Miwa, Operator approach to the Kadomtsev-Petviashvili equation–Transformation groups for soliton equations III, *J. Phys. Soc. Japan* **50** (1981), 3806-3812.

[DLT] P. Deift, F. Lund and E. Trubowitz, Nonlinear wave equations and constrained harmonic motion, *Comm. Math. Phys.* **74** (1980), 141-188.

[D] P. Deligne, Equations différentielles à points singuliers réguliers, *Lecture Notes in Math 163*, Springer, Berlin 1970.

[Do] I. Ya. Dorfman, Dirac structures of integrable evolution equations, *Phys. Lett. A* **125** (1987), 240-246.

[DN] I. Ya. Dorfman and F.W. Nijhoff, On a 2 + 1-dimensional version of the Krichever-Novikov equation, preprint *INS* **167**, Clarkson Univ., 1990.

[DS] V.G. Drinfel'd and V.V. Sokolov, Lie algebras and equations of Korteweg-de Vries type, *J. Soviet Math.* **30** (1985), 1975-2036.

[Du] B.A. Dubrovin, The Kadomcev-Petviašvili equation and the relations between the periods of homomorphic differentials on Riemann surfaces, *Math. USSR-Izv.* **19** (1982), 285-296.

[DV] P. Du Val, Elliptic Functions and Elliptic Curves, LMS Lecture Note Series No. 9, Cambridge Univ. Press, Cambridge, 1973.

[EK] F. Ehlers and H. Knörrer, An algebro-geometric interpretation of the Bäcklund transformation for the Korteweg-de Vries equation, *Comment. Math. Helv.* **57** (1982), 1-10.

[E] N. Engberg, Hirota type equations as Plücker relations in an infinite dimensional Grassmann manifold, MS thesis, Imperial College 1989.

[EF] N.M. Ercolani and H. Flaschka, The geometry of the Hill equation and of the Neumann system, *Philos. Trans. Roy. Soc. London Ser. A* **315** (1985), 405-422.

[F] L.D. Fairbanks, Lax equation representation of certain completely integrable systems, *Compositio Math.* **68** (1988), 31-40.

[Fa] J.D. Fay, Theta functions on Riemann surfaces, *Lecture Notes in Math 352*, Springer, Berlin, 1973.

[FN] H. Flaschka and A.C. Newell, Integrable systems of nonlinear evolution equations, in "Dynamical Systems Theory and Applications," *Proceedings*, Seattle 1974, ed. J. Moser, *Lectures Notes in Phys. 38*, pp. 355-466, Springer-Verlag, Berlin, 1975.

[GP1] R.E. Goldstein and D.M. Petrich, The Korteweg-de Vries hierarchy as dynamics of closed curves in the plane, *Phys. Rev. Lett.* **67** (1991), 3203-3206.

[GP2] R.E. Goldstein and D.M. Petrich, Solitons, Euler's equation, and the geometry of curve motion, preprint 1992, to appear in *Proc. NATO ARW* Heraklion, Greece, 1992.

[HH] G.F. Helminck and A.G. Helminck, The structure of Hilbert flag varieties, preprint 1993.

[KN] I.M. Krichever and S.P. Novikov, Holomorphic fiberings and nonlinear equations, Finite zone solutions of rank 2, *Soviet Math. Dokl.* **20** (1979), 650-654.

[LP] G. Latham and E. Previato, Higher rank Darboux transformations, to appear in *Proc. NATO ARW* Lyon 1991, eds. N.M. Ercolani et al.

[MW] W. Magnus and W. Winkler, *Hill's equation*, Wiley-Interscience, New York, 1966.

[McK] H.P. McKean, Variation on a theme of Jacobi, *Comm. Pure Appl. Math.* **38** (1985), 669-678.

[M] O.I. Mokhov, Commuting differential operators of rank 3, and nonlinear differential equations, *Math. USSR Izv.* **35** (1990), 629-655.

[Mu] D. Mumford, Tata Lectures on Theta II, Birkhäuser, Boston, 1984.

[P1] E. Previato, Hyperelliptic quasi-periodic and soliton solutions of the nonlinear Schrödinger equation, *Duke Math. J.* **52** (1985), 329-377.

[P2] E. Previato, A particle-system model of the sine-Gordon hierarchy, *Phys. D* **18** (1986), 312-314.

[P3] E. Previato, Generalized Weierstrass $\wp$-functions and KP flows in affine space, *Comment. Math. Helv.* **62** (1987), 292-310.

[S] R.J. Schilling, Neumann systems for the algebraic AKNS problem, *Mem. Amer. Math. Soc.* **467**, Providence, RI 1992.

CMcL] A.C. Scott, F.Y.F. Chu and D.W. McLaughlin, The soliton–A new concept in applied science, *Proc. IEEE* **61** (1973), 1443-1483.

[Se] G. Segal, The geometry of the KdV equation, preprint 1990.

[SW] G. Segal, and G. Wilson, Loop groups and equations of KdV type, *Inst. Hautes Etudes Sci. Publ. Math.* **61** (1985), 5-65.

[W1] G. Wilson, Commuting flows and conservation laws for Lax equations, *Math. Proc. Cambridge Philos. Soc.* **86** (1979), 131-143.

[W2] G. Wilson, Habillage et fonctions τ, *C.R. Acad. Sci. Paris Sér. I Math.* **299** (1984), 587-590.

[W3] G. Wilson, On the quasi-hamiltonian formalism of the KdV equation, *Phys. Lett. A* **132** (1988), 445-450.

Integrable Hierarchies and Quantum Gravity

Robbert Dijkgraaf

Department of Mathematics, University of Amsterdam,
Plantage Muidergracht 24, 1018 TV Amsterdam, The Netherlands

1. 2D Quantum Gravity

Euclidean quantum gravity tries to make sense of the following problem. Let M be a smooth manifold, which we will assume to be compact and without boundary, and let Met_M be the space of all Riemannian metrics h on M. We wish to consider the integral

$$Z_M = \int_{\mathrm{Met}_M} [dh] \cdot e^{-S}, \tag{1.1}$$

where $[dh]$ is the natural measure on Met_M and $S[h]$ is some suitably chosen weight function. Since we integrate over all Riemannian structures, Z_M is by definition a topological invariant. Of course, since the integral is infinite-dimensional, it is a highly nontrivial problem to make mathematical sense out of this definition.

As it stands, the expression for Z_M already suffers from severe problems due to the following phenomenon. We have a natural action on Met_M of Diff_M, the group onpf diffeomorphisms of the space M. Elements of Diff_M relate equivalent Riemannian structures. Since the actions S that we will consider are in general invariant under this action, the path-integral Z_M will contain a factor proportional to the volume of Diff_M, which is clearly infinite. So a first step is to define Z_M by integration over the orbit space $\mathrm{Met}_M/\mathrm{Diff}_M$

$$Z_M = \int_{\mathrm{Met}_M/\mathrm{Diff}_M} [dh] \cdot e^{-S}. \tag{1.2}$$

For the action S one usually choses the Einstein-Hilbert action

$$S[h] = \kappa \int_M \sqrt{h}(R+\mu), \tag{1.3}$$

with R the scalar curvature of the metric h, κ Newton's contant, and μ the cosmological constant. The critical points of this action are given by the metrics h that satisfies Einstein's equation in the absence of matter.

Unfortunately, for arbitrary dimensions, this definition of Euclidean quantum gravity suffers from grave difficulties. One simple indication of the many problems surrounding this issue is that in general S is not positive-definite.

However, the situation improves dramatically if we descend to two dimensions and consider a surface Σ. Here the first part of our action S is actually a well-known topological invariant, the Euler characteristic

$$\int_\Sigma \frac{d^2z}{4\pi}\sqrt{h}R = 2-2g, \tag{1.4}$$

where g is the number of handles of the surface, the genus. So we have

$$Z_\Sigma = \lambda^{2g-2} Z_g \tag{1.5}$$

with λ, the so-called string coupling constant, related to Newton's constant κ by

$$\log\lambda \sim \kappa \tag{1.6}$$

The second term gives the total area A of the surface

$$\int_\Sigma \sqrt{h} = A \tag{1.7}$$

So, if we restrict to surfaces of a given topology, *i.e.* fixed genus h, the path-integral simply reduces to an integral over all metrics weighted by their area $A = \int_\Sigma \sqrt{h}$,

$$Z_g = \int_{\mathrm{Met}_\Sigma/\mathrm{Diff}_\Sigma} [dh]\cdot e^{-\mu A}. \tag{1.8}$$

Quite remarkably this path-integral can be exactly evaluated!

Our first observation is, that, although the space $\mathrm{Met}_\Sigma/\mathrm{Diff}_\Sigma$ is infinite dimensional, it can be contracted to the moduli space $\mathcal{M}_g$ of inequivalent complex structures on the surface. This is a finite dimensional space of dimension $6g-6$ (for $g>1$). One understands this fact by the result that any metric can be written as

$$h_{\mu\nu} = e^{2\phi}\cdot \hat{h}_{\mu\nu}, \tag{1.9}$$

where $\hat{h}_{\mu\nu}$ is a fixed representative of a conformal class, for example a metric of constant curvature $R=-1$. A conformal class of metrics uniquely determines a complex structure through

$$J_\mu{}^\nu = \sqrt{h}\epsilon_{\mu\lambda}h^{\lambda\nu}. \tag{1.10}$$

$\mathcal{M}_g$ is a rich mathematical object that plays a crucial role in our understanding of two-dimensional quantum gravity and string theory. Our knowledge of this space has also benefited from the recent developments in string theory [11, 12], see *e.g.* the review [13].

So we expect that the path-integral can be reduced to an integral over the conformal factor ϕ and a residual finite-dimensional integral over the moduli space $\mathcal{M}_g$:

$$Z_g = \int_{\mathcal{M}_g}\int [d\phi]\cdot e^{-S'} \tag{1.11}$$

Of course, the problem here is to determine the action S'. We will state here just the answer. It is given by the so-called Liouville action (we ignore ghosts for a moment)

$$S' = \int \frac{d^2z}{8\pi}\sqrt{\hat{h}}(3\hat{h}^{\mu\nu}\partial_\mu\phi\partial_\nu\phi + \frac{5}{4}R(\hat{h})\phi + \frac{3}{4}\mu e^{2\phi}) \tag{1.12}$$

A simple scaling argument show that the μ-dependence is given by

$$Z_g(\mu) = c_g \cdot \mu^{\frac{5}{2}(1-g)} \tag{1.13}$$

Indeed we can redefine $\phi \to \phi + c$ in the path-integral. This gives the identity

$$Z_g(\mu) = Z_g[\mu \cdot e^{2c}] \cdot \exp\left[-\frac{5c}{4}\int_\Sigma \frac{d^2z}{8\pi}\sqrt{\hat{h}}R\right] = Z_g[\mu \cdot e^{2c}] \cdot e^{\frac{5c}{4}(g-1)}. \tag{1.14}$$

1.1. Matrix models

To go one step further and calculate the absolute normalization c_g, and more importantly, also the correlation functions, one has to make use of the formalism of matrix model [1] and their solutions in the double-scaling limit [2], see *e.g.* the reviews in [3]. These models can be completely solved within the framework of integrable hierarchies [4, 5]. We will not be in a position to explain this development here in any detail. The crucial idea is to approximate the integral over metrics by a summation over triangularizations of the surface, The combinatorics of these summations can be very conveniently be encoded in matrix integrals. One considers these matrix integral for variable size $N \times N$ and then studies the asymptotic expansion in $1/N$. The result of all this is the following answer.

Let $Z(\mu)$ be the partition function where we also sum over topologies

$$Z(\mu) = \sum_{g=0}^{\infty} \lambda^{2g-2} Z_g(\mu), \tag{1.15}$$

(in general this should be regarded as an asymptotic expansion) and let

$$u = \lambda^2 \frac{\partial^2 Z}{\partial \mu^2} \tag{1.16}$$

be the second derivative with respect to μ. Then u satisfies the nonlinear Painlevé equation

$$u^2 + \frac{\lambda^2}{3}u'' = \mu, \tag{1.17}$$

where primes denote differentiation with respect to μ. This can be reformulated as a recursion relation for the coefficients c_g after we normalize, say $c_0 = 1$.

Actually, the solution of the matrix model is much more powerful. One can also consider correlation functions

$$\langle \sigma_{n_1} \ldots \sigma_{n_s} \rangle_g \equiv \int_{\mathrm{Met}_M/\mathrm{Diff}_M} [dh] \cdot \sigma_{n_1} \cdots \sigma_{n_s} \cdot e^{-S}. \tag{1.18}$$

where the expressions σ_n, $n = 0, 1, 2 \ldots$, are certain expressions in the metric h on a surface of genus g. For example

$$\sigma_0 = \int \sqrt{h}, \quad \sigma_1 = \int \sqrt{h} R, \ldots \tag{1.19}$$

All these correlation functions can also be exactly calculated. The final result can be formulated as follows. Let

$$Z(t_0, t_1, \ldots) = \sum_{g=0}^{\infty} \lambda^{2g-2} \langle \exp \sum_{n=0}^{\infty} t_n \sigma_n \rangle_g \tag{1.20}$$

be the generating functional of all correlation functions. We see immediately that $t_0 = \mu$, $t_1 = \kappa$. Consider again the object $u = \lambda^2 Z''$. It is now a function of infinite many variables $t_0, t_1, \ldots$ and satisfies two equations:

1. $u(t)$ is a solution of the KdV hierarchy. That is, we have equations of the type

$$\frac{\partial u}{\partial t_1} = u u' + \frac{\lambda^2}{3} u''', \tag{1.21}$$

or more general

$$\frac{\partial u}{\partial t_n} = \frac{\partial}{\partial t_0} R_{n+1}[u]. \tag{1.22}$$

Here $R_n[u]$ is a polynomial in u and its derivatives $u', u'', \ldots$ with respect to the cosmological constant $t_0 = \mu$ given by

$$R_n[u] = \frac{(n-1)!}{(2n-1)!!} \left[\tfrac{1}{2} \lambda^2 D^2 + u + D^{-1} u D \right]^n \cdot 1, \tag{1.23}$$

with $D = \partial / \partial t_0$. An equivalent statement reads

$$\tau(t) = \exp Z(t) \tag{1.24}$$

is a tau-function of the KdV hierarchy.

2. $u(t)$ satisfies the so-called string equation

$$\sum_{n=o}^{\infty} t_n R_n[u] = t_0 \tag{1.25}$$

which can be seen as a generalization of (1.17).

This beautiful solution has been generalized to more complicated ssytems that also include two-dimensional matter fields. In this way we are naturally led to (noncritical) strings.

2. Non-Critical Strings

Recent years have seen an enormous increase in our understanding of a particular class of string theories – the so-called non-critical bosonic strings. Most of this progress is due to the advent of matrix models.

A special role is played by the so-called $c = 1$ model. In many aspects this is the most physical and richest non-critical model. It also can be solved using matrix model techniques [6], but the determination of its correlation functions has been rather complicated [7, 8, 9]. In [10] the explicit construction of these scattering amplitudes has been clarified and the relation between integrable hierarchies and the $c = 1$ model has been explained, thereby completing in a sense the unification of all solvable non-critical string models. In these notes we will give a more leisurely account of all this.

2.1. Elementary introduction to string theory

Before we plunge into the details of the explicit solutions of these string theories, let us start with some very general remarks to place this work in the relevant context. Recall that from a perturbative point of view a string theory is a quantum theory of excitations (particles) in some space-time X that has an interpretation in terms of surfaces, *i.e.* in terms of maps

$$x : \Sigma \to X, \tag{2.1}$$

with Σ a surface of a particular genus. Indeed the 'central dogma' of string theory (see *fig.* 2.1.) equates (connected) S-matrix elements in space-time with correlation functions in a two-dimensional field theory

$$\langle out|S|in\rangle_c \sim \sum_{g=0}^{\infty} \lambda^{2g-2} \langle \prod V^{out} \prod V^{in} \rangle_g \,. \tag{2.2}$$

Here the object on the LHS is a scattering amplitude in the space-time X whose expansion in terms of the string coupling constant λ is given by correlation functions of vertex operators V^{out}, V^{in} in a two-dimensional quantum field theory defined on a surface Σ of genus g — the world-sheet. Since the correlation functions on the RHS should not depend on intrinsic properties of the surface, like the metric and the position of the vertex operators, the world-sheet field theory should include in some form an integral over all metrics on the surface, and thus we are naturally lead to consider two-dimensional quantum gravity.

The analogue idea in quantum field theory is well-known: amplitudes in field theory can be written as correlation functions in a one-dimensional field theory, *i.e.* quantum mechanics. For example, the usual propagator of a massive scalar field $\phi(x)$

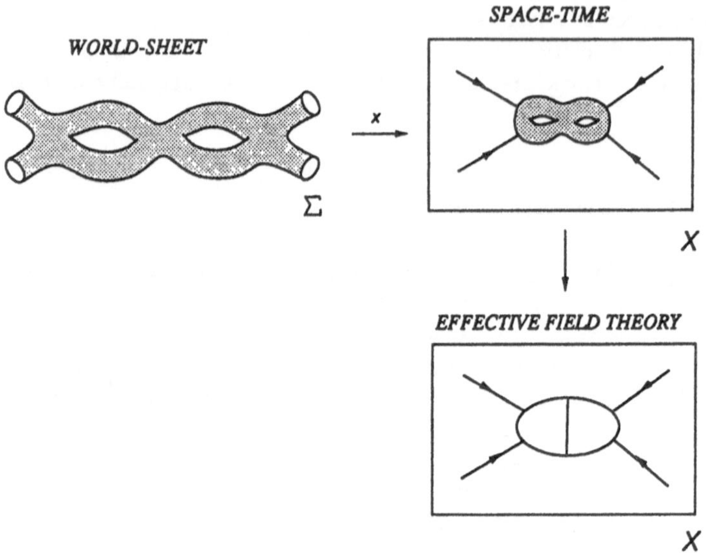

Fig. 2.1: *The 'central dogma' of string theory expresses the space-time scattering amplitudes in terms of correlation functions in a quantum field theory defined on a world-sheet surface. In the effective field theory limit, these string perturbation contributions correspond to (sums of) point-particle Feynman diagrams.*

on $X = \mathbf{R}^d$ with Lagrangian

$$L = (\nabla\phi)^2 + m^2\phi^2 \tag{2.3}$$

is given by the amplitude of a particle to propagate from x to y, and can be written as one-dimensional path-integral over a field $x(t) \in \mathbf{R}^d$ and a metric $g(t)$ on the world-line

$$\begin{aligned} x \longrightarrow y \; &= \; \langle x | \frac{1}{\partial^2 + m^2} | y \rangle \\ &= \int [dx][dg] e^{-\int dt \sqrt{g}(g^{-1}\dot{x}^2 + m^2)} \, . \end{aligned} \tag{2.4}$$

The path-integral is subject to the boundary conditions $x(0) = x$ and $x(1) = y$. The integral over the one-dimensional metric $g(t)$ modulo diffeomorphisms is not very interesting, and reduces to an integral over the total length T of our one-dimensional world. In this way we recover the familiar Schwinger parametrization of the propagator

$$\int [dx][dg] e^{-\int dt \sqrt{g}(g^{-1}\dot{x}^2 + m^2)} = \int_0^\infty dT \, \langle x | e^{-T(\partial^2 + m^2)} | y \rangle \, . \tag{2.5}$$

More complicated Feynman diagrams are treated by considering maps of graphs Γ into space-time, with overlap integrals for the vertices.

In string theory we consider maps $x : S^1 \to X$ instead of points $x \in X$. The analogue of the propagator is the two point function

$x(\sigma)$  $y(\sigma) = \int [dx][dg] e^{-\int d^2z\sqrt{g}(g^{\mu\nu}\partial_\mu x \partial_\nu x + \mu)}\,, \qquad (2.6)$

where $x(\sigma,\tau)$ and $g_{\mu\nu}(\sigma,\tau)$ are now a field and a metric on a two-dimensional surface, which for a propagator has the topology of a cylinder. We also introduced a two-dimensional cosmological constant μ that is in some sense the analogue of the mass of the string field.

This definition can now be very naturally extended to include all interactions, just by considering the two-dimensional field theory on surfaces Σ with more complicated topology

 $_\Sigma = \int [dx][dg] e^{-\int_\Sigma d^2z\sqrt{g}(g^{\mu\nu}\partial_\mu x \partial_\nu x + \mu)} \qquad (2.7)$

So, in string theories the two-dimensional metric $g_{\mu\nu}$ is a fundamental variable and the path-integral should include an integral over all inequivalent Riemannian structures, *i.e.* metrics modulo diffeomorphisms.

The string theories with target space $\mathbf{R}^d$ (with a flat metric) can be most easily quantized in conformal gauge. One finds the action

$$S = \int \frac{d^2z}{2\pi}(\partial x^i \bar{\partial} x^i + \alpha_1 \partial\phi\bar{\partial}\phi + \alpha_2 R_h \phi + \alpha_3 \mu e^{2\phi} + b\bar{\partial}c + \bar{b}\partial\bar{c}) \qquad (2.8)$$

with some appropriate constants α_i [14] chosen such that the total central charge of the x-field and the Liouville field equals 26 and thereby cancels the central charge $c = -26$ of the ghosts b, c. For d arbitrary the Liouville field does not decouple and represents an extra degree of freedom. It is a remarkable effect that the Liouville field forms essentially an extra embedding dimension if $d \neq 26$. Although this effect would be an embarrassment for, say, a four-dimensional string describing QCD flux tubes, it is a very welcome phenomenon in the case $d = 1$ that we will be interested in the subsequent. For $d = 26$ the Liouville field decouples, $\alpha_i = 0$, and we obtain a critical string.

2.2. $c = 1$ string theory

The $c = 1$ string is perhaps the most simple bosonic string theory, and and yet it has a very rich structure. We are naturally led to this model, since the class of bosonic strings that we wrote down in (2.8) with target space $\mathbf{R}^d$ is actually inconsistent if $d > 1$. This is a consequence of the famous tachyon instability. The zero-mode of the string

$$x(\tau) = \int d\sigma\, x(\sigma,\tau) \qquad (2.9)$$

gives rise to a scalar particle whose mass can be calculated to be

$$m^2 = \frac{1-d}{24}. \qquad (2.10)$$

This is indeed a tachyon for dimensions larger than one. The world-sheet interpretation of this instability is conjectured to be the formation of long, thin tree-like surfaces (branched polymers). However we see that this tachyon field is actually massless in dimension $d = 1$ and massive in $d = 0$. So here we stand a chance of constructing consistent bosonic string theories. We will concentrate on the case $d = 1$ where the space-time physics is more clear.

This model consists of one free field x that we couple to two-dimensional gravity. So our action reads

$$S = \int \frac{d^2 z}{2\pi} \sqrt{g} g^{\mu\nu} \partial_\mu x \partial_\nu x. \tag{2.11}$$

Here the field x might be compactified, *i.e.* , we also consider the case where x is a periodic field

$$x \cong x + 2\pi\beta. \tag{2.12}$$

Since one scalar field gives a conformal field theory of central charge $c = 1$, we will refer to this model as the $c = 1$ string.

As explained above, the Liouville field ϕ does not decouple in this case. With the appropriate constants the action reads (not including the ghosts)

$$S = \int \frac{d^2 z}{2\pi} \sqrt{h} \left[\tfrac{1}{2} h^{\mu\nu} (\partial_\mu x \partial_\nu x + \partial_\mu \phi \partial_\nu \phi) + R_h \phi + \mu e^{2\phi} \right] \tag{2.13}$$

This Lagrangian effectively describes a sigma model with a two-dimensional target space, with the Liouville field regarded as a spontaneously generated extra dimension. So we are dealing with a *two-dimensional critical string* (critical in the sense that the total matter central charge is 26). Of course, because of the background charge and the exponential potential, we do not have translation invariance in the ϕ-direction. The Hilbert space of the model is simply given by the tensor products of the state spaces of the free field x, the Liouville field ϕ and the ghosts b, c.

Physical modes in this model are limited. Since for a string in n dimensions we expect $n - 2$ transverse oscillations, naively only the zero-mode of a two-dimensional string theory can be expected to lead to a physical excitation. This field is the massless scalar field known as the tachyon. The world-sheet vertex operators that describe the emission/absorption of tachyons of energy ω simply read (here and in the subsequent $\omega > 0$)

$$T^{\pm}_{\omega} = \frac{\Gamma(\omega)}{\Gamma(-\omega)} \mu^{1-\omega/2} e^{\pm i\omega x} e^{(2-\omega)\phi}. \tag{2.14}$$

With the explicit factors of μ inserted, we see that a genus g correlation function

$$\langle T^{\pm}_{\omega_1} \cdots T^{\pm}_{\omega_n} \rangle_g \tag{2.15}$$

scales simply as μ^{2-2g}. (This can be seen most easily by shifting the Liouville field by $\phi \to \phi - \frac{1}{2} \log \mu$.) Consequently we can identify the string coupling constant λ with the inverse of the cosmological constant

$$\lambda = \frac{1}{\mu}. \tag{2.16}$$

If we compactify x, the energies ω will be quantized

$$\omega_n = \frac{n}{\beta}, \qquad n \in \mathbf{Z}^+. \tag{2.17}$$

We will write T_n instead of T_{ω_n}. The point $\beta = 1$ is very special, since it is the invariant point under the duality transformation

$$\beta \to 1/\beta, \tag{2.18}$$

that interchanges a large and a small radius of compactification.

Clearly the 'holy grail' in $c = 1$ string theory would be the generating functional of all connected scattering amplitudes. For the compactified case it would read

$$F(t,\bar{t}) \sim \sum_{g=0}^{\infty} \lambda^{2g-2} \langle \exp \sum_n (t_n T_n^+ + \bar{t}_n T_n^-) \rangle_g \tag{2.19}$$

In terms of space-time field theory we would have incoming states $|k_1, k_2, \ldots\rangle$ with k_i tachyons of energy ω_i and corresponding outgoing states $\langle \bar{k}_1, \bar{k}_2, \ldots|$, and $\tau = e^F$ would be the generating function of all matrix elements

$$\tau(t,\bar{t}) = \sum_{\{k_i,\bar{k}_i\}} \prod_{i,j} \frac{t_i^{k_i}\, \bar{t}_i^{\bar{k}_i}}{k_i!\, \bar{k}_i!} \langle \bar{k}_1, \bar{k}_2, \ldots |S| k_i | k_1, k_2, \ldots \rangle\ . \tag{2.20}$$

Actually there are more degrees of freedom than just the tachyons, but these so-called discrete states appear at very particular energy and momentum [15]. Consequently, they do not have an interpretation as propagating particles, but are more analogous to quantum mechanical degrees of freedom.

2.3. Minimal models

Before we go on and discuss the solution of the $c = 1$ string theory, we want to turn briefly to another interesting class of models, the so-called $c < 1$ models. These models are the unique CFTs with a finite number of irreducible representations of the Virasoro algebra. They are labeled by two integers (p,q) which should be coprime. It is rather difficult to write a Lagrangian for these models, but roughly they can be described as a $c = 1$ model with an extra background charge for the scalar field x

$$S = \int \partial x \bar{\partial} x + QRx. \tag{2.21}$$

So we loose all translation invariance. Furthermore, the occurrence of screening operators and an extra BRST charge make the analysis of the correct physical states much more complicated. Nevertheless, these string theories have been completely solved, again due to the magic of matrix models. We will just give a brief account of the nature of the solution as found in [4, 5].

First of all, the spectrum is extremely simple and universal for all (p,q) models and the massive deformation that can be obtained of them. For every positive integer we have a physical state, so we have vertex operators

$$V = \mathcal{O}_1, \mathcal{O}_2, \mathcal{O}_3, \ldots \tag{2.22}$$

(The $(2,3)$ model corresponds to the trvial matter theory and thus gives pure quantum gravity as discussed in §1. The operators σ_n of (1.19) correspond to $\mathcal{O}_{2n+1}$.)

According to our discussion above, the string theory correlation functions of these operators are on general grounds given by a (complicated and in general unknown) integral over moduli space, where the volume form is a correlator in a CFT consisting of the minimal model coupled to ghosts and Liouville theory

$$\langle \mathcal{O}_{n_1} \cdots \mathcal{O}_{n_s} \rangle_g = \int_{\mathcal{M}_{g,s}} \langle \ldots \rangle_{CFT} \tag{2.23}$$

The matrix model techniques give an alternative route to the calculation of these correlators, and this result can be summarized as follows. The generating functional of correlation functions for the (p,q) model

$$\tau_{p,q}(t) = \exp \sum_{g=0}^{\infty} \lambda^{2-2g} \langle \exp \sum_n t_n \mathcal{O}_n \rangle_g \tag{2.24}$$

is a tau-function of the KP hierarchy and all (p,q)-generating functions for fixed p lie on one orbit. More precisely, relative to a convenient choice of origin, the (p,q) model is obtained as

$$\tau_{p,q}(t_k) = \tau_p(t_k + \delta_{k,p+q}). \tag{2.25}$$

Furthermore, the KP hierarchy reduces to the pth KdV hierarchy, which implies that all correlation functions of the operators $\mathcal{O}_n$ with $n \equiv 0 \pmod p$ vanish.

At the point $(p,1)$ we also have a very explicit representation of the τ-function due to Kontsevich [12] (for more details see *e.g.* [13, 16]). Let Z be an $N \times N$ Hermitian matrix, and choose particular values for the KP times t_n given by

$$t_n = \frac{\lambda}{n} Tr Z^{-n} \, . \tag{2.26}$$

With this parametrization, the τ-function is given by the following matrix integral, where Y is also an $N \times N$ Hermitian matrix

$$\tau_{p,1}(t) = c^{-1} \cdot \int dY \cdot e^{i\mathrm{Tr}(Z^p Y - \frac{Y^{p+1}}{p+1})/\lambda} \, . \tag{2.27}$$

Here the normalization constant is given by

$$c = (2\pi i/\lambda)^{\frac{N^2}{2}} \det(Z^{\frac{1-p}{2}}) \frac{\Delta(z)}{\Delta(z^p)} e^{i\frac{p}{p+1} Tr Z^{p+1}/\lambda} \, . \tag{2.28}$$

and $\Delta(z)$ is the Vandermonde determinant

$$\Delta(z) = \det z_i^{j-1} = \prod_{i>j}(z_i - z_j). \tag{2.29}$$

This rather formidable looking expression for the normalization constant is actually very natural. The integral (2.27) can be expanded around the critical point $Y = Z$ of the action

$$\mathrm{Tr}\,(Z^pY - \frac{Y^{p+1}}{p+1}). \tag{2.30}$$

The normalization constant is now just the classical value of the action plus the gaussian (one-loop) contribution to the integral

$$c = e^{i\frac{p}{p+1}TrZ^{p+1}/\lambda} \cdot c', \tag{2.31}$$

where c'_p is defined by the Gaussian integral

$$\begin{aligned} c' = \int dY \cdot e^{iS(Y,Z)/\lambda} &= (2\pi i/\lambda)^{\frac{N^2}{2}} \prod_{i,j} \sqrt{\frac{z_i - z_j}{z_i^p - z_j^p}} \\ &= (2\pi i/\lambda)^{\frac{N^2}{2}} \det Z^{\frac{1-p}{2}} \frac{\Delta(z)}{\Delta(z^p)}, \end{aligned} \tag{2.32}$$

and

$$S(Y,Z) = \sum_{k=0}^{p-1} \mathrm{Tr}\,[YZ^kYZ^{p-1-k}]. \tag{2.33}$$

This representation makes clear that $\tau_{p,1}(t)$ can indeed have an asymptotic expansion in the variables t_n.

3. Integrable Hierarchies and Quantum Field Theory

In order to understand better the integrable hierarchies that naturally appear in non-critical string theory and in particular the remarkable integral representation (2.27), we will now review some relations between quantum field theory and integrable systems. We follow the standard exposition of the KP hierarchy [17].

3.1. 2d Chiral quantum field theory

Consider a two-dimensional free chiral scalar field $\varphi(z)$, with the usual mode expansion

$$\partial\varphi(z) = \sum_n \alpha_n z^{-n-1}\,. \tag{3.1}$$

Here $z = e^{ix-\tau}$ is a coordinate on a cylinder, with x the periodic space variable and τ Euclidean time. We have standard commutation relations $[\alpha_n, \alpha_m] = n\delta_{n+m}$, and a vacuum $|0\rangle$ satisfying $\alpha_n|0\rangle = 0$, $n \geq 0$. The Hilbert space $\mathcal{H}$ can be considered as the completion of the Fock space spanned by states $\alpha_{-n_1}\ldots\alpha_{-n_s}|0\rangle$. The reader is encouraged to think about this two-dimensional scalar field as the tachyon field living in the space-time with coordinates (x,ϕ). Here we roughly have $\phi=\tau$.

As in the case of a harmonic oscillator one can consider coherent states,

$$|t\rangle = \exp\left(\sum_{n=1}^{\infty} it_n\alpha_{-n}\right)|0\rangle, \tag{3.2}$$

and their Hermitian conjugates

$$\langle t| = \langle 0|\exp\left(\sum_{n=1}^{\infty} -it_n\alpha_n\right). \tag{3.3}$$

Now to any state $|W\rangle$ in the Hilbert space $\mathcal{H}$ we can associate a coherent state wave-function $\tau_W(t)$ by considering the inner product

$$\tau_W(t) = \langle t|W\rangle\ . \tag{3.4}$$

This function is a tau-function of the KP hierarchy if and only if the state $|W\rangle$ lies in the so-called Grassmannian.

To explain the concept of the Grassmannian we have to turn to the alternative description of the chiral boson in terms of chiral Weyl fermions $\psi(z)$, $\overline{\psi}(z)$ by means of the well-known bosonization formulas

$$i\partial\varphi = \overline{\psi}\psi, \quad \psi = e^{i\varphi}, \quad \overline{\psi} = e^{-i\varphi}\ . \tag{3.5}$$

These free fermions have mode expansions

$$\psi(z) = \sum_{a\in\mathbf{Z}+\frac{1}{2}} \psi_a z^{-a-\frac{1}{2}}, \tag{3.6}$$

$$\overline{\psi}(z) = \sum_{a\in\mathbf{Z}+\frac{1}{2}} \overline{\psi}_a z^{-a-\frac{1}{2}}, \tag{3.7}$$

and canonical anti-commutation relations

$$[\psi_a,\overline{\psi}_b]_+ = \delta_{a+b}, \qquad [\psi_a,\psi_b]_+ = [\overline{\psi}_a,\overline{\psi}_b]_+ = 0. \tag{3.8}$$

Loosely speaking, the Grassmannian can be defined as the collection of all *fermionic* Bogoliubov transforms of the vacuum $|0\rangle$. That is, the state $|W\rangle$ belongs to the Grassmannian if it is annihilated by particular linear combinations of the fermionic creation and annihilation operators.

$$(\psi_{n+\frac{1}{2}} - \sum_{m=1}^{\infty} A_{nm}\psi_{-m+\frac{1}{2}})|W\rangle = 0, \qquad n\geq 0\ , \tag{3.9}$$

or equivalently,

$$|W\rangle = S|0\rangle, \qquad S = \exp\left(\sum_{n,m} A_{nm}\overline{\psi}_{-n-\frac{1}{2}}\psi_{-m+\frac{1}{2}}\right). \tag{3.10}$$

Note that the operator S can be considered as an element of the infinite-dimensional linear group, $S \in GL(\infty, \mathbf{C})$.

By replacing the vacuum $|0\rangle$ by the state $|W\rangle$, we simply made another decomposition into positive and negative energy states, and filled these new negative energy states. The positive energy wave-functions are no longer given by

$$z^0, z^1, z^2, \ldots \tag{3.11}$$

but are now replaced by the functions

$$v_0(z), v_1(z), v_2(z), \ldots \tag{3.12}$$

with

$$v_n(z) = z^n - \sum_{m=1}^{\infty} A_{nm} z^{-m} . \tag{3.13}$$

If one prefers the language of semi-infinite differential forms, we have a formula

$$|W\rangle = v_0 \wedge v_1 \wedge \ldots \tag{3.14}$$

which should be contrasted with

$$|0\rangle = z^0 \wedge z^1 \wedge \ldots \tag{3.15}$$

We want to mention at this point one important generalization. In the above fashion one generates solutions to the KP hierarchy. This construction can be extended to give a tau-function for the two-dimensional Toda Lattice hierarchy by considering a second set of times $\bar{t}_k$, as discussed in detail in [18]. In terms of our conformal field theory, the *Toda tau-function* is simply obtained as

$$\tau(t, \bar{t}) = \langle t|S|\bar{t}\rangle , \tag{3.16}$$

with $|\bar{t}\rangle$ and $\langle t|$ the coherent states (3.2) and (3.3) and S a general $GL(\infty, \mathbf{C})$ element, *i.e.* an exponentiated fermion bilinear of type (3.10). We will return to the Toda hierarchy in §3.

Instead of taking the inner product of the state $|W\rangle$ with a coherent bosonic state, one can also consider *fermionic* N-point functions. In fact, one finds in this way a simple expression in terms of an $N \times N$ determinant of the wave-functions (3.13) [13, 16]

$$\langle N|\psi(z_1)\ldots\psi(z_N)|W\rangle = \det v_{j-1}(z_i) . \tag{3.17}$$

Using the bosonization formulas, one recognizes this correlation function as a special coherent state wave-function $\langle t|W\rangle$ where the parameters t_n are given by

$$t_n = \sum_{i=1}^{N} \frac{1}{n} z_i^{-n} \,. \tag{3.18}$$

With this choice of parameterization, and after taking into account a normal ordering contribution, the tau-function can be written as

$$\tau(t) \doteq \frac{\det v_{j-1}(z_i)}{\Delta(z)} \,, \tag{3.19}$$

with $\Delta(z)$ the Vandermonde determinant.

3.2. Kontsevich integrals and the $c < 1$ models

We are now in a position to understand the integral representation that we gave for the minimal models partition functions $\tau_{p,q}(t)$. The state $|W\rangle$ corresponding to this orbit is most simply described at the $(p,1)$ point, where a description in terms of topological field theory can be given. In the Lax operator description of the KdV hierarchies [17], this point can be very efficiently characterized by the following initial value for the pth order linear differential operator L

$$L = D^p + x. \tag{3.20}$$

For an explicit basis for the state $|W\rangle$ one can take the wave-functions

$$v_n(z) = \sqrt{\frac{ipz^{p-1}}{2\pi\lambda}}\, e^{\frac{ip}{p+1} z^{p+1}/\lambda} \int_{-\infty}^{\infty} dy \cdot y^n \cdot e^{i(z^p y - \frac{y^{p+1}}{p+1})/\lambda} \,, \tag{3.21}$$

where the normalization is chosen such that we have the appropriate asymptotic expansion

$$v_n(z) = z^n (1 + O(z^{-1})) \,. \tag{3.22}$$

Since the wave-functions are moments in a Fourier transform, the fermionic determinant formula can be explicitly evaluated. This is due to the following identity, first given by Harish-Chandra, for the integral over the unitary group $U(N)$ [19]

$$\int dU \, e^{i \mathrm{Tr}[UXU^\dagger Y]} = c \cdot \frac{\det e^{i x_i y_j}}{\Delta(x)\Delta(y)}, \tag{3.23}$$

with dU the Haar measure and x_i, y_i the eigenvalues of the Hermitian matrices X and Y and c some normalization constant. As a special application of the above equation consider the matrix Fourier transform, *i.e.* the integral over a Hermitian matrix Y in an external field X, both $N \times N$ matrices, of the form

$$\tau = \int dY \, e^{i \mathrm{Tr}[XY + V(Y)]}. \tag{3.24}$$

By conjugation invariance, this is only a function of the eigenvalues $x_1, \ldots, x_N$. We can use (3.23) to integrate out the angular variables U in the decomposition

$$Y = U \cdot \mathrm{diag}(y_1, \ldots, y_N) \cdot U^\dagger, \tag{3.25}$$

which also introduces a Jacobian

$$dY = \Delta(y)^2 \cdot dU \cdot [dy]. \tag{3.26}$$

This leaves us with an integral over the eigenvalues y_i of the form

$$\tau = \int [dy]\, \Delta(y)\Delta(x)^{-1} \exp \sum_j i[x_j y_j + V(y_j)] \tag{3.27}$$

Now the Vandermonde determinant $\Delta(y)$ is a sum of terms of the form

$$\pm\, y_1^{i_1} \cdots y_N^{i_N}, \tag{3.28}$$

and for each of these terms the integral τ factorizes in separate integrals over the individual eigenvalues y_i. If we introduce the function

$$w(x) = \int dy \cdot e^{ixy + iV(y)} \tag{3.29}$$

and its derivatives

$$w_k(x) = \int dy \cdot y^k \cdot e^{ixy + iV(y)}, \tag{3.30}$$

the contribution of (3.28) is simply

$$\pm\, w_{i_1}(x_1) \cdots w_{i_N}(x_N). \tag{3.31}$$

So we can evaluate $\tau(X)$ straightforwardly as

$$\tau = \frac{\det(w_{j-1}(x_i))}{\Delta(x)}. \tag{3.32}$$

If we apply this method to the determinant (3.19) of the wave-functions (3.21), we find the matrix integral (2.27).

This result can be generalized to the 'generalized Kontsevich model' [13, 16] which features an arbitrary potential $V(z)$

$$\tau(t) = c \cdot \int DY \cdot e^{iTr(V'(Z)Y - V(Y))/\lambda}\,. \tag{3.33}$$

with

$$c = (2\pi i/\lambda)^{-\frac{N^2}{2}} \cdot \det V''(Z) \cdot \frac{\Delta(V'(z))}{\Delta(z)} \cdot e^{iTr(V(Z) - V'(Z)Z)/\lambda}\,. \tag{3.34}$$

It has been noticed by many authors that the case $p = -1$ (*i.e.* a logarithmic potential $V(z) = \log z$) is likely to be associated with the $c = 1$ model. In the next section we will proceed to show that this is indeed the case.

4. The $c = 1$ String

We will now return to the $c = 1$ string and discuss its solution within the matrix model framework.

4.1. The $c = 1$ matrix model

Matrix models have been introduced in two-dimensional quantum gravity to describe discretizations of smooth surfaces [1]. We will not explain this method here in any detail. The basic idea is that the path-integral over all metrics on the surface can be approximated by a summation over triangularizations. The combinatorics of these tessellations of surfaces by regular polyhedra is organized very efficiently by considering the perturbative expansions of Hermitian matrix integrals.

The coupling of matter fields to theories of quantum gravity, or equivalently random surface models, proceeds by making the matrix variables depend on other (discrete or continuous) variables. In this fashion the $c = 1$ matrix model describes the quantum mechanics of an $N \times N$ hermitian matrix $\Lambda(x)$. The real variable x should be considered as (Euclidean) time. In particular we can opt to compactify x with period $2\pi\beta$. This essentially means we consider matrix quantum mechanics at some finite temperature $1/\beta$. As our Lagrangian we choose a simple kinetic term plus a potential $\mathrm{Tr}\, U(\Lambda)$ with U some polynomial. The partition function is thus given by

$$Z(N,\beta) = \int d\Lambda \, \exp\left\{ -N \int_0^{2\pi\beta} dx \, \mathrm{Tr} \left[\left(\frac{d\Lambda}{dx} \right)^2 + U(\Lambda) \right] \right\}. \tag{4.1}$$

This model has a nontrivial double scaling limit $N \to \infty$ if we fine-tune the potential U to a critical point [6]. The model can consistently be reduced to the eigenvalues λ_i of the matrix Λ. The wonderful simple result is that in the double scaling limit all the interesting dynamics is reproduced by considering *free* fermions $\psi(\lambda)$ moving in a potential [6]

$$U(\lambda) = -\lambda^2. \tag{4.2}$$

This is the famous inverted harmonic oscillator. The physical model is obtained by filling, say, the right-hand side of the Fermi sea to a level μ from the top of the potential (see *fig.* 4.1.). This parameter has naturally the interpretation of the cosmological constant. All tunneling effects (leaking through the barrier) are thus $e^{-\mu}$ effects. When we recall that for the two-dimensional string $\mu = 1/\lambda$, we see that all these effects are also non-perturbative in the string coupling constant. They can consequently not be given an interpretation in terms of surfaces.

Of course, the Hamiltonian

$$H = -\frac{d^2}{d\lambda^2} - \lambda^2 \tag{4.3}$$

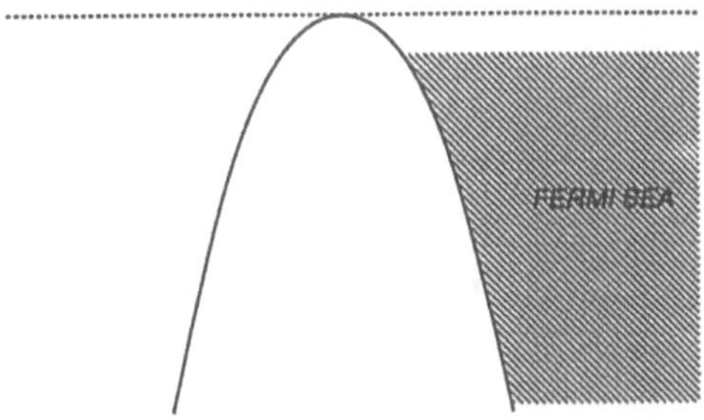

Fig. 4.1: *The $c = 1$ matrix model reduces to a collection of non-interacting fermions in an inverted harmonic potential*

has a continuous spectrum. Its eigenfunctions

$$H\psi_\omega = \omega\psi_\omega \tag{4.4}$$

can be expressed in parabolic cylinder functions [7]. These functions have the following interesting asymptotic behaviour. In the limit $\lambda \to \infty$ they behave as

$$\psi_\omega(\lambda) \sim \psi_0(\lambda)e^{i\omega(\pm\tau - x)} \tag{4.5}$$

where the new variable τ is defined as

$$\lambda = 2\sqrt{\mu}\cosh\tau. \tag{4.6}$$

That is, in (x, τ)-coordinates we find at spatial infinity (up to a factor) *relativistic* fermions.

Of course, the incoming and outgoing wave functions are not independent. They are related by a one-particle scattering process. That is, in general we will have

$$\psi_\omega^{out} = R_\omega \cdot \psi_\omega^{in} \tag{4.7}$$

where R_ω is the reflection factor that is fully determined by the shape of the potential. In this case the potential is the inverted quadratic, and the factor R_ω can be determined by a straightforward, but non-trivial, calculation. The outcome of the calculation certainly depends on how we treat the LHS of the potential. In the symmetric case we find [9]

$$R_\omega = (-i\mu)^{-\omega}\frac{\Gamma(\frac{1}{2} - i\mu + \omega)}{\Gamma(\frac{1}{2} - i\mu)}. \tag{4.8}$$

In some sense, this function is all one should know about the $c = 1$ matrix model in order to understand the scattering of tachyons. One verifies easily perturbative unitarity

$$R_\omega R^*_{-\omega} = 1 + O(e^{-\mu}). \tag{4.9}$$

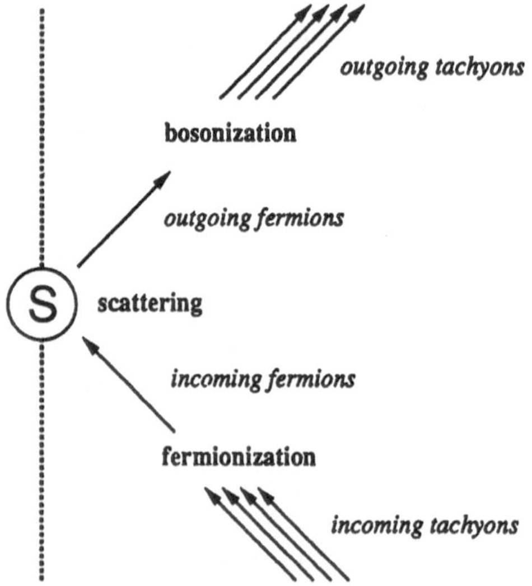

Fig. 4.2: *The calculating of tachyon scattering amplitudes proceeds through fermionization and bosonization.*

In the Euclidean case we cannot really speak about scattering, since there are no separate past and future null infinities. We simply have a mixing of left and right-movers, given again by the factor R_ω. In the compactified case, to which we will restrict for the remainder, the energies will be furthermore quantized and are given by (recall the fermions are naturally anti-periodic)

$$\omega_n = \frac{1}{\beta}(n + \tfrac{1}{2}), \qquad n \in \mathbf{Z}. \tag{4.10}$$

The fermion S-matrix is now defined by

$$\psi^{in}(z) = S\psi^{out}(z)S^{-1}, \qquad \overline{\psi}^{in}(z) = S\overline{\psi}^{out}(z)S^{-1}, \tag{4.11}$$

with

$$S =: exp\Big[\sum_{m\in\mathbf{Z}} \log R_{\omega_m} \psi^{out}_{-(m+\frac{1}{2})} \overline{\psi}^{out}_{m+\frac{1}{2}}\Big] : \tag{4.12}$$

In order to calculate the tachyon scattering amplitudes we can now adopt the following simple strategy described in [9] (see *fig.* 4.2.). At spatial infinity the incoming and outgoing tachyons can be expressed in fermions through the standard formulas of relativistic fermionization and bosonization in two dimensions. The fermions have very simple scattering properties. Individual fermions just scatter with a reflection factor R_ω. In particular the number of fermions (and not only the fermion number) is conserved. After the scattering process the fermions propagate to infinity where they can be reassembled into bosons.

Thus we may write the full generating functional for connected Green's functions in terms of a single free boson

$$\partial\varphi(z) = \sum_n \alpha_n z^{-n-1} . \tag{4.13}$$

This scalar field should be thought of as the asymptotic limit of the tachyon field at τ, or equivalently $\phi, \to \infty$, where the Liouville interaction can be ignored. We now have a very simple representation for the generating function of all amplitudes [10]

$$\tau(t,\bar{t}) = \langle t|S|\bar{t}\rangle, \tag{4.14}$$

where the S-matrix is given by (4.12) and the coherent states now include an extra factor of μ

$$|\bar{t}\rangle = \exp\left(\sum_{n=1}^{\infty} i\mu\bar{t}_n\alpha_{-n}\right)|0\rangle, \qquad \langle t| = \langle 0|\exp\left(\sum_{n=1}^{\infty} -i\mu t_n\alpha_n\right). \tag{4.15}$$

This formula is an enormous simplification over previous expressions for $c=1$ amplitudes. It emphasizes the analogy with the $c<1$ results which we recall were of the form

$$\tau(t) = \langle t|W\rangle = \langle t|S|0\rangle. \tag{4.16}$$

In particular we observe immediately from (4.14) that the $c=1$ string forms a realization of the Toda Lattice hierarchy.

4.2. The Kontsevich-Penner model

As we have seen the partition function of the $c=1$ string is a solution of the Toda Lattice hierarchy. For fixed values of the incoming times $\bar{t}_k$, the partition function $\tau(t,\bar{t})$ becomes a solution of the KP hierarchy. We want to determine in more detail the element $W_{\bar{t}}$ in the Grassmannian that parametrizes this particular orbit of the KP flows. To this end we have to consider the state

$$|W_{\bar{t}}\rangle = S\cdot U_{\bar{t}}|0\rangle, \qquad U_{\bar{t}} = \exp\sum_{n=1}^{\infty} i\mu\bar{t}_n\alpha_{-n} . \tag{4.17}$$

We will describe $|W_{\bar{t}}\rangle$ by giving a basis $v_k(z;\bar{t})$, $k\geq 0$, of one-particle wave-functions. First we observe that the operator $U_{\bar{t}}$ acts on the wave-functions z^n by simple multiplication

$$U_{\bar{t}}:\ z^n \to \exp\left(\sum i\mu\bar{t}_k z^{-k}\right)\cdot z^n . \tag{4.18}$$

Similarly we have for the action of S a multiplication

$$S:\ z^n \to R_n\cdot z^n . \tag{4.19}$$

We have already seen that the reflection factors R_n contain all the relevant information of the $c = 1$ matrix model. At radius β they can be chosen to be

$$R_n = (-i\mu)^{-\frac{n+\frac{1}{2}}{\beta}} \frac{\Gamma(\frac{1}{2} - i\mu + \frac{n+\frac{1}{2}}{\beta})}{\Gamma(\frac{1}{2} - i\mu)} . \tag{4.20}$$

(Recall, we are only interested in the perturbative part in μ^{-1} of this expression.) The usual vacuum $|0\rangle$ is spanned by the non-negative powers z^k. Therefore the basis elements $v_k(z;\bar{t})$ of $W_{\bar{t}}$ are simply determined as

$$v_k(z;\bar{t}) = c_k \cdot S\, U_{\bar{t}} z^k , \tag{4.21}$$

with a normalization constant c_k such that $v_k(z;0) = z^k$. (This corresponds to the normal ordering of the S-matrix in (4.12).) Since the reflection factor is basically a gamma function, the result can be expressed as a Laplace transform

$$v_k(z;\bar{t}) = c'(z) \cdot \int_0^\infty dy \cdot y^k \cdot y^{-i\mu\beta+(\beta-1)/2} e^{i\mu(y/z)^\beta} \exp\left(\sum i\mu\bar{t}_k y^{-k}\right) \tag{4.22}$$

Here the constant $c'(z)$ is given by

$$c'(z) = \beta \frac{(-i\mu/z^\beta)^{\frac{1}{2}-i\mu}}{\sqrt{z}\Gamma(\frac{1}{2} - i\mu)} . \tag{4.23}$$

These integral representations are of Kontsevich type if and only if $\beta = 1$, that is, only at the self-dual radius. Indeed in that case we have

$$v_k(z;\bar{t}) = c'(z) \cdot \int_0^\infty dy \cdot y^k \cdot \exp i\mu \left(y/z - \log y + \sum \bar{t}_k y^{-k}\right) . \tag{4.24}$$

Therefore, following the procedure of §2.2 we can write the following matrix integral representation for the generating functional. Define the integral

$$\sigma(Z,\bar{t}) = \int dY e^{i\mu Tr[YZ^{-1}+V(Y)]} , \tag{4.25}$$

where

$$V(Y) = -\log Y + \sum \bar{t}_k Y^{-k} , \tag{4.26}$$

and we integrate over positive definite matrices Y. Then we have

$$\tau(t,\bar{t}) = \frac{\sigma(Z,\bar{t})}{\sigma(Z,0)} , \tag{4.27}$$

with the parameterization

$$t_n = \frac{\mu^{-1}}{n} Tr Z^{-n} . \tag{4.28}$$

Note that with this normalization $\tau(t,0) = 1$, which is appropriate since we consider normalized correlation functions. In order to write down the result we had to treat the incoming and outgoing tachyons very differently, parametrizing the outgoing states through whereas the coupling coefficients to the incoming states enter the matrix integral in a much more straightforward fashion.

References

[1] V. Kazakov, Phys. Lett. **159B** (1985) 303; F. David, Nucl. Phys. **B257** (1985) 45; V. Kazakov, I. Kostov, and A. Migdal, Phys. Lett. **157B** (1985) 295; J. Fröhlich, *The statistical mechanics of surfaces,* in *Applications of Field Theory to Statistical Mechanics*, L. Garrido ed. (Springer, 1985).

[2] E. Brézin and V. Kazakov, *Exactly solvable field theories of closed strings,* Phys. Lett. **B236** (1990) 144.
M. Douglas and S. Shenker, *Strings in less than one dimension,* Nucl. Phys. **B335** (1990) 635.
D.J. Gross and A. Migdal, *Nonperturbative two dimensional quantum gravity,* Phys. Rev. Lett. **64** (1990) 127.

[3] *Random Surfaces, Quantum Gravity and Strings,* Proceedings of the 1990 Cargèse workshop, O. Alvarez, E. Marinari, and P. Windey eds. (Plenum Press, 1991).
Two Dimensional Quantum Gravity and Random Surfaces, Proceedings of the Jerusalem Winter School for Theoretical Physics, D. Gross, T. Piran, and S. Weinberg eds. (World Scientific, 1992)
String Theory and Quantum Gravity '91, Proceedings of the Trieste Spring School and Workshop 1991, J. Harvey *et al* eds. (World Scientific, 1992).

[4] T. Banks, M.R. Douglas, N. Seiberg, and S. Shenker, *Microscopic and macroscopic loops in nonperturbative two dimensional gravity,* Phys. Lett. **238B** (1990) 279.
D.J. Gross and A. Migdal, *A nonperturbative treatment of two dimensional quantum gravity,* Nucl. Phys. **B340** (1990) 333.

[5] M. Douglas, *Strings in less than one dimension and the generalized KdV hierarchies,* Phys. Lett. **238B** (1990) 176.

[6] E. Brézin, V. Kazakov, and Al.B. Zamolodchikov, *Scaling violation in a field theory of closed strings in one physical dimension,* Nucl. Phys. **B338** (1990).
D.J. Gross and N. Miljković, *A nonperturbative solution of $D = 1$ string theory,* Phys. Lett. **238B** (1990) 217.
P. Ginsparg and J. Zinn-Justin, *2-d Gravity and 1-d matter,* Phys. Lett. **240B** (1990) 333.
D.J. Gross and I.R. Klebanov, *One-dimensional string theory on a circle,* Nucl. Phys. **B344** (1990) 475.

[7] G. Moore, *Double-scaled field theory at $c = 1$,* Nucl. Phys. **B368** (1992) 557.

[8] I.R. Klebanov and D. Lowe, *Correlation functions in two-dimensional gravity coupled to a compact scalar field*, Nucl. Phys. **B363** (1991) 543.

[9] G. Moore, R. Plesser, and S. Ramgoolam, *Exact S-matrix for 2D string theory*, Nucl. Phys. **B377**(1992)143.
G. Moore and R. Plesser, *Classical scattering in 1+1 dimensional string theory*, Yale preprint YCTP-P7-92.

[10] R. Dijkgraaf, G. Moore and R. Plesser, *The partition function of 2d string theory*, to be published in Nucl. Phys. B.

[11] E. Witten, *Two dimensional gravity and intersection theory on moduli space*, Surveys In Diff. Geom. **1** (1991) 243-310.

[12] M. Kontsevich, *Intersection theory on the moduli space of curves and the matrix Airy function*, Commun. Math. Phys. **147** (1992) 1–23.

[13] R. Dijkgraaf, *Intersection theory, integrable hierarchies and topological field theory*, in Proceedings of the Cargèse Summer School on "New Symmetry Principles in Quantum Field Theory" (Plenum Press).

[14] T. Curtright and C. Thorn, Phys. Rev. Lett. **48** (1982) 1309; Ann. Phys. **147** (1983) 365; Ann. Phys. **153** (1984) 147.

[15] U. Danielsson and D.J. Gross, *On the correlation functions of the special operators in $c = 1$ quantum gravity*, Nucl. Phys. **B366** (1991) 3.
E. Witten, *Ground ring of two dimensional string theory*, Nucl. Phys. **B373** (1992) 187.
I.R. Klebanov and A.M. Polyakov, *Interaction of discrete states in two-dimensional string theory*, Mod. Phys. Lett. **A6** (1991) 3273.
D. Kutasov, E. Martinec, and N. Seiberg, *Ground rings and their modules in 2d gravity with $c \leq 1$ matter*, Phys. Lett. **276B** (1992) 437.
I.R. Klebanov, *Ward identities in two-dimensional string theory*, Mod. Phys. Lett. **A7** (1992) 723.

[16] S. Kharchev, A. Marschakov, A. Mironov, A. Morozov, A. Zabrodin, *Unification of all string models with $c \leq 1$*, Phys. Lett. **275B** (1992) 311.*Towards unified theory of 2d gravity*, Lebedev Institute preprint FIAN/TD-10/91, hepth 9201013.
A. Marshakov, *On string field theory at $c \leq 1$*, Lebedev Institute preprint FIAN/TD-8/92, hepth 9208022.

[17] G. Segal and G. Wilson, *Loop groups and equations of KdV type*, Publ. Math. I.H.E.S. **61** (1985) 1.
E. Date, M. Jimbo, M. Kashiwara, T. Miwa, *Transformation groups for soliton equations*, RIMS Symp. *Nonlinear Integrable Systems—Classical Theory and Quantum Theory* (World scientific, Singapore, 1983).

[18] K. Ueno and K. Takasaki, *Toda lattice hierarchy*, in *Group Representations and Systems of Differential Equations*, H. Morikawa Ed., Advanced Studies in Pure Mathematics 4.

[19] Harish-Chandra, *Differential operators on a semisimple Lie algebra,* Am. J. Math. **79** (1987) 87.
C. Itzykson and J.-B. Zuber, *The planar approximation II,* J. Math. Phys. **21** (1980) 411.
M.L. Mehta, *A method of integration over matrix variables,* Commun. Math. Phys. **79** (1981) 327.

A geometric construction of solutions of the Toda lattice hierarchy

G.F. Helminck

University of Twente, Faculty of Applied Mathematics, P.O.Box 217, 7500 AE Enschede

Abstract. In this paper we present an analytic and geometric framework for the construction of solutions of the Toda lattice hierarchy.

Introduction. In [UT] , Ueno and Takasaki introduced the Toda lattice hierarchy, a system of nonlinear differential difference equations. Their approach is of a formal character and does not consider convergence questions for the objects they are dealing with. Here we will describe a convergent setting in which one can construct solutions of the system mentioned above. A more detailed description of the different sections is as follows: the first section contains the algebraic formulation of the hierarchy in Lax-form and the reduction of the system to a set of equations for so-called "wavematrices". In the second section we give the geometric setting for the construction of the convergent solutions. The final section gives the construction of the wavematrices and the proof that they satisfy the equations discussed in the first section.

§1 The equations.

1.1 Since the Toda lattice hierarchy is defined as a system of differential equations for the matrixcoefficients of a number of $\mathbb{Z} \times \mathbb{Z}$-matrices, we first introduce some notations.
Let R be a commutative ring. Then we write $M_{\mathbb{Z}}(R)$ for the R-module consisting of $\mathbb{Z} \times \mathbb{Z}$-matrices with coefficients from R. If $A = (A_{ij})$ and $B = (B_{ij})$ belong to $M_{\mathbb{Z}}(R)$, then the product $A \cdot B$ in $M_{\mathbb{Z}}(R)$, where

$$(A \cdot B)_{ik} = \sum_{j \in \mathbb{Z}} A_{ij} B_{jk},$$

is only defined for special A and B. It always exists if A or B belongs to the diagonalmatrices

$$\mathcal{D}(R) = \{A | A \in M_{\mathbb{Z}}(R),\ A_{ij} = 0 \text{ if } i \neq j\}.$$

Moreover an element A of $\mathcal{D}(R)$ is invertible if and only if all the A_{ii} are invertible in R. In the sequel, an important role is played by the element Λ of $M_{\mathbb{Z}}(R)$ given by

$$\Lambda_{ij} = \begin{cases} 1 \text{ if } j = i - 1 \\ 0 \text{ if } j \neq i - 1. \end{cases}$$

It has an inverse Λ^{-1} with entries given by

$$(\Lambda^{-1})_{ij} = \begin{cases} 1 \text{ if } j = i+1 \\ 0 \text{ if } j \neq i+1. \end{cases}$$

The matrix Λ acts on $\mathcal{D}(R)$ by conjugation: $(\Lambda d \Lambda^{-1})_{jj} = d_{j-1j-1}$ for all $d \in \mathcal{D}(R)$ and all $j \in \mathbb{Z}$. Moreover, for each $k \in \mathbb{Z}$ and each $d \in \mathcal{D}(R)$, the element $d\Lambda^k$, has only on the "kth-diagonal" possibly non-zero entries, i.e.

$$(d\Lambda^k)_{ij} = \begin{cases} d_{j+kj+k} \text{ if } i = j+k \\ 0 \text{ if } i \neq j+k. \end{cases}$$

Therefore we can write each element A in $M_{\mathbb{Z}}(R)$ uniquely as

1.1.1 $$A = \sum_{k \in \mathbb{Z}} a_k \Lambda^k, \text{ with } a_k \in \mathcal{D}(R).$$

To the decomposition (1.1.1) we link some notations: if A is as in (1.1.1) then we write

1.1.2 $$A_+ = \sum_{k \geq 0} a_k \Lambda^k \text{ and } A_- = \sum_{k<0} a_k \Lambda^k.$$

Inside $M_{\mathbb{Z}}(R)$ we consider 2 subspaces that form an algebra w.r.t. the product.

1.1.3 Definition. An element $A \in M_{\mathbb{Z}}(R)$ is called *uppertriangular of level k*, if it can be written as

$$A = \sum_{l \geq k} a_l \Lambda^l, \text{ with } a_l \in \mathcal{D}(R).$$

The collection of all these elements we denote by UT_k. It is a direct verification that $UT := \bigcup_{k \in \mathbb{Z}} UT_k$ is an algebra with the product introduced above and that an element of UT is invertible if and only if the leading diagonalcomponent is invertible. Likewise one can introduce

1.1.4 Definition An element $A \in M_{\mathbb{Z}}(R)$ is called *lowertriangular of level k*, if it can be written as

$$A = \sum_{l \leq k} a_l \Lambda^l, \text{ with } a_l \in \mathcal{D}(R).$$

The collection of all these elements we denote by LT_k. If we put again $LT := \bigcup_{k \in \mathbb{Z}} LT_k$, then LT is an algebra and an element is invertible in LT if and only if its leading diagonal part is invertible.

1.1.5 Remark The product of an element in LT and an element in UT is in general not defined and requires some convergence conditions on both factors in order to make sense.

1.2 The hierarchy we are interested in can conveniently be formulated in terms of a matrix $\tilde{L}$ in LT_1 and a matrix $\tilde{M}$ in UT_{-1} of the form

1.2.1 $$\tilde{L} = \Lambda + \sum_{j\geq 0} \tilde{l}_j \Lambda^{-j} \text{ and } \tilde{M} = \sum_{i\geq -1} \tilde{m}_i \Lambda^i \text{ with } \tilde{m}_{-1} \text{ invertible.}$$

For all relevant j and s, we will write $\tilde{l}_j(s)$ and $\tilde{m}_j(s)$ instead of $(\tilde{l}_j)_{ss}$ and $(\tilde{m}_j)_{ss}$. To a pair $(\tilde{L}, \tilde{M})$ as in 1.2.1 we associate for all $n \geq 1$ the elements B_n and C_n in $UT \cap LT$ defined by

$$B_n = (\tilde{L}^n)_+ \quad \text{and} \quad C_n = (\tilde{M}^n)_-.$$

The equations of the hierarchy will consist of a system of differential difference equations for the elements $\{\tilde{l}_j(s), \tilde{m}_i(s)\}$. In order to give the algebraic description of the hierarchy, we consider the $\{\tilde{l}_j(s), \tilde{m}_i(s)\}$ first as free commuting variables and we take for R the algebra $B = \mathbb{C}[\tilde{l}_j(s), \tilde{m}_i(s), \tilde{m}_{-1}(s)^{-1}]$, where $j \in \mathbb{N}, i \in \mathbb{Z}, i \geq -1$, and $s \in \mathbb{Z}$. Clearly, one can define a $\mathbb{C}$-linear derivation of B by simply prescribing the image of all the $\tilde{l}_j(s)$ and all the $\tilde{m}_i(s)$. In $M_{\mathbb{Z}}(B)$ the following commutators exist and belong to the indicated subsets

$$[B_n, \tilde{L}] = [-(\tilde{L})^n_-, \tilde{L}] \in LT_0$$

$$[C_n, \tilde{M}] = [-(\tilde{M}^n)_+, \tilde{M}] \in UT_{-1}$$

$$[B_n, \tilde{M}] \in UT_{-1} \text{ and } [C_n, \tilde{L}] \in LT_0$$

In the light of the foregoing remark, we can define $\mathbb{C}$-linear derivations $\tilde{\partial}_{t_n}$ and $\tilde{\partial}_{s_n}$ of B by the equations

1.2.2 $$\tilde{\partial}_{t_n} \tilde{L} = [B_n, \tilde{L}],\ \tilde{\partial}_{t_n} \tilde{M} = [B_n, \tilde{M}],$$

1.2.3 $$\tilde{\partial}_{s_n} \tilde{L} = [C_n, \tilde{L}],\ \tilde{\partial}_{s_n} \tilde{M} = [C_n, \tilde{M}].$$

Let R be a general $\mathbb{C}$-algebra and let (L, M) be a pair of elements in $M_{\mathbb{Z}}(R)$ of the form 1.2.1. These data are the same as giving a $\mathbb{C}$-algebra homomorphism $\alpha : B \to R$ by the prescription

1.2.4 $$\alpha(\tilde{l}_j(s)) = (l_j)_{ss} \text{ and } \alpha(\tilde{m}_i(s)) = (m_i)_{ss}.$$

The map α defines by coefficientwise action a $\mathbb{C}$-linear map $\alpha : M_{\mathbb{Z}}(B) \to M_{\mathbb{Z}}(R)$ and we have $L = \alpha(\tilde{L})$ and $M = \alpha(\tilde{M})$. Now we would like to have derivations ∂_{t_n} and ∂_{s_n}, $n \geq 1$, of R that are prolongations of the derivations $\tilde{\partial}_{t_n}$ and $\tilde{\partial}_{s_n}$ defined by 1.2.2 and 1.2.3, i.e. they should satisfy for all $n \geq 1$,

1.2.5 $$\alpha \circ \tilde{\partial}_{t_n} = \partial_{t_n} \circ \alpha \text{ and } \alpha \circ \tilde{\partial}_{s_n} = \partial_{s_n} \circ \alpha.$$

One directly verifies that condition 1.2.5 is equivalent to showing that the pair (L, M) and the derivations $\{\partial_{t_n}, \partial_{s_n}, n \geq 1\}$ satisfy the equations

1.2.6 $$\partial_{t_n} L = [(L)^n_+, L], \partial_{t_n} M = [(L^n)_+, M],$$

1.2.7 $$\partial_{s_n} M = [(M)^n_-, M] \text{ and } \partial_{s_n} L = [(M)^n_-, L].$$

The equations 1.2.6 and 1.2.7 are called the equations of the Toda lattice hierarchy, since the simplest non-linear equations contained in it are that of the generalized Toda lattice, see [UT]. The data $(R, \partial_{t_n}, \partial_{s_n}, \alpha(\tilde{L}), \alpha(\tilde{M}))$ we call a solution of this hierarchy.
¿From now on, we assume that R is a $\mathbb{C}$-algebra of functions in the parameters $\{t_n | n \geq 1\}$ and $\{s_n | n \geq 1\}$, that it is stable under taking the partial derivative ∂_{t_n} resp. ∂_{s_n} w.r.t. t_n resp. s_n, and that R contains $\mathbb{C}[\approx_\kappa, \sim_\kappa, \kappa \geq \mathbb{K}]$. We will now comment on the linear problem associated to (1.2.6) and (1.2.7) in [UT]. There they considered "operators" $W^{(\infty)}$ and $W^{(0)}$ of the form

$$W^{(\infty)} = \widehat{W}^{(\infty)} \cdot \exp(\sum_{i>0} t_i \Lambda^i) := \{ \text{Id} + \sum_{i<0} \omega_i^{(\infty)} \Lambda^i \} \exp(\sum_{i>0} t_i \Lambda^i) \text{ and}$$
$$W^{(0)} = \widehat{W}^{(0)} \cdot \exp(\sum_{i>0} s_i \Lambda^{-i}) := \{ \sum_{i\geq 0} \omega_i^{(0)} \Lambda^i \} \exp(\sum_{i>0} s_i \Lambda^{-i}), \text{ with } \omega_0^{(0)} \text{ invertible,}$$

where all the $\omega_i^{(\infty)}$ and $\omega_i^{(o)}$ belong to $\mathcal{D}(R)$ and such that $\omega_0^{(o)}$ is invertible in $\mathcal{D}(R)$.
Since $W^{(\infty)}$ and $W^{(0)}$ are products of an element in UT and in LT, these products make in general no sense in $M_{\mathbb{Z}}(R)$ and one must give some convergent context such that $W^{(\infty)}$ and $W^{(0)}$ can be seen as elements of $M_{\mathbb{Z}}(R)$. This will be done in the second and third section. In that case one notes that both $W^{(\infty)}$ and $W^{(0)}$ are invertible in $M_{\mathbb{Z}}(R)$.
If W^{∞} and W^0 are well-defined elements of $M_{\mathbb{Z}}(R)$, then we will call them wavematrices in $M_{\mathbb{Z}}(R)$. The linear system associated with (L, M) consists of the following equations that couple the wavematrices to the pair (L, M),

1.2.8 $$LW^{(\infty)} = W^{(\infty)}\Lambda \quad \text{and } MW^{(0)} = W^{(0)}\Lambda^{-1},$$

1.2.9 $$\partial_{t_n} W^{(\infty)} = B_n W^{(\infty)} \quad \text{and } \partial_{t_n} W^{(0)} = B_n W^{(0)} \text{ for all } n \geq 1,$$

1.2.10 $$\partial_{s_n} W^{(\infty)} = C_n W^{(\infty)} \quad \text{and } \partial_{s_n} W^{(0)} = C_n W^{(0)} \text{ for all } n \geq 1.$$

Since $W^{(\infty)}$ and $W^{(0)}$ are invertible, equation (1.2.8) lead to

1.2.11 $$L = \widehat{W}^{(\infty)} \Lambda \widehat{W}^{(\infty)^{-1}} \text{ and } M = \widehat{W}^{(0)} \Lambda^{-1} \widehat{W}^{(0)^{-1}}.$$

By differentiating the equations in (1.2.8) w.r.t. the variable t_n resp. s_n and by substituting (1.2.9) resp. (1.2.10), we get that L and M defined by (1.2.11) satisfy the equations (1.2.3). In the rest of this paper we present a geometric context from which one can construct wave matrices $W^{(\infty)}$ and $W^{(0)}$ in $M_{\mathbb{Z}}(R)$ that satisfy (1.2.9) and (1.2.10). The operators L and M defined by (1.2.11) are then the solutions of the Toda lattice hierarchy.

§2The geometric setting.

2.1 Let H be a complex Hilbert space with orthonormal basis $\{e_i | i \in \mathbb{Z}\}$ and innerproduct $< \cdot | \cdot >$. The space of bounded linear operators from H to H, we denote by $B(H)$ and we assume it to be equiped with the operator norm. Its group of invertible elements is denoted by $Gl(H)$. The group $Gl(H)$ is an open part of $B(H)$ and as such, it is a Banach Lie group with Lie algebra $B(H)$.

2.1.1 Notation To each operator g in $B(H)$ we associate a $\mathbb{Z} \times \mathbb{Z}$-matrix $[g] = ([g]_{ij})$ in $M_{\mathbb{Z}}(\mathbb{C})$ by putting $[g]_{ij} =< g(e_j)|e_i >$, $i, j \in \mathbb{Z}$.

Next we introduce some Lie subgroups of $Gl(H)$ and their corresponding Lie algebras. First of all we have the Borel subgroup B_+ and the "opposite" Borel subgroup B_- given by

$$B_+ = \{g | g \in Gl(H), [g]_{ij} = 0 = [g^{-1}]_{ij} \text{ for all } i < j\},$$
$$B_- = \{g | g \in Gl(H), [g]_{ij} = 0 = [g^{-1}]_{ij} \text{ for all } i > j\}.$$

Their Lie algebras $L(B_+)$ and $L(B_-)$ satisfy

$$L(B_+) = \{b | b \in B(H), [b]_{ij} = 0 \text{ for all } i < j\}$$
$$L(B_-) = \{b | b \in B(H), [b]_{ij} = 0 \text{ for all } i > j\}.$$

As in the finite-dimensional case, B_+ and B_- are the semi-direct product of a diagonal group

$$D = \{g | g \in Gl(H), [g]_{ij} = 0 \text{ for all } i \neq j\}$$

and the unipotent subgroups U_+ respectively U_- given by

$$U_+ = \{g|g \in B_+, [g]_{ii} = 1 \text{ for all } i \in \mathbb{Z}\} \text{ and } U_- = \{g|g \in B_-, [g]_{ii} = 1 \text{ for all } i \in \mathbb{Z}\}.$$

Their Lie algebras are denoted respectively by $L(D), L(U_+)$ and $L(U_-)$. Since $B(H)$ decomposes as

$$B(H) = L(U_-) \oplus L(D) \oplus L(U_+)$$

and exp is a local diffeomorphism around zero, we see that

$$O_1 = B_+U_- \quad \text{and } O_2 = U_-B_+$$

are open subsets of $Gl(H)$. We will give another characterization of O_1 and O_2. For each $n \in \mathbb{Z}$, let H_n be the topological span of the $\{e_i|i \geq n\}$ and let p_n be the orthogonal projection onto H_n. If one decomposes an operator $g \in O_1$ and $h \in O_2$ w.r.t. $H = H_n \oplus H_n^\perp$, then one computes directly that for all $n \in \mathbb{Z}$

2.1.2
$$g = \begin{pmatrix} g_{11}(n) & g_{12}(n) \\ g_{21}(n) & g_{22}(n) \end{pmatrix} \text{ with } g_{22}(n) \in Gl(H_n^\perp) \text{ and}$$

2.1.3
$$h = \begin{pmatrix} h_{11}(n) & h_{12}(n) \\ h_{21}(n) & h_{22}(n) \end{pmatrix} \text{ with } h_{11}(n) \in Gl(H_n).$$

Reversely, these properties characterize the sets O_1 and O_2, for we have

2.1.4 Proposition The sets O_1 resp. O_2 consist of all g resp. h in $Gl(H)$ satisfying (2.1.2) resp. (2.1.3) for all $n \in \mathbb{Z}$.

Proof We give the proof for O_2, the one for O_1 is similar. Take any $n \in \mathbb{Z}$ then one has

$$h = \begin{pmatrix} \mathrm{Id} & 0 \\ h_{21}(n)h_{11}(n)^{-1} & \mathrm{Id} \end{pmatrix} \begin{pmatrix} h_{11}(n) & h_{12}(n) \\ 0 & h_{22}(n) - h_{21}(n)h_{11}(n)^{-1}h_{12}(n) \end{pmatrix}.$$

Hence we may assume $h_{21}(n) = 0$. With respect to the decomposition $H_n =< e_n > \oplus H_{n+1}$ we have that

$$h_{11}(n) = \begin{pmatrix} h_{11}(n+1) & \beta \\ \gamma & \delta \end{pmatrix} = \begin{pmatrix} \mathrm{Id} & 0 \\ \gamma h_{11}(n+1)^{-1}1 & \mathrm{Id} \end{pmatrix} \begin{pmatrix} h_{11}(n+1) & \\ 0 & * \end{pmatrix}.$$

Continuing in this fashion we can find an u_1 in $Gl(H_n)$ and a b_1 in $Gl(H_n)$ such that $h_{11}(n) = u_1b_1$ and their matrices w.r.t. the $\{e_k|k \geq n\}$ have the form

$$[u_1] = \begin{pmatrix} \ddots & & 0 \\ & 1 & \\ & & \ddots \\ * & & 1 \end{pmatrix} \text{ and } [b_1] = \begin{pmatrix} \ddots & \ddots & & * \\ & 0 & * & \\ & & \ddots & \ddots \\ \cdots & 0 & \cdots & 0 & * \end{pmatrix}.$$

Since h belongs to O_2 and since we may assume $h_{21}(n)=0$, one sees that $h_{22}(n)$ decomposes w.r.t. $H_n^\perp = < e_{n-1} > \oplus H_{n-1}^\perp$ as

$$h_{22}(n) = \begin{pmatrix} \alpha_1 & \beta_1 \\ \gamma_1 & h_{22}(n-1) \end{pmatrix} \text{ with } \alpha_1 \neq 0.$$

Hence we can solve by a step by step procedure that there is a u_2 and a b_2 in $Gl(H_n^\perp)$ such that $h_{22}(n)=u_2 b_2$ and such that their matrices w.r.t. the $\{e_k | k < n\}$ have the form

$$[u_2] = \begin{pmatrix} 1 & & & & 0 \\ * & \ddots & & & \\ \vdots & \ddots & 1 & & \\ & & * & \ddots & \\ & & & \ddots & \ddots \end{pmatrix} \text{ and } [b_2] = \begin{pmatrix} * & \cdots & * & \cdots \\ 0 & \ddots & & \\ \vdots & \ddots & * & \\ 0 & & 0 & \ddots \\ & & & \ddots \end{pmatrix}.$$

By combining the u_1, u_2, b_1, b_2 and $h_{12}(n)$ and $h_{21}(n)$, one finds the desired decomposition of h. This concludes the proof of the proposition. □

2.2 Next we introduce the flows that play a role in the Toda lattice hierarchy. For each open neighbourhood U of $\{\lambda | \lambda \in \mathbb{C},\ |\lambda| = 1\}$, consider

$$\Gamma(U) = \left\{ \sum_{i\in\mathbb{Z}} a_i \lambda^i \;\middle|\; \begin{array}{l} a_i \in \mathbb{C}, \sum_{j\in\mathbb{Z}} a_j \lambda^j \text{ is a holomorphic} \\ \text{function: } U \to \mathbb{C}^* \end{array} \right\}$$

with the topology of uniform convergence on compact subsets. It is a group w.r.t. point wise multiplication. Let $\tilde{\Gamma}$ be the direct limit of the $\Gamma(U)$ with the corresponding topology. Inside $\tilde{\Gamma}$ we have the subgroups $\tilde{\Gamma}_+$ and $\tilde{\Gamma}_-$ given by

$$\tilde{\Gamma}_+ = \{\exp(\sum_{i>0} t_i \lambda^i) \in \tilde{\Gamma}\} \text{ and } \tilde{\Gamma}_- = \{\exp(\sum_{j>0} s_j \lambda^{-j}) \in \tilde{\Gamma}\}.$$

According to [HP], each element γ in $\tilde{\Gamma}$ can be decomposed uniquely as follows

2.3.1 $$\gamma = \gamma_+ \gamma_- \lambda^k a, \text{ with } \gamma_+ \in \tilde{\Gamma}_+, \gamma_- \in \tilde{\Gamma}_-, k \in \mathbb{Z} \text{ and } a \in \mathbb{C}^*.$$

Let $\tilde{\Gamma}(o)$ be the subgroup of $\tilde{\Gamma}$ consisting of all γ in $\tilde{\Gamma}$ with k equal to zero. The group $\tilde{\Gamma}$ maps continuously into $Gl(H)$. For, let $\underline{\Lambda} : H \to H$ be the shift operator defined by

$$\underline{\Lambda}(\sum_i \alpha_i e_i) = \sum \alpha_i e_{i+1}.$$

Then we define a continuous embedding $M : \tilde{\Gamma} \to Gl(H)$ by

$$M(\gamma) = M(\sum_{i\in\mathbb{Z}} a_i \lambda^i) = \sum_{i\in\mathbb{Z}} a_i \underline{\Lambda}^i.$$

The image under M of $\tilde{\Gamma}, \tilde{\Gamma}_+, \tilde{\Gamma}_-$ and $\tilde{\Gamma}(\circ)$, we denote respectively by $\Gamma, \Gamma_+, \Gamma_-$ and $\Gamma(\circ)$. Note that $\Gamma(\circ)$ corresponds exactly with the intersection of Γ with the open sets O_1 and O_2.

2.2.2 Remark Since the $\mathbb{Z} \times \mathbb{Z}$-matrix of $\underline{\Lambda}$ is exactly the matrix Λ from section (1.1), it will be clear that the matrices of elements of Γ_+ and Γ_- are exactly the matrices occurring in the wavematrices of section (1.2).

§3 The construction of the solutions.

3.1 Inside $Gl(H)$ we consider the open subset Ω defined by

$$\Omega = \Gamma(\circ) O_2 \Gamma(\circ).$$

Since $\Gamma(\circ)$ is the union of the $a\Gamma_+\Gamma_-$, with $a \in \mathbb{C}^*$, and since the open set O_2 can be written as U_-B_+, we see that

$$\Omega = \Gamma_+ O_2 \Gamma_-.$$

The set Ω is not equal to $Gl(H)$, for if one considers for example the orbit $\Gamma_+\Lambda^k\Gamma_-$, with $k \neq 0$, then it has empty intersection with O_2. For, each g in $\Gamma_+\Lambda^k\Gamma_-$ decomposes w.r.t. $H = H_n \oplus H_n^\perp$ as

$$g = \begin{pmatrix} g_{11} & g_{12} \\ g_{21} & g_{22} \end{pmatrix}, \text{ with } g_{11} \text{ a Fredholm operator of index } -k.$$

Hence, by lemma (2.1.4) these elements do not belong to O_2. Now we take elements γ_+ in Γ_+ and γ_- in Γ_- such that

$$\gamma_+ = \gamma_+(t) = \exp(\sum_{i>0} t_i \underline{\Lambda}^i) \text{ and } \gamma_- = \gamma_-(s) = \exp(\sum_{j>0} s_j \underline{\Lambda}^{-j}),$$

and we consider the left Γ_+-flow and the right Γ_--flow in Ω. That is to say, we choose a g in Ω and we look at

$$G(t,s) = \gamma_+ g \gamma_-^{-1}.$$

Clearly G is a holomorphic map from $\tilde{\Gamma}_+ \times \tilde{\Gamma}_-$ to $Gl(H)$ and since g belongs to Ω, we have that $G^{-1}(O_2)$ is a non-empty open subset of $\tilde{\Gamma}_+ \times \tilde{\Gamma}_-$. For the ring R in the first section we take now the ring of holomorphic functions on $G^{-1}(O_2)$. For all $(\gamma_+(t), \gamma_-(s))$ in $G^{-1}(O_2)$, we have a decomposition

3.1.1
$$G(t,s) = \widehat{\mathcal{W}}^{(\infty)}(t,s)^{-1} \widehat{\mathcal{W}}^{(0)}(t,s),$$

with $\widehat{\mathcal{W}}^{(\infty)}(t,s) \in U_-$ and $\widehat{\mathcal{W}}^{(0)}(t,s) \in B_+$. If we write $\widehat{W}^{(\infty)}$ resp. $\widehat{W}^{(0)}$ for the matrices of $\widehat{\mathcal{W}}^{(\infty)}$ resp. $\widehat{\mathcal{W}}^{(0)}$, then the coefficients of these matrices belong to R, and they decompose as

$$\widehat{W}^{(\infty)} = \Lambda^0 + \sum \underline{\omega}_j^{(\infty)} \Lambda^j, \text{ with } \underline{\omega}_j^{(\infty)} \in \mathcal{D}(R) \text{ and}$$
$$\widehat{W}^{(0)} = \sum_{i \geq 0} \underline{\omega}_i^{(0)} \Lambda^i, \text{ with } \underline{\omega}_i^{(0)} \in \mathcal{D}(R) \text{ and } \underline{\omega}_0^{(0)} \text{ invertible in } \mathcal{D}(R).$$

Next we introduce the operators $\mathcal{W}^{(\infty)}$ and $\mathcal{W}^{(0)}$ in $Gl(H)$ by

$$\mathcal{W}^{(\infty)} = \widehat{\mathcal{W}}^{(\infty)} \cdot \gamma_+ \text{ and } \mathcal{W}^{(0)} = \widehat{\mathcal{W}}^{(0)} \cdot \gamma_-.$$

Denote the matrices of $\mathcal{W}^{(\infty)}$ and $\mathcal{W}^{(0)}$ by $W^{(\infty)}$ resp. $W^{(0)}$. The set-up has been chosen such that in $M_{\mathbb{Z}}(R)$ the products of $\widehat{W}^{\infty)}$ and $[\gamma_+]$ and of $\widehat{W}^{(0)}$ and $[\gamma_-]$ are well-defined and that they are equal to $W^{(\infty)}$ resp. $W^{(0)}$. Again the coefficients of $W^{(\infty)}$ and $W^{(0)}$ belong to R. Moreover they have the form of a "wavematrix" as considered in section (1.2).
To $W^{(\infty)}$ and $W^{(0)}$ we associate the Lax-matrices L and M according to

3.1.2 $L_g = W^{(\infty)}\Lambda W^{(\infty)^{-1}} = \widehat{W}^{(\infty)}\Lambda \widehat{W}^{(\infty)^{-1}}$ and $M_g = W^{(0)}\Lambda W^{(0)^{-1}} = \widehat{W}^{(0)}\Lambda \widehat{W}^{(0)^{-1}}$.

We are now ready to prove the main result.

3.1.3 Theorem (a) If g belongs to Ω, then the matrices L_g and M_g constructed in (3.1.2) are solutions of the Toda lattice hierarchy.
(b) If $\delta_- \in \Gamma_-, \delta_+ \in \Gamma_+$ and $a \in \mathbb{C}^*$ then the constructed solutions of the Toda lattice hierarchy corresponding to g and $\delta_- g a \delta_+$ are the same, i.e.

$$L_{\delta_- g a \delta_+} = L_g \text{ and } M_{\delta_- g a \delta_+} = M_g.$$

Proof First of all we note that $\mathcal{W}^{(0)}$ and $\mathcal{W}^{(\infty)}$ are constructed in such a way that $W^{(\infty)}[g] = W^{(0)}$. Hence, if we can proof

$$\partial_{t_n} W^{(\infty)} = B_n W^{(\infty)} \text{ for all } n \geq 1,$$

then the same equations holds for $W^{(0)}$. Analogously, it suffices to prove

$$\partial_{s_n} W^{(0)} = C_n W^{(0)} \text{ for all } n \geq 1,$$

to get these equations for $W^{(\infty)}$. Consider first the equations (1.2.9). On one hand we have

$$\begin{aligned}\partial_{t_n} W^{(\infty)} &= \partial_{t_n}(\widehat{W}^{(\infty)}) \exp(\sum_{i>0} t_i \Lambda^i) + \widehat{W}^{(\infty)} \Lambda^n \exp(\sum_{i>0} t_i \Lambda^i) \\ &= \{\partial_{t_n}(\widehat{W}^{(\infty)}) + \widehat{W}^{(\infty)} \Lambda^n\} \widehat{W}^{(\infty)^{-1}} W^{(\infty)} \\ &= \{\partial_{t_n}(\widehat{W}^{(\infty)}) \widehat{W}^{(\infty)^{-1}} + L^n\} W^{(\infty)}.\end{aligned}$$

On the other hand, if we differentiate $W^{(0)}[g]^{-1}$, w.r.t. t_n, we get

$$\begin{aligned}\partial_{t_n} W^{(0)}[g]^{-1} &= \partial_{t_n}(\widehat{W}^{(0)})\exp(\sum_{j>0} s_j\Lambda^{-j})[g]^{-1}\\ &= \{\partial_{t_n}(\widehat{W}^{(0)})\widehat{W}^{(0)^{-1}}\}\widehat{W}^{(\infty)}\\ &= \{\sum_{k\geq 0} b_k\Lambda^k\}\widehat{W}^{(\infty)}.\end{aligned}$$

In particular we may conclude that

$$\sum_{k\geq 0} b_k\Lambda^k = \{\partial_{t_n}(\widehat{W}^{(\infty)})\widehat{W}^{(\infty)^{-1}} + L^n\}_+ = (L^n)_+ = B_n.$$

The second equality in this equation follows from the fact that $\partial_{t_n}(\widehat{W}^{(\infty)})$ is lowertriangular of level -1. The equations (1.2.10) are also obtained by differentiating once $W^{(0)}$ and once $W^{(\infty)}[g]$:

$$\begin{aligned}\partial_{s_n}(W^{(0)}) &= \{\partial_{s_n}(\widehat{W}^{(0)}) + \widehat{W}^{(0)}\Lambda^{-1}\}\exp(\sum_{j>0} s_j\Lambda^{-j})\\ &= \{\partial_{s_n}(\widehat{W}^{(0)})\widehat{W}^{(0)^{-1}} + M^n\}W^{(0)}\\ \partial_{s_n}(\widehat{W}^{(\infty)}[g]) &= \partial_{s_n}(\widehat{W}^{(\infty)})\exp(\sum_{i>0} t_i\Lambda^i)\\ &= \{\partial_{s_n}(\widehat{W}^{(\infty)})\widehat{W}^{(\infty)^{-1}}\}\widehat{W}^{(\infty)}[g].\end{aligned}$$

Since $\partial_{s_n}(\widehat{W}^{(\infty)})$ is lowertriangular of level -1 and $\partial_{s_n}(\widehat{W}^{(0)})$ is uppertriangular of level 0, we get that

$$\{\partial_{s_n}(\widehat{W}^{(\infty)})\widehat{W}^{(\infty)^{-1}}\} = \{\partial_{s_n}(\widehat{W}^{(0)})\widehat{W}^{(0)^{-1}} + M^n\}_- = (M^n)_- = C_n.$$

This proves the equations (1.2.9) and (1.2.10) for L_g and M_g and the first part of the theorem. Next we consider $\tilde{g} = \delta_- g a \delta_+$, with $a \in \mathbb{C}^*, \delta_- \in \lneqq_-, \delta_+ \in \lneqq_+$ and $g \in \Omega$. It belongs again to Ω and, since $\delta_-, \gamma_+, \delta_+$ and γ_- commute the corresponding wavematrices are easily seen to have the form $W^{(\infty)}[\delta_-]^{-1}$ and $W^{(0)}[\delta_+]a$, where $W^{(\infty)}$ and $W^{(0)}$ are the wavematrices belonging to g. This implies that $L_{\tilde{g}}$ and $M_{\tilde{g}}$ are given by

$$\begin{aligned}L_{\tilde{g}} &= \widehat{W}^{(\infty)}[\partial_-]^{-1}\Lambda[\partial_-]\widehat{W}^{(\infty)^{-1}} = L_g \text{ and}\\ M_{\tilde{g}} &= \widehat{W}^{(0)}[\partial_+]a\Lambda^{-1}a^{-1}[\partial_+]^{-1}\widehat{W}^{(0)^{-1}} = \widehat{W}^{(0)}\Lambda\widehat{W}^{(0)^{-1}} = M_g.\end{aligned}$$

This concludes the proof of the theorem. □

References

[UT] K. Ueno and K. Takasaki, Toda Lattice Hierarchy, Publication 423, RIMS Kyoto 1983.

[PS] A. Pressley and G. Segal, Loop groups, Oxford Mathematical Monographs, Clarendon Press, Oxford, 1986.

[HP] G.F. Helminck and G.F. Post, The geometry of differential difference equations, Memorandum 999, University of Twente, september 1991.

Introduction to Random Matrices

Craig A. Tracy
Department of Mathematics and Institute of Theoretical Dynamics, University of California, Davis, CA 95616, USA

Harold Widom
Department of Mathematics, University of California, Santa Cruz, CA 95064, USA

These notes provide an introduction to the theory of random matrices. The central quantity studied is $\tau(a) = \det(1 - K)$ where K is the integral operator with kernel

$$\frac{1}{\pi}\frac{\sin\pi(x-y)}{x-y}\,\chi_I(y)\,.$$

Here $I = \bigcup_j (a_{2j-1}, a_{2j})$ and $\chi_I(y)$ is the characteristic function of the set I. In the Gaussian Unitary Ensemble (GUE) the probability that no eigenvalues lie in I is equal to $\tau(a)$. Also $\tau(a)$ is a tau-function and we present a new simplified derivation of the system of nonlinear completely integrable equations (the a_j's are the independent variables) that were first derived by Jimbo, Miwa, Môri, and Sato in 1980. In the case of a single interval these equations are reducible to a Painlevé V equation. For large s we give an asymptotic formula for $E_2(n;s)$, which is the probability in the GUE that exactly n eigenvalues lie in an interval of length s.

I. INTRODUCTION

These notes provide an introduction to that aspect of the theory of random matrices dealing with the distribution of eigenvalues. To first orient the reader, we present in Sec. II some numerical experiments that illustrate some of the basic aspects of the subject. In Sec. III we introduce the invariant measures for the three "circular ensembles" involving unitary matrices. We also define the level spacing distributions and express these distributions in terms of a particular Fredholm determinant. In Sec. IV we explain how these measures are modified for the orthogonal polynomial ensembles. In Sec. V we discuss the universality of these level spacing distribution functions in a particular scaling limit. The discussion up to this point (with the possible exception of Sec. V) follows the well-known path pioneered by Hua, Wigner, Dyson, Mehta and others who first developed this theory (see, e.g., the reprint volume of Porter [39] and Hua [18]). This, and much more, is discussed in Mehta's book [27]—*the classic reference* in the subject.

An important development in random matrices was the discovery by Jimbo, Miwa, Môri, and Sato [22] (hereafter referred to as JMMS) that the basic Fredholm determinant mentioned above is a τ-function in the sense of the Kyoto School. Though it has been some twelve years since [22] was published, these results are not widely appreciated by the practitioners of random matrices. This is due no doubt to the complexity of their paper. The methods of JMMS are methods of discovery; but now that we know the result, simpler proofs can be constructed. In Sec. VI we give such a proof of the JMMS equations. Our proof is a simplification and generalization of Mehta's [29] simplified proof of the single interval case. Also our methods build on the earlier work of Its, Izergin, Korepin, and Slavnov [19] and Dyson [13]. We include in this section a discussion of the connection between the JMMS equations and the integrable Hamiltonian systems that appear in the geometry of quadrics and spectral theory as developed by Moser [35]. This section concludes with a discussion of the case of a single interval (viz., probability that exactly n eigenvalues lie in a given interval). In this case the JMMS equations can be reduced to a single ordinary differential equation—the Painlevé V equation.

Finally, in Sec. VII we discuss the asymptotics in the case of a large single interval of the various level spacing distribution functions [4, 43, 31]. In this analysis both the Painlevé representation and new results in Toeplitz/Wiener-Hopf theory are needed to produce these asymptotics. We also give an approach based on the asymptotics of the eigenvalues of the basic linear integral operator [15, 27, 40]. These results are then compared with the continuum model calculations of Dyson [13].

II. NUMERICAL EXPERIMENTS

The Gaussian orthogonal ensemble (GOE) consists of $N \times N$ real symmetric matrices whose elements (subject to the symmetric constraint) are independent and identically distributed Gaussian random variables of mean zero and variance one. Pioneers in the simulation of random matrices were Porter and Rosenzweig (see, e.g., pgs. 235–299 in [39]). Today one can easily use a Gaussian random number generator to produce a "typical" such matrix. Given this matrix we can diagonalize it to produce our "random eigenvalues." Using the software MATHEMATICA, 25 such 100×100 GOE matrices were generated and Fig. 1 is a histogram of the density of eigenvalues where the x-axis has been normalized so that all eigenvalues lie in $[-1, 1]$. Also shown is the *Wigner semicircle law*

$$\rho_W(x) = \frac{2}{\pi}\sqrt{1 - x^2}. \tag{2.1}$$

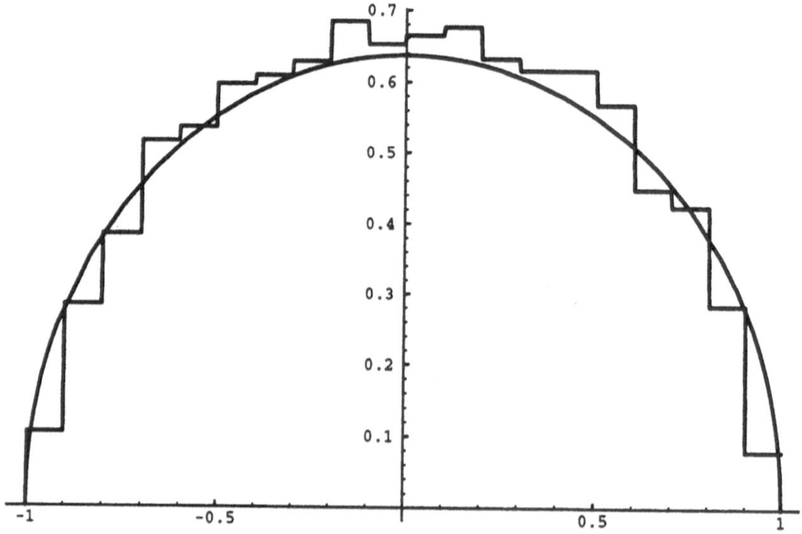

FIG. 1. Density of eigenvalues histogram for 25, 100×100 GOE matrices. Also plotted is the Wigner semicircle which is known to be the limiting distribution.

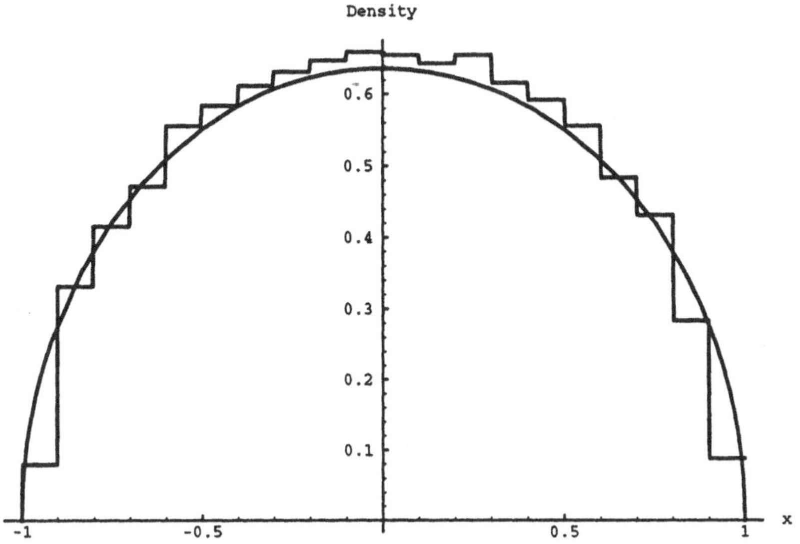

FIG. 2. Density of eigenvalues histogram for 25, 100×100 symmetric matrices whose elements are uniformly distributed on $[-1,1]$. Also plotted is the Wigner semicircle distribution.

Given any such distribution (or density) function, one can ask to what extent is it "universal." In Fig. 2 we plot the same density histogram except we change the distribution of matrix elements to the uniform distribution on $[-1,1]$. One sees that the same semicircle law is a good approximation to the density of eigenvalues. See [27] for further discussion of the Wigner semicircle law.

A fundamental quantity of the theory is the (conditional) probability that given an eigenvalue at a, the next eigenvalue lies between b and $b+db$: $p(0;a,b)\,db$. In measuring this quantity it is usually assumed that the system is well approximated by a translationally invariant system of constant eigenvalue density. This density is conveniently normalized to one. In the translationally invariant system $p(0;a,b)$ will depend only upon the difference $s := b-a$. When there is no chance for confusion, we denote this probability density simply by $p(s)$. Now as Figs. 1 and 2 clearly show, the eigenvalue density in the above examples is not constant. However, since we are mainly interested in the case of large matrices (and ultimately $N \to \infty$), we take for our data the eigenvalues lying in an interval in which the density does not change significantly. For this data we compute a histogram of *spacings* of eigenvalues. That is to say, we order the eigenvalues E_i and compute the level spacings $S_i := E_{i+1} - E_i$. The scale of both the x-axis and y-axis are fixed once we require that the integrated density is one and the mean spacing is one. Fig. 3 shows the resulting histogram for 20, 100×100 GOE matrices where the eigenvalues were taken from the middle half of the entire eigenvalue spectrum. The important aspect of this data is it shows *level repulsion of eigenvalues* as indicated by the vanishing of $p(s)$ for small s. Also plotted is the *Wigner surmise*

$$p_W(s) = \frac{\pi}{2} s \exp\left(-\frac{\pi}{4}s^2\right) \tag{2.2}$$

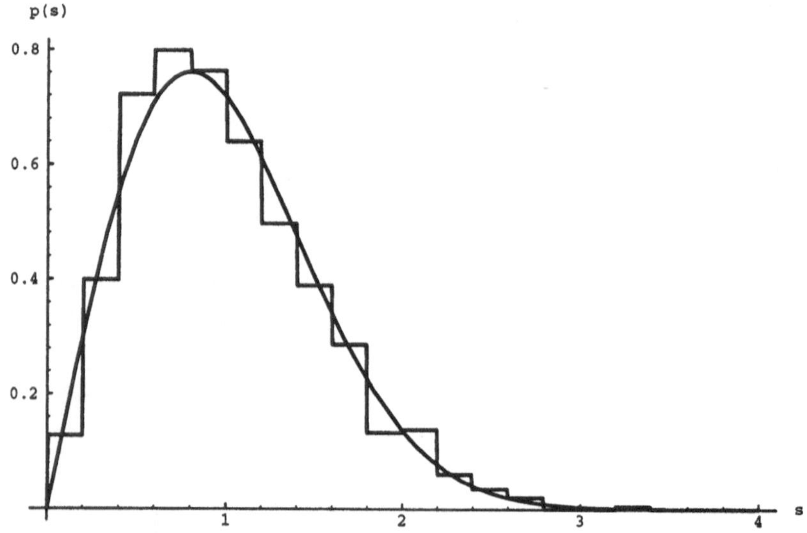

FIG. 3. Level spacing histogram for 20, 100×100 GOE matrices. Also plotted is the Wigner surmise (2.2).

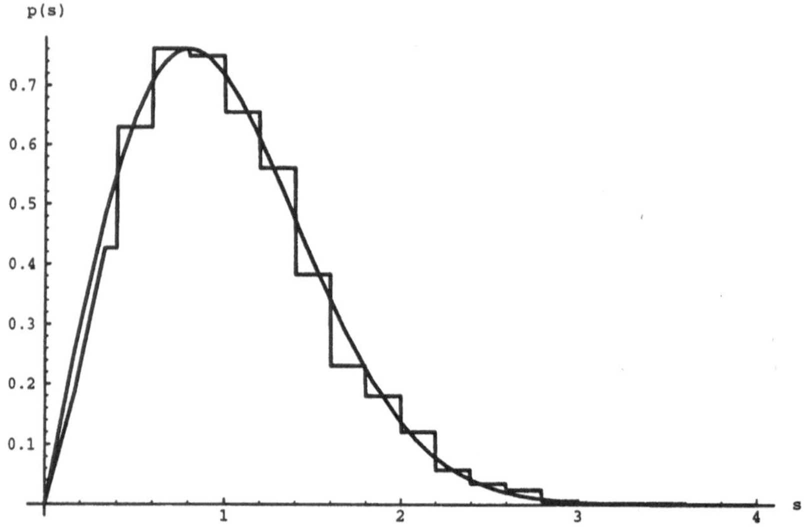

FIG. 4. Level spacing histogram for 50, 100×100 symmetric matrices whose elements are uniformly distributed on $[-1, 1]$. Also plotted is the Wigner surmise (2.2).

which for these purposes numerically well approximates the exact result (to be discussed below) in the range $0 \leq s \leq 3$. In Fig. 4 we show the same histogram except now the data are from 50, 100×100 real symmetric matrices whose elements are iid random variables with uniform distribution on $[-1, 1]$. In computing these histograms, the spacings were computed for each realization of a random matrix, and then the level spacings of several experiments were lumped together to form the data. If one first forms the data by mixing

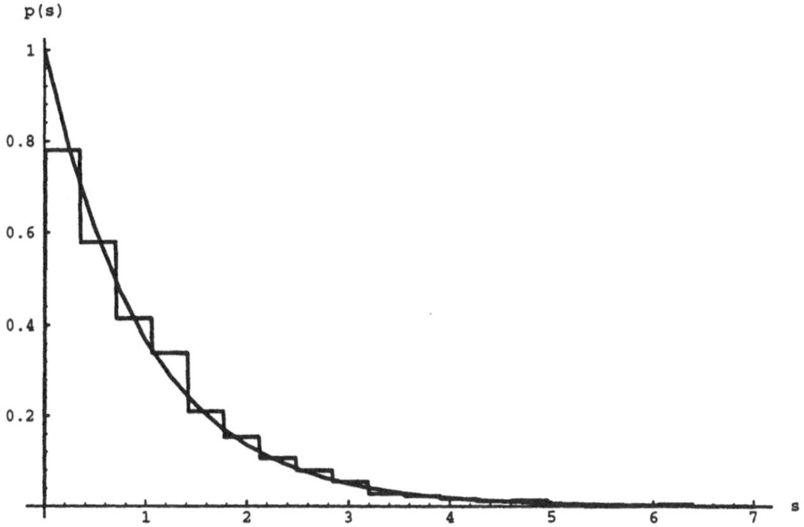

FIG. 5. Level spacing histogram for mixed data sets. Also plotted is the Poisson distribution.

together the eigenvalues from the random matrices, and then computes the level spacing histogram the results are completely different. This is illustrated in Fig. 5 where the resulting histogram is well approximated by the Poisson density $\exp(-s)$ (see Appendix 2 in [27] for further discussion of the superposition of levels). There are other numerical experiments that can be done, and the reader is invited to discover various aspects of random matrix theory by devising ones own experiments.

III. INVARIANT MEASURES AND LEVEL SPACING DISTRIBUTIONS

A. Preliminary Remarks

In classical statistical mechanics, the microcanonical ensemble is defined by the measure that assigns equal a priori probability to all states of the given system (which in turn is defined by specifying a Hamiltonian on phase space) of fixed energy E and volume V. The motivation for this measure is that after specifying the energy of the system, every point in phase space lying on the energy surface should be equally likely since we have "no further macroscopic information." In the applications of random matrix theory to quantum systems, the Hamiltonian is modeled by a matrix H. However, in this case we give up knowledge of the system, i.e. H itself is unknown. Depending upon the symmetries of the system, the sets of possible H's are taken to be real symmetric matrices, Hermitian matrices, or self-dual Hermitian matrices ("quaternion real") [10, 11, 27]. The question then is what measure do we choose for each of these sets of H's?

What we would intuitively like to do is to make each H equally likely to reflect our "total ignorance" of the system. Because these spaces of H's are noncompact, it is of course impossible to have a probability measure that assigns equal a priori probability. It is useful to recall the situation on the real line $\mathbf{R}$. If we confine ourselves to a finite interval $[a, b]$, then the unique translationally invariant probability measure is the normalized

Lebesgue measure. Another characterization of this probability measure is that it is the unique density that maximizes the information entropy

$$S[p] = -\int_a^b p(x)\log p(x)\,dx. \tag{3.1}$$

On $\mathbf{R}$ the maximum entropy density subject to the constraints $E(1) = 1$ and $E(x^2) = \sigma^2$ is the Gaussian density of variance σ^2. The Gaussian ensembles of random matrix theory can also be characterized as those measures that have maximum information entropy subject to the constraint of a fixed value of $E(H^*H)$ [1, 41]. The well-known explicit formulas are given below in Sec. IV.

Another approach, first taken by Dyson [10] and the one we follow here, is to consider unitary matrices rather than hermitian matrices. The advantage here is that the space of unitary matrices is compact and the eigenvalue density is constant (translationally invariant distributions).

B. Haar Measure for $U(N)$

We denote by $G = U(N)$ the set of $N \times N$ unitary matrices and recall that $\dim_{\mathbf{R}} U(N) = N^2$. One can think of $U(N)$ as an N^2-dimensional submanifold of $\mathbf{R}^{2N^2}$ under the identification of a complex $N \times N$ matrix with a point in $\mathbf{R}^{2N^2}$. The group G acts on itself by either left or right translations, i.e. fix $g_0 \in G$ then

$$L_{g_0} : g \to g_0 g \quad \text{and} \quad R_{g_0} : g \to g g_0 \,.$$

The normalized Haar measure μ_2 is the unique probability measure on G that is both left- and right-invariant:

$$\mu_2(gE) = \mu_2(Eg) = \mu_2(E) \tag{3.2}$$

for all $g \in G$ and every measurable set E (the reason for the subscript 2 will become clear below). Since for compact groups a measure that is left-invariant is also right-invariant, we need only construct a left-invariant measure to obtain the Haar measure. The invariance (3.2) reflects the precise meaning of "total ignorance" and is the analogue of translational invariance picking out the Lebesgue measure.

To construct the Haar measure we construct the matrix of left-invariant 1-forms

$$\Omega_g = g^{-1}dg, \quad g \in G, \tag{3.3}$$

where Ω_g is anti-Hermitian since g is unitary. Choosing N^2 linearly independent 1-forms ω_{ij} from Ω_g, we form the associated volume form obtained by taking the wedge product of these ω_{ij}'s. This volume form on G is left-invariant, and hence up to normalization, it is the desired probability measure.

Another way to contruct the Haar measure is to introduce the standard Riemannian metric on the space of $N \times N$ complex matrices $Z = (z_{ij})$:

$$(ds)^2 = \operatorname{tr}(dZ\,dZ^*) = \sum_{j,k=1}^{N} |dz_{ij}|^2 \,.$$

We now restrict this metric to the submanifold of unitary matrices. A simple computation shows the restricted metric is

$$(ds)^2 = \operatorname{tr}\left(\Omega_g \Omega_g^*\right). \tag{3.4}$$

Since Ω_g is left-invariant, so is the metric $(ds)^2$. If we use the standard Riemannian formulas to construct the volume element, we will arrive at the invariant volume form.

We are interested in the induced probability density on the eigenvalues. This calculation is classical and can be found in [18, 42]. The derivation is particularly clear if we make use of the Riemannian metric (3.4). To this end we write

$$g = X\,D\,X^{-1}\,, \quad g \in G, \tag{3.5}$$

where X is a unitary matrix and D is a diagonal matrix whose diagonal elements we write as $\exp(i\varphi_k)$. Up to an ordering of the angles φ_k the matrix D is unique. We can assume that the eigenvalues are distinct since the degenerate case has measure zero. Thus X is determined up to a diagonal unitary matrix. If we denote by $T(N)$ the subgroup of diagonal unitary matrices, then to each $g \in G$ there corresponds a unique pair (X, D), $X \in G/T(N)$ and $D \in T(N)$. The Haar measure induces a measure on $G/T(N)$ (via the natural projection map). Since

$$X^* dgX = \Omega_X D - D\Omega_X + dD\,,$$

we have

$$\begin{aligned}(ds)^2 &= \mathrm{tr}\left(\Omega_g \Omega_g^*\right) = \mathrm{tr}\,(dg\,dg^*) = \mathrm{tr}\,(X^* dgXX^* dg^* X)\\ &= \mathrm{tr}\,([\Omega_X, D]\,[\Omega_X, D]^*) + \mathrm{tr}\,(dD\,dD^*)\\ &= \sum_{k,\ell=1}^{N} |\delta X_{k\ell}\,(\exp(i\varphi_k) - \exp(i\varphi_\ell))|^2 + \sum_{k=1}^{N} (d\varphi_k)^2\end{aligned}$$

where $\Omega_X = (\delta X_{k\ell})$. Note that the diagonal elements δX_{kk} do not appear in the metric. Using the Riemannian volume formula we obtain the volume form

$$\omega_g = \omega_X \prod_{j<k} |\exp(i\varphi_j) - \exp(i\varphi_k)|^2\, d\varphi_1 \cdots d\varphi_N \tag{3.6}$$

where $\omega_X = \mathrm{const} \prod_{j>k} \delta X_{jk}$.

We now integrate over the entire group G subject to the condition that the elements have their angles between φ_k and $\varphi_k + d\varphi_k$ to obtain

Theorem 1 *The volume of that part of the unitary group $U(N)$ whose elements have their angles between φ_k and $\varphi_k + d\varphi_k$ is given by*

$$P_{N2}(\varphi_1, \ldots, \varphi_N)\, d\varphi_1 \cdots d\varphi_N = C_{N2} \prod_{j<k} |\exp(i\varphi_j) - \exp(i\varphi_k)|^2\, d\varphi_1 \cdots d\varphi_N \tag{3.7}$$

where C_{N2} is a normalization constant.

We mention that there exist algorithms [8] to generate unitary matrices that are Haar distributed. The algorithm of Diaconis and Shahshahani [8] is of order N^3 and is easily implemented on a computer.

C. Orthogonal and Symplectic Ensembles

Dyson [11], in a careful analysis of the implications of time-reversal invariance for physical systems, showed that (a) systems having time-reversal invariance and rotational symmetry or having time-reversal invariance and integral spin are characterized by symmetric unitary matrices; and (b) those systems having time-reversal invariance and half-integral spin but no rotational symmetry are characterized by self-dual unitary matrices (see also

Chp. 9 in [27]). For systems without time-reversal invariance there is no restriction on the unitary matrices. These three sets of unitary matrices along with their respective invariant measures, which we denote by $E_\beta(N)$, $\beta = 1, 4, 2$, respectively, constitute the *circular ensembles*. We denote the invariant measures by μ_β, e.g. μ_2 is the normalized Haar measure discussed in Sec. III B.

A symmetric unitary matrix S can be written as

$$S = V^T V, \quad V \in U(N).$$

Such a decomposition is not unique since

$$V \to RV, \quad R \in O(N),$$

leaves S unchanged. Thus the space of symmetric unitary matrices can be identified with the coset space

$$U(N)/O(N).$$

The group $G = U(N)$ acts on the coset space $U(N)/O(N)$, and we want the invariant measure μ_1. If π denotes the natural projection map $G \to G/H$, then the measure μ_1 is just the induced measure:

$$\mu_1(B) = \mu_2(\pi^{-1}(B)),$$

and hence $E_1(N)$ can be identified with the pair $(U(N)/O(N), \mu_1)$.

The space of self-dual unitary matrices can be identified (for even N) with the coset space

$$U(N)/Sp(N/2)$$

where $Sp(N)$ is the symplectic group. Similarly, the circular ensemble $E_4(N)$ can be identified with $(U(N)/Sp(N/2), \mu_4)$ where μ_4 is the induced measure.

As in Sec. III B we want the the probability density $P_{N\beta}$ on the eigenvalue angles that results from the measure μ_β for $\beta = 1$ and $\beta = 4$. This calculation is somewhat more involved and we refer the reader to the original literature [10, 18] or to Chp. 9 in [27]. The basic result is

Theorem 2 *In the ensemble $E_\beta(N)$ ($\beta = 1, 2, 4$) the probability of finding the eigenvalues $\exp(i\varphi_j)$ of S with an angle in each of the intervals $(\theta_j, \theta_j + d\theta_j)$ ($j = 1, \ldots, N$) is given by*

$$P_{N\beta}(\theta_1, \ldots, \theta_N)\, d\theta_1 \cdots d\theta_N = C_{N\beta} \prod_{1 \le \ell < j \le N} |\exp(i\theta_\ell) - \exp(i\theta_j)|^\beta \, d\theta_1 \cdots d\theta_N \tag{3.8}$$

where $C_{N\beta}$ is a normalization constant.

The normalization constant follows from

Theorem 3 *Define for $N \in \mathbf{N}$ and $\beta \in \mathbf{C}$*

$$\Psi_N(\beta) = (2\pi)^{-N} \int_0^{2\pi} \cdots \int_0^{2\pi} \prod_{j<k} |\exp(i\theta_j) - \exp(i\theta_k)|^\beta \, d\theta_1 \cdots d\theta_N \tag{3.9}$$

then

$$\Psi_N(\beta) = \frac{\Gamma(1 + \beta N/2)}{(\Gamma(1 + \beta/2))^N}. \tag{3.10}$$

The integral (3.9) has an interesting history, and it is now understood to be a special case of Selberg's integral (see, e.g., the discussion in [27]).

D. Physical Interpretation of the Probability Density $P_{N\beta}$

The 2D Coulomb potential for N unit like charges on a circle of radius one is

$$W_N(\theta_1,\ldots,\theta_N) = -\sum_{1\le j<k\le N} \log|\exp(i\theta_j) - \exp(i\theta_k)| \,. \tag{3.11}$$

Thus the (positional) equilibrium Gibbs measure at inverse temperature $0 < \beta < \infty$ for this system of N charges with energy W_N is

$$\frac{\exp\left(-\beta W_N(\theta_1,\ldots,\theta_N)\right)}{\Psi_N(\beta)} \,. \tag{3.12}$$

For the special cases of $\beta = 1, 2, 4$ this Gibbs measure is the probability density of Theorem 2. Thus in this mathematically equivalent description, the term "level repulsion" takes on the physical meaning of repulsion of charges. This Coulomb gas description is due to Dyson [10], and it suggests various physically natural approximations that would not otherwise be so clear.

E. Level Spacing Distribution Functions

For large matrices there is too much information in the probability densities $P_{N\beta}(\theta_1,\ldots,\theta_N)$ to be useful in comparison with data. Thus we want to integrate out some of this information. Since the probability density $P_{N\beta}(\theta_1,\ldots,\theta_N)$ is a completely symmetric function of its arguments, it is natural to introduce the *n-point correlation functions*

$$R_{n\beta}(\theta_1,\ldots,\theta_n) = \frac{N!}{(N-n)!}\int_0^{2\pi}\cdots\int_0^{2\pi} P_{N\beta}(\theta_1,\ldots,\theta_N)\, d\theta_{n+1}\cdots d\theta_N \,. \tag{3.13}$$

The case $\beta = 2$ is significantly simpler to handle and we limit our discussion here to this case.

Lemma 1

$$P_{N2}(\theta_1,\ldots,\theta_N) = \frac{1}{N!}\det\left(K_N(\theta_j,\theta_k)\Big|_{j,k=1}^N\right) \tag{3.14}$$

where

$$K_N(\theta_j,\theta_k) = \frac{1}{2\pi}\frac{\sin\left(N(\theta_j-\theta_k)/2\right)}{\sin\left((\theta_j-\theta_k)/2\right)} \,. \tag{3.15}$$

Proof:

Recalling the Vandermonde determinant we can write

$$\prod_{j<k}|\exp(i\theta_j) - \exp(i\theta_k)|^2 = \det(M^T)\det(\overline{M}) \tag{3.16}$$

where $M_{jk} = \exp\left(i(j-1)\theta_k\right)$. A simple calculation shows

$$\left(M^T\overline{M}\right)_{jk} = 2\pi D_{jj} K_N(\theta_j,\theta_k)\bar{D}_{kk} \tag{3.17}$$

where D is the diagonal matrix with entries $\exp\left(i(N-1)\theta_j/2\right)$. Except for the normalization constant the lemma now follows. Getting the correct normalization constant requires a little more work (see, e.g., Chp. 5 in [27]). ∎

From this lemma and the combinatoric Theorem 5.2.1 of [27] follows

Theorem 4 *Let $R_{n2}(\theta_1,\ldots,\theta_n)$ be the n-point function defined by (3.13) for the circular ensemble $E_2(N)$; then*

$$R_{n2}(\theta_1,\ldots,\theta_n) = \det\left(K_N(\theta_j,\theta_k)\Big|_{j,k=1}^{n}\right) \tag{3.18}$$

where $K_N(\theta_j,\theta_k)$ is given by (3.15).

We now discuss the behavior for large N. The 1-point correlation function

$$R_{1,2}'(\theta_1) = \frac{N}{2\pi} \tag{3.19}$$

is just the density, ρ, of eigenvalues with mean spacing $D = 1/\rho$. As the size of the matrices goes to infinity so does the density. Thus if we are to construct a meaningful limit as $N \to \infty$ we must scale the angles θ_j. This motivates the definition of the *scaling limit*

$$\rho \to \infty, \quad \theta_j \to 0, \quad \text{such that} \quad x_j := \rho\theta_j \in \mathbf{R} \quad \text{is fixed.} \tag{3.20}$$

We will abbreviate this scaling limit by simply writing $N \to \infty$. In this limit

$$R_{n2}(x_1,\ldots,x_n)\,dx_1\cdots dx_n := \lim_{N\to\infty} R_n(\theta_1,\ldots,\theta_n)\,d\theta_1\cdots d\theta_n \tag{3.21}$$

where we used the slightly confusing notation of denoting the scaling limit of the n-point functions by the same symbol. From Theorem 4 follows

Theorem 5 *In the scaling limit (3.20) the n-point functions become*

$$R_{n2}(x_1,\ldots,x_n) = \det\left(K(x_j,x_k)\Big|_{j,k=1}^{n}\right) \tag{3.22}$$

where the kernel $K(x,y)$ is given by

$$K(x,y) = \frac{1}{\pi}\,\frac{\sin\pi(x-y)}{x-y}\,. \tag{3.23}$$

The three sets of correlation functions

$$\mathcal{E}_\beta := \{R_{n\beta}(x_1,\ldots,x_n); x_j \in \mathbf{R}\}_{n=1}^{\infty}\,, \quad \beta = 1,2,4, \tag{3.24}$$

define three different statistics (called the *orthogonal ensemble, unitary ensemble, symplectic ensemble*, respectively) of an infinite sequence of eigenvalues (or as we sometimes say, *levels*) on the real line.

We now have the necessary machinary to discuss the level spacing correlation functions in the ensemble $\mathcal{E}_2$. We denote by $\mathcal{I}$ the union of m disjoint sub-intervals of the unit circle:

$$\mathcal{I} = \mathcal{I}_1 \cup \cdots \cup \mathcal{I}_m. \tag{3.25}$$

We begin with the probability of finding exactly n_1 eigenvalues in interval $\mathcal{I}_1,\ldots,\ n_m$ eigenvalues in interval $\mathcal{I}_m$ in the ensemble $E_2(N)$. We denote this probability by $E_{2N}(n_1,\ldots,n_m;\mathcal{I})$ and we will let $N \to \infty$ at the end.

If χ_A denotes the characteristic function of the set A and $n := n_1 + \cdots + n_m$, then the probability we want is

$$
\begin{aligned}
E_{2N}(n_1,\ldots,n_m;\mathcal{I}) = \binom{N}{n_1 \cdots n_m \; N-n} \int_0^{2\pi} d\theta_1 \cdots \int_0^{2\pi} d\theta_N \, P_{N2}(\theta_1,\ldots,\theta_N) \\
\times \prod_{j_1=1}^{n_1} \chi_{\mathcal{I}_1}(\theta_{j_1}) \prod_{j_2=n_1+1}^{n_1+n_2} \chi_{\mathcal{I}_2}(\theta_{j_2}) \cdots \prod_{j_m=n_1+\cdots+1}^{n_1+\cdots+n_m} \chi_{\mathcal{I}_m}(\theta_{j_m}) \\
\times \prod_{j=n+1}^{N} (1-\chi_{\mathcal{I}}(\theta_j)) \, . \qquad (3.26)
\end{aligned}
$$

We define the quantities

$$
\begin{aligned}
r_{n_1\cdots n_m} = \int_0^{2\pi} d\theta_1 \cdots \int_0^{2\pi} d\theta_n \, R_{n2}(\theta_1,\ldots,\theta_n) \\
\times \prod_{j_1=1}^{n_1} \chi_{\mathcal{I}_1}(\theta_{j_1}) \prod_{j_2=n_1+1}^{n_1+n_2} \chi_{\mathcal{I}_2}(\theta_{j_2}) \cdots \prod_{j_m=n_1+\cdots+1}^{n_1+\cdots+n_m} \chi_{\mathcal{I}_m}(\theta_{j_m}) \, . \qquad (3.27)
\end{aligned}
$$

The idea is to expand the last product term in (3.26) involving the characteristic function $\chi_{\mathcal{I}}$ and to regroup the terms according to the number of times a factor of $\chi_{\mathcal{I}}$ appears. Doing this one can then integrate out those θ_k variables that do not appear as arguments of any of the characteristic functions and express the result in terms of the quantities $r_{n_1\cdots n_m}$. To recognize the resulting terms we define the Fredholm determinant

$$
D_N(\mathcal{I};\lambda) = \det\left(1 - \sum_{j=1}^{m} \lambda_j K_N(\theta,\theta')\chi_{\mathcal{I}_j}(\theta')\right)
$$

where "$\sum_{j=1}^{m} \lambda_j K_N(\theta,\theta')\chi_{\mathcal{I}_j}(\theta')$" means the operator with that kernel and λ is the m-tuple $(\lambda_1,\ldots,\lambda_m)$. A slight rewriting of the Fredholm expansion gives

$$
D_N(\mathcal{I};\lambda) = 1 + \sum_{j=1}^{\infty}(-1)^j \sum_{\substack{j_k \geq 0 \\ j_1+\cdots+j_m=j}} \frac{\lambda^{j_1}\ldots\lambda^{j_m}}{j_1!\cdots j_m!} r_{n_1\cdots n_m} \, .
$$

The expansion above is then recognized to be proportional to

$$
\left.\frac{\partial^n D_N(\mathcal{I};\lambda)}{\partial\lambda_1^{n_1}\cdots\partial\lambda_m^{n_m}}\right|_{\lambda_1=\cdots=\lambda_m=1} .
$$

The scaling limit $N\to\infty$ can be taken with the result:

Theorem 6 *Given m disjoint open intervals $I_k = (a_{2k-1}, a_{2k}) \subset$ **R**, let*

$$
I := I_1 \cup \cdots \cup I_m . \qquad (3.28)
$$

The probability $E_2(n_1,\ldots,n_m;I)$ in the ensemble $\mathcal{E}_2$ that exactly n_k levels occur in interval I_k $(k=1,\ldots,m)$ is given by

$$
E_2(n_1,\ldots,n_m,I) = \frac{(-1)^n}{n_1!\cdots n_m!}\left.\frac{\partial^n D(I;\lambda)}{\partial\lambda^{n_1}\cdots\partial\lambda^{n_m}}\right|_{\lambda_1=\cdots=\lambda_m=1} \qquad (3.29)
$$

where

$$
D(I;\lambda) = \det\left(1 - \sum_{j=1}^{m} \lambda_j K(x,y)\chi_{I_j}(y)\right), \qquad (3.30)
$$

$K(x,y)$ is given by (3.23), and $n := n_1 + \cdots + n_m$.

In the case of a single interval $I = (a,b)$, we write the probability in ensemble $\mathcal{E}_\beta$ of exactly n eigenvalues in I as $E_\beta(n;s)$ where $s := b-a$. Mehta [27, 28] has shown that if we define

$$D_\pm(s;\lambda) = \det(1-\lambda K_\pm) \tag{3.31}$$

where $K_\pm$ are the operators with kernels $K(x,y) \pm K(-x,y)$, and

$$E_\pm(n;s) = \frac{(-1)^n}{n!} \frac{\partial^n D_\pm(s;\lambda)}{\partial \lambda^n}\Big|_{\lambda=1}, \tag{3.32}$$

then $E_1(0;s) = E_+(0;s)$,

$$E_+(n;s) = E_1(2n;s) + E_1(2n-1;s), \quad n > 0\,, \tag{3.33}$$

$$E_-(n;s) = E_1(2n;s) + E_1(2n+1;s), \quad n \geq 0\,, \tag{3.34}$$

and

$$E_4(n;s) = \frac{1}{2}(E_+(n;2s) + E_-(n;2s)), \quad n \geq 0\,. \tag{3.35}$$

Using the Fredholm expansion, small s expansions can be found for $E_\beta(n;s)$. We quote [27, 28] here only the results for $n = 0$:

$$\begin{aligned} E_1(0;s) &= 1 - s + \frac{\pi^2 s^3}{36} - \frac{\pi^4 s^5}{1200} + O(s^6)\,, \\ E_2(0;s) &= 1 - s + \frac{\pi^2 s^4}{36} - \frac{\pi^4 s^6}{675} + O(s^8)\,, \\ E_4(0;s) &= 1 - s + \frac{8\pi^4 s^6}{2025} + O(s^8)\,. \end{aligned} \tag{3.36}$$

The conditional probability in the ensemble $\mathcal{E}_\beta$ of an eigenvalue between b and $b+db$ given an eigenvalue at a, is given [27] by $p_\beta(0;s)\,ds$ where

$$p_\beta(0;s) = \frac{d^2 E_\beta(0;s)}{ds^2}\,. \tag{3.37}$$

Using this formula and the expansions (3.36) we see that $p_\beta(0;s) = O(s^\beta)$, making connection with the numerical results discussed in Sec. II. Note that the Wigner surmise (2.2) gives for small s the correct power of s for $p_1(0;s)$, but the slope is incorrect.

IV. ORTHOGONAL POLYNOMIAL ENSEMBLES

Orthogonal polynomial ensembles have been studied since the 1960's, e.g. [14], but recently interest has revived because of their application to the matrix models of 2D quantum gravity [7, 9, 16, 17]. Here we give the main results that generalize the previous sections.

The orthogonal polynomial ensemble associated to V assigns a probability measure on the space of $N \times N$ hermitian matrices proportional to

$$\exp(-\mathrm{Tr}\,(V(M)))\,dM \tag{4.1}$$

where $V(x)$ is a real-valued function such that

$$w_V(x) = \exp(-V(x))$$

defines a weight function in the sense of orthogonal polynomial theory. The quantity dM denotes the product of Lebesgue measures over the independent elements of the hermitian matrix M. Since $\mathrm{Tr}\,(V(M))$ depends only upon the eigenvalues of M, we may diagonalize M

$$M = XDX^*,$$

and as before integrate over the "X" part of the measure to obtain a probability measure on the eigenvalues. Doing this gives the density

$$P_N(x_1,\dots,x_N) = \prod_{1\le k<\ell\le N} (x_k - x_\ell)^2 \exp\left(-\sum_{j=1}^{N} V(x_j)\right). \tag{4.2}$$

If we introduce the orthogonal polynomials

$$\int_{\mathbf{R}} p_m(x)p_n(x)w_V(x)\,dx = \delta_{mn}, \quad m,n = 0,1,\dots \tag{4.3}$$

and associated functions

$$\varphi_m(x) = \exp\left(-V(x)/2\right) p_m(x),$$

then the probability density (4.2) becomes

$$\begin{aligned} P_N(x_1,\dots,x_N) &= \frac{1}{N!}\left(\det\left(\varphi_{j-1}(x_k)\right)\Big|_{j,k=1}^{N}\right)^2 \\ &= \frac{1}{N!}\det\left(K_N(x_j,x_k)\right) \end{aligned} \tag{4.4}$$

where

$$\begin{aligned} K_N(x,y) &= \sum_{j=0}^{N-1} \varphi_j(x)\varphi_j(y) \\ &= \frac{k_{N-1}}{k_N}\,\frac{\varphi_N(x)\varphi_{N-1}(y) - \varphi_{N-1}(x)\varphi_N(y)}{x-y} \quad \text{for } x \ne y \\ &= \frac{k_{N-1}}{k_N}\left(\varphi_N'(x)\varphi_{N-1}(x) - \varphi_{N-1}'(x)\varphi_N(x)\right) \quad \text{for } x = y. \end{aligned} \tag{4.5}$$

The last two equalities follow from the Christoffel-Darboux formula and the k_n are defined by

$$p_n(x) = k_n x^n + \cdots, \quad k_n > 0.$$

Using the orthonormality of the $\varphi_j(x)$'s one shows exactly as in [27] that

$$\begin{aligned} R_n(x_1,\dots,x_n) &:= \frac{N!}{(N-n)!}\int_{\mathbf{R}}\cdots\int_{\mathbf{R}} P_N(x_1,\dots,x_N)\,dx_{n+1}\cdots dx_N \\ &= \det\left(K_N(x_j,x_k)\Big|_{j,k=1}^{n}\right). \end{aligned} \tag{4.6}$$

In particular, the density of eigenvalues is given by

$$\rho_N(x) = K_N(x,x). \tag{4.7}$$

Arguing as before, we find the probability that an interval I contains no eigenvalues is given by the determinant

$$\det\left(1 - K_N\right) \tag{4.8}$$

where K_N denotes the operator with kernel

$$K_N(x,y)\chi_I(y)$$

and $K_N(x,y)$ given by (4.5). Analogous formulas hold for the probability of exactly n eigenvalues in an interval. We remark that the size of the matrix N has been kept fixed throughout this discusion. The reader is referred to the work of Mahoux and Mehta [30, 32] for further discussion of integration over matrix variables.

V. UNIVERSALITY

We now consider the limit as the size of the matrices tends to infinity in the orthogonal polynomial ensembles of Sec. IV. Recall that we defined the scaling limit by introducing new variables such that in these scaled variables the mean spacing of eigenvalues was unity. In Sec. III the system for finite N was translationally invariant and had constant mean density $N/2\pi$. The orthogonal polynomial ensembles are not translationally invariant (recall (4.7) so we now take a *fixed* point x_0 in the support of $\rho_N(x)$ and examine the local statistics of the eigenvalues in some small neighborhood of this point x_0. Precisely, the scaling limit in the orthogonal polynomial ensemble that we consider is

$$N \to \infty, \quad x_j \to x_0 \quad \text{such that} \quad \xi_j := \rho_N(x_0)(x_j - x_0) \quad \text{is fixed.} \tag{5.1}$$

The problem is to compute $K_N(x,y)\,dy$ in this scaling limit. From (4.5) one sees this is a question of asymptotics of the associated orthogonal polynomials. For weight functions $w_V(x)$ corresponding to classical orthogonal polynomials such asymptotic formulas are well known and using these it has been shown [14, 36] that

$$K_N(x,y)\,dy \to \frac{1}{\pi}\,\frac{\sin\pi(\xi-\xi')}{\xi-\xi'}\,d\xi'\,. \tag{5.2}$$

Note this result is independent of x_0 and any of the parameters that might appear in the weight function $w_V(x)$. Moore [34] has given heuristic semiclassical arguments that show that we can expect (5.2) in great generality (see also [23]).

There is a growing literature on asymptotics of orthogonal polynomials when the weight function $w_V(x)$ has polynomial V, see [25] and references therein. For example, for the case of $V(x) = x^4$ Nevai [37] has given rigorous asymptotic formulas for the associated polynomials. It is straightforward to use Nevai's formulas to verify that (5.2) holds in this non-classical case (see also [30, 32, 38]).

There are places where (5.2) will fail and these will correspond to choosing the x_0 at the "edge of the spectrum" [6, 34] (in these cases x_0 varies with N). This will correspond to the double scaling limit discovered in the matrix models of 2D quantum gravity [7, 9, 16, 17]. Different multicritical points will have different limiting $K_N(x,y)\,dy$ [6, 34].

VI. JIMBO-MIWA-MÔRI-SATO EQUATIONS

A. Definitions and Lemmas

In this section we denote by a the $2m$-tuple $(a_1,\ldots,a_{2m})$ where the a_j are the endpoints of the intervals given in Theorem 6 and by d_a exterior differentiation with respect to the a_j $(j = 1,\ldots,2m)$. We make the specialization

$$\lambda_j = \lambda \quad \text{for} \quad j$$

which is the case considered in [22]. It is not difficult to extend the considerations of this section to the general case.

We denote by K the operator that has kernel

$$\lambda K(x,y)\chi_I(y) \tag{6.1}$$

where $K(x,y)$ is given by (3.23). It is convenient to write

$$\lambda K(x,y) = \frac{A(x)A'(y) - A'(x)A(y)}{x-y} \tag{6.2}$$

where

$$A(x) = \frac{\sqrt{\lambda}}{\pi}\sin \pi x\,.$$

The operator K acts on $L^2(\mathbf{R})$, but can be restricted to act on a dense subset of smooth functions. From calculus we get the formula

$$\frac{\partial}{\partial a_j}K = (-1)^j K(x,a_j)\delta(y-a_j) \tag{6.3}$$

where $\delta(x)$ is the Dirac delta function. Note that in the right hand side of the above equation we are using the shorthand notation that "$A(x,y)$" means the operator with that kernel. We will continue to use this notation throughout this section.

We introduce the functions

$$Q(x;a) = (1-K)^{-1}A(x) = \int_{\mathbf{R}} \rho(x,y)A(y)\,dy \tag{6.4}$$

and

$$P(x;a) = (1-K)^{-1}A'(x) = \int_{\mathbf{R}} \rho(x,y)A'(y)\,dy \tag{6.5}$$

where $\rho(x,y)$ denotes the distributional kernel of $(1-K)^{-1}$. We will sometimes abbreviate these to $Q(x)$ and $P(x)$, respectively. It is also convenient to introduce the resolvent kernel

$$R = (1-K)^{-1}K\,.$$

In terms of kernels, these are related by

$$\rho(x,y) = \delta(x-y) + R(x,y)\,.$$

We define the fundamental 1-form

$$\omega(a) := d_a \log D(I;\lambda)\,. \tag{6.6}$$

Since the integral operator K is trace-class and depends smoothly on the parameters a, we have the well known result

$$\omega(a) = -\mathrm{Tr}\left((1-K)^{-1}d_a K\right)\,. \tag{6.7}$$

Using (6.3) this last trace can be expressed in terms of the resolvent kernel:

$$\omega(a) = -\sum_{j=1}^{2m}(-1)^j R(a_j,a_j)\,da_j \tag{6.8}$$

which shows the importance of the quantities $R(a_j,a_j)$. A short calculation establishes

$$\frac{\partial}{\partial a_j}(1-K)^{-1} = (-1)^j R(x,a_j)\rho(a_j,y)\,. \tag{6.9}$$

We will need two commutators which we state as lemmas.

Lemma 2 *If $D = \frac{d}{dx}$ denotes the differentiation operator, then*

$$\left[D,(1-K)^{-1}\right] = -\sum_{j=1}^{2m}(-1)^j R(x,a_j)\rho(a_j,y)\,.$$

Proof:
Since

$$\left[D,(1-K)^{-1}\right] = (1-K)^{-1}\,[D,K]\,(1-K)^{-1}\,,$$

we begin by computing $[D,K]$. An integration by parts shows that

$$[D,K] = -\sum_j (-1)^j K(x,a_j)\delta(y-a_j)$$

where we used the property

$$\frac{\partial K(x,y)}{\partial x} + \frac{\partial K(x,y)}{\partial y} = 0$$

satisfied by our $K(x,y)$ and the well known formula for the derivative of $\chi_I(x)$. The lemma now follows from the fact that

$$R(x,y) = \int_{\mathbf{R}} \rho(x,z)K(z,y)\,dz = \int_{\mathbf{R}} K(x,z)\rho(z,y)\,dz\,.$$

■

Lemma 3 *If M_x denotes the multiplication by x operator, then*

$$\left[M_x,(1-K)^{-1}\right] = Q(x)\,\left(1-K^t\right)^{-1} A'\chi_I(y) - P(x)\,\left(1-K^t\right)^{-1} A\chi_I(y)$$

where K^t denotes the transpose of K, and also

$$\left[M_x,(1-K)^{-1}\right] = (x-y)R(x,y)\,.$$

Proof:
We have

$$[M_x,K] = (A(x)A'(y) - A'(x)A(y))\,\chi_I(y)\,(1-K)^{-1}\,.$$

From this last equation the first part of the lemma follows using the definitions of Q and P. The alternative expression for the commutator follows directly from the definition of $\rho(x,y)$ and its relationship to $R(x,y)$. ■

This lemma leads to a representation for the kernel $R(x,y)$ [19]:

Lemma 4 *If $R(x,y)$ is the resolvent kernel of (6.2) and Q and P are defined by (6.4) and (6.5), respectively, then for $x,y \in I$ we have*

$$R(x,y) = \frac{Q(x;a)P(y;a) - P(x;a)Q(y;a)}{x-y}\,, \quad x \neq y\,,$$

and

$$R(x,x) = \frac{dQ}{dx}(x;a)\,P(x;a) - \frac{dP}{dx}(x;a)\,Q(x;a)\,.$$

Proof:

Since $K(x,y) = K(y,x)$ we have, on I,

$$(1-K^t)^{-1}A\chi_I = (1-K)^{-1}A\chi_I = (1-K)^{-1}A$$

(the last since the kernel of K vanishes for $y \notin I$). Thus $(1-K^t)^{-1}A\chi_I = Q$ on I, and similarly $(1-K^t)^{-1}A'\chi_I = P$ on I. The first part of the lemma then follows from Lemma 3. The expression for the diagonal follows from Taylor's theorem. ∎

We remark that Lemma 4 used only the property that the kernel $K(x,y)$ can be written as (6.2) and not the specific form of $A(x)$. Such kernels are called "completely integrable integral operators" by Its et. al. [19] and Lemma 4 is central to their work.

B. Derivation of the JMMS Equations

We set

$$q_j = q_j(a) = \lim_{\substack{x\to a_j \\ x\in I}} Q(x;a) \quad \text{and} \quad p_j = p_j(a) = \lim_{\substack{x\to a_j \\ x\in I}} P(x;a), \quad j = 1,\ldots,2m. \tag{6.10}$$

Specializing Lemma 4 we obtain immediately

$$R(a_j,a_k) = \frac{q_j p_k - p_j q_k}{a_j - a_k}\,, \quad j \neq k\,. \tag{6.11}$$

Referring to (6.9) we easily deduce that

$$\frac{\partial q_j}{\partial a_k} = (-1)^k R(a_j,a_k) q_k\,, \quad j \neq k\,, \tag{6.12}$$

and

$$\frac{\partial p_j}{\partial a_k} = (-1)^k R(a_j,a_k) p_k\,, \quad j \neq k\,. \tag{6.13}$$

Now

$$\begin{aligned}\frac{dQ}{dx} &= D\,(1-K)^{-1}\,A(x)\\ &= (1-K)^{-1}\,DA(x) + \left[D,(1-K)^{-1}\right]A(x)\\ &= (1-K)^{-1}\,A'(x) - \sum_k (-1)^k R(x,a_k) q_k\,.\end{aligned}$$

Thus

$$\frac{dQ}{dx}(a_j;a) = p_j - \sum_k (-1)^k R(a_j,a_k) q_k\,. \tag{6.14}$$

Similarly,

$$\begin{aligned}\frac{dP}{dx} &= (1-K)^{-1}\,A''(x) + \left[D,(1-K)^{-1}\right]A'(x)\\ &= -\pi^2\,(1-K)^{-1}\,A(x) - \sum_k (-1)^k R(x,a_k) p_k\,.\end{aligned}$$

Thus

$$\frac{dP}{dx}(a_j;a) = -\pi^2 q_j - \sum_k (-1)^k R(a_j,a_k) p_k \,. \tag{6.15}$$

Using (6.14) and (6.15) in the expression for the diagonal of R in Lemma 4, we find

$$R(a_j,a_j) = \pi^2 q_j^2 + p_j^2 + \sum_k (-1)^k R(a_j,a_k) R(a_k,a_j)(a_j - a_k) \,. \tag{6.16}$$

Using

$$\frac{\partial q_j}{\partial a_j} = \frac{dQ(x;a)}{dx}\bigg|_{x=a_j} + \frac{\partial Q(x;a)}{\partial a_j}\bigg|_{x=a_j} ,$$

(6.12) and (6.14) we obtain

$$\frac{\partial q_j}{\partial a_j} = p_j - \sum_{k \neq j} (-1)^k R(a_j,a_k) q_k \,. \tag{6.17}$$

Similarly,

$$\frac{\partial p_j}{\partial a_j} = -\pi^2 q_j - \sum_{k \neq j} (-1)^k R(a_j,a_k) p_k \,. \tag{6.18}$$

Equations (6.11)–(6.13) and (6.16)–(6.18) are the JMMS equations. We remark that they appear in slightly different form in [22] due to the use of sines and cosines rather than exponentials in the definitions of Q and P.

C. Hamiltonian Structure of the JMMS Equations

To facilitate comparison with [22, 35] we introduce

$$\begin{aligned} q_{2j} &= -\frac{i}{2} x_{2j} , \quad q_{2j+1} = \frac{1}{2} x_{2j+1} , \\ p_{2j} &= -i y_{2j} , \quad p_{2j+1} = y_{2j+1} , \\ M_{jk} &:= \frac{1}{2}(x_j y_k - x_k y_j) , \\ G_j(x,y) &:= \frac{\pi^2}{4} x_j^2 + y_j^2 - \sum_{\substack{k=1 \\ k \neq j}}^{2m} \frac{M_{jk}^2}{a_j - a_k} . \end{aligned} \tag{6.19}$$

In this notation,

$$\omega(a) = \sum_j G_j(x,y)\, da_j \,.$$

If we introduce the canonical symplectic structure

$$\{x_j, x_k\} = \{y_j, y_k\} = 0, \quad \{x_j, y_k\} = \delta_{jk} , \tag{6.20}$$

then as shown in Moser [35]

Theorem 7 *The integrals $G_j(x,y)$ are in involution; that is, if we define the symplectic structure by (6.20) we have*

$$\{G_j, G_k\} = 0 \quad \text{for all} \quad j,k = 1, \ldots, 2m.$$

Furthermore as can be easily verified, the JMMS equations take the following form [22]:

Theorem 8 *If we define the Hamiltonian*

$$\omega(a) = \sum_{j=1}^{2m} G_j(x,y)da_j \,,$$

then Eqs. (6.12), (6.13), (6.17) and (6.18) are equivalent to Hamilton's equations

$$d_a x_j = \{x_j, \omega(a)\} \quad \text{and} \quad d_a y_j = \{y_j, \omega(a)\} \,.$$

In words, the flow of the point (x,y) in the "time variable" a_j is given by Hamilton's equations with Hamiltonian G_j.

The (Frobenius) complete integrability of the JMMS equations follows immediately from Theorems 7 and 8. We must show

$$d_a \{x_j, \omega\} = 0 \quad \text{and} \quad d_a \{y_j, \omega\} = 0.$$

Now

$$d_a \{x_j, \omega\} = \{d_a x_j, \omega\} + \{x_j, d_a\omega\} \,,$$

but $d_a\omega = 0$ since the G_k's are in involution. And we have

$$\{d_a x_j, \omega\} = \sum_{k<\ell} (\{\{x_j, G_k\}, G_\ell\} - \{\{x_j, G_\ell\}, G_k\})\, da_k \wedge da_\ell$$

which is seen to be zero from Jacobi's identity and the involutive property of the G_j's.

D. Reduction to Painlevé V in the One Interval Case

1. The $\sigma(x;\lambda)$ differential equation

We consider the case of one interval:

$$m = 1, \quad a_1 = -t, \quad a_2 = t, \quad \text{with} \quad s := 2t \,. \tag{6.21}$$

Since $\rho(x,y)$ is both symmetric and even for $x, y \in I$, we have $q_2 = -q_1$ and $p_2 = p_1$. Introducing the quantity

$$r_1 = \int_{-t}^{t} \rho(-t,x) \exp(-i\pi x)\, dx \,,$$

we write

$$q_1 = \frac{\sqrt{\lambda}}{2\pi i}(\overline{r_1} - r_1) \quad \text{and} \quad p_1 = \frac{\sqrt{\lambda}}{2}(\overline{r_1} + r_1) \,.$$

Specializing the results of Sec. VI B to $m = 1$ we have

$$\omega(a) = -2R(t,t)\, dt \,, \tag{6.22}$$

$$R(-t,t) = -\frac{1}{t} q_1 p_1 = \frac{\lambda}{4\pi i t}(r_1^2 - \overline{r}_1^2) \,, \tag{6.23}$$

$$\frac{dq_1}{dt} = -\frac{\partial q_1}{\partial a_1} + \frac{\partial q_1}{\partial a_2} = -p_1 + 2R(-t,t) q_1 \,,$$

$$\frac{dp_1}{dt} = \pi^2 q_1 + 2R(-t,t) p_1 \,,$$

$$\frac{dr_1}{dt} = i\pi r_1 + 2R(-t,t)\overline{r}_1 \,, \tag{6.24}$$

$$
\begin{aligned}
R(t,t) &= \pi^2 q_1^2 + p_1^2 - 2tR(-t,t)^2 \\
&= \lambda \bar{r}_1 r_1 + \frac{\lambda^2}{8\pi^2 t}\left(\bar{r}_1^2 - r_1^2\right)^2 .
\end{aligned} \tag{6.25}
$$

A straightforward computation from (6.23)-(6.25) shows

$$
\frac{d}{dt}\left(tR(-t,t)\right) = \lambda \Re(r_1^2)\,, \tag{6.26}
$$

$$
\frac{d}{dt}\left(tR(t,t)\right) = \lambda |r_1|^2\,, \tag{6.27}
$$

$$
\frac{d}{dt}R(t,t) = 2\left(R(-t,t)\right)^2 . \tag{6.28}
$$

Eq. (6.28) is known as *Gaudin's relation* and Eqs. (6.26) and (6.27) are identities derived by Mehta [29] (see also Dyson[13]) in his proof of the one interval JMMS equations. Here we made the JMMS equations central and derived (6.26)-(6.28) as consequences.

These equations make it easy to derive a differential equation for

$$
\sigma(x;\lambda) := -2tR(t,t) = x\frac{d}{dx}\log D(\frac{x}{\pi};\lambda), \quad \text{where} \quad x = 2\pi t. \tag{6.29}
$$

We start with the identity

$$
|r_1|^4 = \left(\Re(r_1^2)\right)^2 + \left(\Im(r_1^2)\right)^2 ,
$$

and define temporarily $a(t) := tR(t,t)$ and $b(t) := tR(-t,t)$; then (6.23), (6.26), and (6.27) imply

$$
\left(\frac{da}{dt}\right)^2 = \left(\frac{db}{dt}\right)^2 + 4\pi^2 b^2 .
$$

Using (6.28) and its derivative to eliminate b and db/dt, we get an equation for $a(t)$ and hence $\sigma(x;\lambda)$:

Theorem 9 *In the case of a single interval $I = (-t,t)$ with $s = 2t$, the Fredholm determinant*

$$
D(s;\lambda) = \det\left(1 - \lambda K\right)
$$

is given by

$$
D(s;\lambda) = \exp\left(\int_0^{\pi s} \frac{\sigma(x;\lambda)}{x}\,dx\right) , \tag{6.30}
$$

where $\sigma(x;\lambda)$ satisfies the differential equation

$$
\left(x\sigma''\right)^2 + 4\left(x\sigma' - \sigma\right)\left(x\sigma' - \sigma + (\sigma')^2\right) = 0 \tag{6.31}
$$

with boundary condition as $x \to 0$

$$
\sigma(x;\lambda) = -\frac{\lambda}{\pi}x - (\frac{\lambda}{\pi})^2 x^2 - \cdots . \tag{6.32}
$$

Proof: Only (6.32) needs explanation. The small x expansion of $\sigma(x;\lambda)$ is fixed from the small s expansion of $D(s;\lambda)$ which can be computed from the Neumann expansion. ■

The differential equation (6.31) is the "σ representation" of the Painlevé V equation. This is discussed in [22], and in more detail in Appendix C of [21]. In terms of the monodromy parameters θ_i ($i = 0, 1, \infty$) of [21], (6.31) corresponds to $\theta_0 = \theta_1 = \theta_\infty = 0$ which is the case of no local monodromy. For an introduction to Painlevé functions see [20, 24].

2. The $\sigma_\pm(x;\lambda)$ equations

Recalling the discussion following Eq. (3.31), we see we need the determinants $D_\pm(s;\lambda)$ to compute $E_\beta(n;s)$ for $\beta = 1$ and 4. Let $R_\pm$ denote the resolvent kernels for the operators

$$K_\pm := \frac{1 \pm J}{2} K = K \frac{1 \pm J}{2},$$

where $(Jf)(x) = f(-x)$ and the last equality of the above equation follows from the evenness of K. Thus

$$R_\pm := (1 - K_\pm)^{-1} K_\pm = \frac{1}{2}(1 \pm J) R,$$

which in terms of kernels is

$$R_\pm(x,y) = \frac{1}{2}\left(R(x,y) \pm R(-x,y)\right).$$

Thus

$$\left(R_-(t,t) - R_+(t,t)\right)^2 = \left(R(-t,t)\right)^2 = \frac{1}{2}\frac{d}{dt} R(t,t).$$

Introducing the analogue of $\sigma(x;\lambda)$, i.e.

$$\sigma_\pm(x;\lambda) := x \frac{d}{dx} \log D_\pm\left(\frac{x}{\pi};\lambda\right),$$

the above equation becomes

$$\left(\frac{\sigma_-(x;\lambda) - \sigma_+(x;\lambda)}{x}\right)^2 = -\frac{d}{dx}\frac{\sigma(x;\lambda)}{x}. \tag{6.33}$$

Of course, $\sigma_\pm(x;\lambda)$ also satisfy

$$\sigma_+(x;\lambda) + \sigma_-(x;\lambda) = \sigma(x;\lambda). \tag{6.34}$$

Using (6.33) and (6.34) and integrating $\sigma_\pm(x;\lambda)/x$ we obtain (the square root sign ambiguity can be fixed from small x expansions)

Theorem 10 *Let $D_\pm(s;\lambda)$ be the Fredholm determinants defined by (3.31), then*

$$\log D_\pm(s;\lambda) = \frac{1}{2}\log D(s;\lambda) \pm \frac{1}{2}\int_0^s \sqrt{-\frac{d^2}{dx^2}\log D(x;\lambda)}\, dx. \tag{6.35}$$

VII. ASYMPTOTICS

A. Asymptotics via the Painlevé V Representation

In this section we explain how one derives asymptotic formulas for $E_\beta(n;s)$ as $s \to \infty$ starting with the Painlevé V representations of Theorems 9 and 10. This section follows Basor, Tracy and Widom [4] (see also [31]). We remark that the asymptotics of $E_\beta(0;s)$ as $s \to \infty$ was first derived by Dyson [12] by a clever use of inverse scattering methods.

Referring to Theorems 9 and 10 one sees that the basic problem from the differential equation point of view is to derive large x expansions for

$$\sigma_0(x) := \sigma(x;1) \tag{7.1}$$

$$\sigma_n(x) := \frac{\partial^n \sigma}{\partial \lambda^n}(x;1), \quad n = 1,2,\ldots \tag{7.2}$$

$$\sigma_{\pm,n}(x) := \frac{\partial^n \sigma_\pm}{\partial \lambda^n}(x;1), \quad n = 1,2,\ldots . \tag{7.3}$$

We point out the sensitivity of these results to the parameter λ being set to one. This dependence is best discussed in terms of the differential equation (6.31) where it is an instance of the general problem of *connection formulas*, see e.g. [24] and references therein. In this context the problem is: given the small x boundary condition, find asymptotic formulas as $x \to \infty$ where all constants not determined by a local analysis at ∞ are given as functions of the parameter λ. If we assume an asymptotic solution for large x of the form $\sigma(x) \sim a x^p$, then (6.31) implies either $p = 1$ or 2 and if $p = 2$ then necessarily $a = -\frac{1}{4}$. The connection problem for (6.31) has been studied by McCoy and Tang [26] who show that for $0 < \lambda < 1$ one has

$$\sigma(x,\lambda) = a(\lambda)x + b(\lambda) + o(1)$$

as $x \to \infty$ with

$$a(\lambda) = \frac{1}{\pi}\log(1-\lambda) \quad \text{and} \quad b(\lambda) = \frac{1}{2}a^2(\lambda).$$

Since these formulas make no sense at $\lambda = 1$, it is reasonable to guess that

$$\sigma(x;1) \sim -\frac{1}{4}x^2. \tag{7.4}$$

For a rigorous proof of this fact see [43]. It should be noted that in Dyson's work he too "guesses" this leading behavior to get his asymptotics to work (this leading behavior is not unexpected from the continuum Coulomb gas approximation). Given (7.4), and only this, it is a simple matter using (6.31) to compute recursively the correction terms to this leading asymptotic behavior:

$$\sigma_0(x) = -\frac{1}{4}x^2 - \frac{1}{4} + \sum_{n=1}^{\infty} \frac{c_{2n}}{x^{2n}}, \quad x \to \infty \tag{7.5}$$

($c_2 = -\frac{1}{4}$, $c_4 = -\frac{5}{2}$, etc.). Using (7.5) in (6.30) and (6.35) one can efficiently generate the large s expansions for $D(s;1)$ and $D_\pm(s;1)$ except for overall multiplicative constants. In this instance other methods fix these constants (see discussion in [4, 27]). In general this overall multiplicative constant in a τ-function is quite difficult to determine (for an example of such a determination see [2, 3]). We record here the result:

$$\log D_\pm(s;1) = -\frac{1}{16}\pi^2 s^2 \mp \frac{1}{4}\pi s - \frac{1}{8}\log \pi s \pm \frac{1}{4}\log 2 + \frac{1}{6}\log 2 + \frac{3}{2}\zeta'(-1) + o(1) \tag{7.6}$$

as $s \to \infty$ where ζ is the Riemann zeta function. We mention that for $0 < \lambda < 1$ the asymptotics of $D(s;\lambda)$ as $s \to \infty$ are known [5, 26].

One method to determine the asymptotics of $\sigma_n(x)$ as $x \to \infty$ is to examine the variational equations of (6.31), i.e. simply differentiate (6.31) with respect to λ and then set $\lambda = 1$. These linear differential equations can be solved asymptotically given the asymptotic solution (7.5). In carrying this out one finds there are two undetermined constants corresponding to the two linearly independent solutions to the first variational equation of (6.31). One constant does not affect the asymptotics of $\sigma_1(x)$ (assuming the other is nonzero!). Determining these constants is part of the general connection problem and it has not been solved in the context of differential equations. In [4, 43] Toeplitz and Wiener-Hopf methods are employed to fix these constants. The Toeplitz arguments of [4] depend upon some unproved assumptions about scaling limits, but the considerations of [43] are completely rigorous and we deduce the following result [4] for $\sigma_n(x)$ for all $n = 3, 4, \ldots$:

$$\sigma_n(x) = -\frac{n!}{(2^3\pi)^{n/2}}\frac{\exp(nx)}{x^{n/2-1}}\left[1 + \frac{1}{8}(7n-4)\frac{1}{x} + \frac{7}{128}(7n^2+12n-16)\frac{1}{x^2} + \mathrm{O}(\frac{1}{x^3})\right] \tag{7.7}$$

as $x \to \infty$. For $n = 1, 2$ the above is correct for the leading behavior but for $n = 1$ the correction terms have coefficients $\frac{5}{8}$ and $\frac{65}{128}$, respectively, and for $n = 2$ the above formula gives the coefficient for $1/x$ but the coefficient for $1/x^2$ is $\frac{65}{32}$. See [4] for asymptotic formulas for $\sigma_{\pm,n}(x)$.

Since the asymptotics of $E_\beta(0;s)$ are known, it is convenient to introduce

$$r_\beta(n;s) := \frac{E_\beta(n;s)}{E_\beta(0;s)}\,.$$

Here we restrict our discussion to $\beta = 2$ (see [4] for other cases). Using (7.7) in (3.29) one discovers that there is a great deal of cancellation in the terms which go into the asymptotics of $r_2(n;s)$. To prove a result for all $n \in \mathbf{N}$ by this method we must handle all the correction terms in (7.7)—this was not done in [4] and so the following result was proved only for $1 \le n \le 10$:

$$r_2(n;s) = B_{2,n}\frac{\exp(n\pi s)}{s^{n^2/2}}\left[1 + \frac{n}{8}(2n^2+7)\frac{1}{\pi s} + \frac{n^2}{128}(4n^4+48n^2+229)\frac{1}{(\pi s)^2} + \mathrm{O}(\frac{1}{s^3})\right] \tag{7.8}$$

where

$$B_{2,n} = 2^{-n^2-n/2}\pi^{-(n^2+n)/2}\,(n-1)!\,(n-2)!\cdots 2!\,1!\,.$$

In the next section we derive the leading term of (7.8) for all $n \in \mathbf{N}$. Asymptotic formulas for $r_\beta(n;s)$ ($\beta = 1, 4, \pm$) can be found in [4].

B. Asymptotics of $r_2(n;s)$ from Asymptotics of Eigenvalues

The asymptotic formula (7.8) can also be derived by a completely different method (as was briefly indicated in [4]). If we denote the eigenvalues of the integral operator K by $\lambda_0 > \lambda_1 > \cdots > 0$, then

$$\det(1-\lambda K) = \prod_{i=0}^{\infty}(1-\lambda\lambda_i)\,,$$

and so it follows immediately from (3.29) that

$$r_2(n;s) = \sum_{i_1<\cdots<i_n} \frac{\lambda_{i_1}\cdots\lambda_{i_n}}{(1-\lambda_{i_1})\cdots(1-\lambda_{i_n})}. \tag{7.9}$$

(This is formula (5.4.30) in [27].) Thus the asymptotics of the eigenvalues λ_i as $s \to \infty$ can be expected to give information on the asymptotics of $r_2(n;s)$ as $s \to \infty$.

It is a remarkable fact that the integral operator K, acting on the interval $(-t,t)$, commutes with the differential operator $\mathcal{L}$ defined by

$$\mathcal{L}f = \frac{d}{dx}(x^2-t^2)\frac{df}{dx} + t^2x^2f, \quad s = 2t; \tag{7.10}$$

the boundary condition here is that f be continuous at $\pm t$. Thus the integral operator and the differential operator have precisely the same eigenfunctions—the so-called prolate spheroidal wave functions, see e.g. [33]. Now Fuchs [15], by an application of the WKB method to the differential equation, and using a connection between the eigenvalues λ_i and the values of the normalized eigenfunctions at the end-points, derived the asymptotic formula

$$1-\lambda_i \sim \pi^{i+1}2^{2i+3/2}s^{i+1/2}e^{-\pi s}/i! \tag{7.11}$$

valid for fixed i as $s \to \infty$. Further terms of the asymptotic expansion for the ratio of the two sides were obtained by Slepian [40].

If one looks at the asymptotics of the individual terms on the right side of (7.9), then we see from (7.11) that they all have the exponential factor $e^{n\pi s}$ and that the powers of s that occur are

$$s^{-n/2-(i_1+\cdots+i_n)}.$$

Thus the term corresponding to $i_1 = 0$, $i_2 = 1$, ..., $i_n = n-1$ dominates each of the others. In fact we claim this term dominates the sum of all the others, and so

$$r_2(n;s) \sim 1!\,2!\cdots(n-1)!\,\pi^{-n(n+1)/2}2^{-n^2-n/2}s^{-n^2/2}e^{n\pi s}, \tag{7.12}$$

in agreement with (7.8).

To prove this claim, we write

$$r_2(n;s) = \frac{\lambda_{i_1^0}\cdots\lambda_{i_n^0}}{(1-\lambda_{i_1^0})\cdots(1-\lambda_{i_n^0})} + \sum{}' \frac{\lambda_{i_1}\cdots\lambda_{i_n}}{(1-\lambda_{i_1})\cdots(1-\lambda_{i_n})}$$

where $i_1^0 = 0$, $i_1^0 = 1$, ..., $i_n^0 = n-1$ and the sum here is taken over all $(i_1,\ldots,i_n) \neq (i_1^0,\ldots,i_n^0)$ with $i_1 < \cdots < i_n$. We have to show that

$$\sum{}' \frac{\lambda_{i_1}\cdots\lambda_{i_n}}{(1-\lambda_{i_1})\cdots(1-\lambda_{i_n})} \Big/ \frac{e^{n\pi s}}{s^{n^2/2}} \longrightarrow 0$$

as $s \to \infty$ and we know that this would be true if the sum were replaced by any summand. Write $\sum' = \sum_1 + \sum_2$ where in $\sum_1$ we have $i_j < N$ for all j (N to be determined later) and $\sum_2$ is the rest of $\sum'$. Since $\sum_1$ is a finite sum we have

$$\sum{}_1 \Big/ \frac{e^{n\pi s}}{s^{n^2/2}} \longrightarrow 0$$

so we need consider only $\sum_2$. In any summand of this we have $i_j \geq N$ for some j, and so for this j

$$\frac{1}{1-\lambda_{i_j}} \le \frac{1}{1-\lambda_N}$$

and by (7.11) this is at most

$$a_N\, s^{-N-1/2}\, e^{\pi s}$$

for some constant a_N. The product of all other factors

$$\frac{1}{1-\lambda_{i_j}}$$

appearing in this summand is at most

$$\left(\frac{1}{1-\lambda_0}\right)^{n-1}$$

and so by (7.11) with $i=0$ at most

$$b_n\, s^{-(n-1)/2}\, e^{(n-1)\pi s}$$

for another constant b_n. So we have the estimate

$$\sum\nolimits_2 \le a_N b_n\, s^{-N-n/2}\, e^{n\pi s} \sum \lambda_{i_1} \cdots \lambda_{i_n}$$

where the sum on the right may be taken over all n-tuples $(i_1, \ldots, i_n)$. This sum is precisely equal to $(\operatorname{tr} K)^n = s^n$. Hence

$$\sum\nolimits_2 \le a_N b_n s^{-N+n/2} e^{n\pi s}\,.$$

If we choose $N > (n^2+n)/2$ then we have

$$\sum\nolimits_2 \Big/ \frac{e^{n\pi s}}{s^{n^2/2}} \longrightarrow 0$$

as desired.

C. Dyson's Continuum Model

In [13] Dyson constructs a continuum Coulomb gas model [27] for $E_\beta(n;s)$. In this continuum model,

$$E_\beta(n;s) = \exp\left(-\beta W - (1-\frac{\beta}{2})S\right)$$

where

$$W = -\frac{1}{2}\int\int \hat\rho(x)\hat\rho(y)\log|x-y|\, dx dy$$

is the total energy,

$$S = \int \rho(x)\log\rho(x)\, dx$$

is the entropy, $\hat\rho(x) = \rho(x) - 1$ and $\rho(x)$ is a continuum charge distribution on the line satisfying $\rho(x) \to 1$ as $x \to \pm\infty$ and $\rho(x) \ge 0$ everywhere. The distribution $\rho(x)$ is chosen to minimize the free energy subject to the condition

$$\int_{-s/2}^{s/2} \rho(x)\,dx = n\,.$$

Analyzing his solution in the limit $1 << n << s$, Dyson finds $E_\beta(n;s) \sim \exp(-\beta W_c)$ where

$$W_c = \frac{\pi^2 s^2}{16} - \frac{\pi s}{2}(n+\delta) + \frac{1}{4}n(n+\delta) + \frac{1}{4}n(n+2\delta)\left[\log(\frac{4\pi s}{n}) + \frac{1}{2}\right] \qquad (7.13)$$

with $\delta = 1/2 - 1/\beta$.

We now compare these predictions of the continuum model with the exact results. First of all, this continuum prediction does not get the $s^{-1/4}$ (for $\beta = 2$) or the $s^{-1/8}$ (for $\beta = 1,4$) present in all $E_\beta(n;s)$ that come from the $\log \pi s$ term in (7.6). Thus it is better to compare with the continuum prediction for $r_\beta(n;s)$. We find that the continuum model gives both the correct exponential behavior *and* the correct power of s for all three ensembles. Tracing Dyson's arguments shows that the power of s involving the n^2 exponent is an *energy effect* and the power of s involving the n exponent is an *entropy effect*. Finally, the continuum model also makes a prediction (for large n) for the the $B_{\beta,n}$'s ($B_{2,n}$ is given above). Here we find that the ratio of the exact result to the continuum model result is approximately $n^{-1/12}$ for $\beta = 2$ and $n^{-1/24}$ for $\beta = 1,4$. This prediction of the continuum model is better than it first appears when one considers that the constants themselves are of order $n^{n^2/2}$ ($\beta = 2$) and n^{n^2} ($\beta = 1,4$).

ACKNOWLEDGMENTS

It is a pleasure to acknowledge E. L. Basor, P. Diaconis, F. J. Dyson, P. J. Forrester, P. B. Kahn, M. L. Mehta, and P. Nevai for their many helpful comments and their encouragement. We also thank F. J. Dyson and M. L. Mehta for sending us their preprints prior to publication. The first author thanks the organizers of the 8th Scheveningen Conference, August 16–21, 1992, for the invitation to attend and speak at this conference. These notes are an expanded version of the lectures presented there. This work was supported in part by the National Science Foundation, DMS–9001794 and DMS–9216203, and this support is gratefully acknowledged.

REFERENCES

[1] R. Balian, *Random matrices and information theory*, Nuovo Cimento **B57** (1968) 183–193.

[2] E. L. Basor and C. A. Tracy, *Asymptotics of a tau-function and Toeplitz determinants with singular generating functions*, Int. J. Modern Physics A **7**, Suppl. 1A (1992) 83–107.

[3] E. L. Basor and C. A. Tracy, *Some problems associated with the asymptotics of τ-functions*, Surikagaku (Mathematical Sciences) **30**, no. 3 (1992) 71–76 [English translation appears in RIMS–845 preprint].

[4] E. L. Basor, C. A. Tracy, and H. Widom, *Asymptotics of level spacing distribution functions for random matrices*, Phys. Rev. Letts. **69** (1992) 5–8.

[5] E. L. Basor and H. Widom, *Toeplitz and Wiener-Hopf determinants with piecewise continuous symbols*, J. Func. Analy. **50** (1983) 387–413.

[6] M. J. Bowick and E. Brézin, *Universal scaling of the tail of the density of eigenvalues in random matrix models*, Phys. Letts. **B268** (1991) 21–28.

[7] E. Brézin and V. A. Kazakov, *Exactly solvable field theories of closed strings*, Phys. Letts. **B236** (1990) 144–150.

[8] P. Diaconis and M. Shahshahani, *The subgroup algorithm for generating uniform random variables*, Prob. in the Engin. and Inform. Sci. **1** (1987) 15–32.

[9] M. Douglas and S. H. Shenker, *Strings in less than one dimension*, Nucl. Phys. **B335** (1990) 635–654.

[10] F. J. Dyson, *Statistical theory of energy levels of complex systems, I, II, and III*, J. Math. Phys. **3** (1962) 140–156; 157–165; 166–175.

[11] F. J. Dyson, *The three fold way. Algebraic structure of symmetry groups and ensembles in quantum mechanics*, J. Math. Phys. **3** (1962) 1199–1215.

[12] F. J. Dyson, *Fredholm determinants and inverse scattering problems*, Commun. Math. Phys. **47** (1976) 171–183.

[13] F. J. Dyson, *The Coulomb fluid and the fifth Painlevé transcendent*, IASSNSS-HEP-92/43 preprint.

[14] D. Fox and P. B. Kahn, *Identity of the n-th order spacing distributions for a class of Hamiltonian unitary ensembles*, Phys. Rev. **134** (1964) 1151–1155.

[15] W. H. J. Fuchs, *On the eigenvalues of an integral equation arising in the theory of band-limited signals*, J. Math. Anal. and Applic. **9** (1964) 317–330.

[16] D. J. Gross and A. A. Migdal, *Nonperturbative two-dimensional quantum gravity*, Phys. Rev. Letts. **64** (1990) 127–130; *A nonperturbative treatment of two-dimensional quantum gravity*, Nucl. Phys. **B340** (1990) 333–365.

[17] D. J. Gross, T. Piran, and S. Weinberg, eds., *Two Dimensional Quantum Gravity and Random Surfaces* (World Scientific, Singapore, 1992).

[18] L. K. Hua, *Harmonic analysis of functions of several complex variables in the classical domains* [Translated from the Russian by L. Ebner and A. Koranyi] (Amer. Math. Soc., Providence, 1963).

[19] A. R. Its, A. G. Izergin, V. E. Korepin, and N. A. Slavnov, *Differential equations for quantum correlation functions*, Int. J. Mod. Physics B **4** (1990) 1003–1037.

[20] K. Iwasaki, H. Kimura, S. Shimomura, and M. Yoshida, *From Gauss to Painlevé: A Modern Theory of Special Functions* (Vieweg, Braunschweig, 1991).

[21] M. Jimbo and T. Miwa, *Monodromy preserving deformation of linear ordinary differential equations with rational coefficients. II*, Physica **2D** (1981) 407–448.

[22] M. Jimbo, T. Miwa, Y. Môri, and M. Sato, *Density matrix of an impenetrable Bose gas and the fifth Painlevé transcendent*, Physica **1D** (1980) 80–158.

[23] R. D. Kamien, H. D. Politzer, and M. B. Wise, *Universality of random-matrix predictions for the statistics of energy levels*, Phys. Rev. Letts. **60** (1988) 1995–1998.

[24] D. Levi and P. Winternitz, eds., *Painlevé Transcendents: Their Asymptotics and Physical Applications* (Plenum Press, New York, 1992).

[25] D. S. Lubinsky, *A survey of general orthogonal polynomials for weights on finite and infinite intervals*, Acta Applic. Math. **10** (1987) 237–296.

[26] B. M. McCoy and S. Tang, *Connection formulae for Painlevé V functions*, Physica **19D** (1986) 42–72; **20D** (1986) 187–216.

[27] M. L. Mehta, *Random Matrices*, 2nd edition (Academic, San Diego, 1991).

[28] M. L. Mehta, *Power series for level spacing functions of random matrix ensembles*, Z. Phys. B **86** (1992) 285–290.

[29] M. L. Mehta, *A non-linear differential equation and a Fredholm determinant*, J. de Phys. I France, **2** (1992) 1721–1729.

[30] M. L. Mehta and G. Mahoux, *A method of integration over matrix variables: III*, Indian J. pure appl. Math., **22**(7) (1991) 531–546.

[31] M. L. Mehta and G. Mahoux, *Level spacing functions and non-linear differential equations*, preprint.

[32] G. Mahoux and M. L. Mehta, *A method of integration over matrix variables: IV*, J. Phys. I France **1** (1991) 1093–1108.

[33] J. Meixner and F. W. Schäfke, *Mathieusche Funktionen und Sphäroidfunktionen* (Springer Verlag, Berlin, 1954).

[34] G. Moore, *Matrix models of 2D gravity and isomonodromic deformation*, Prog. Theor. Physics Suppl. No. **102** (1990) 255–285.

[35] J. Moser, *Geometry of quadrics and spectral theory*, in *Chern Symposium 1979* (Springer, Berlin, 1980), 147–188.

[36] T. Nagao and M. Wadati, *Correlation functions of random matrix ensembles related to classical orthogonal polynomials*, J. Phys. Soc. Japan **60** (1991) 3298–3322.

[37] P. Nevai, *Asymptotics for orthogonal polynomials associated with* $\exp(-x^4)$, SIAM J. Math. Anal. **15** (1984) 1177–1187.

[38] L. A. Pastur, *On the universality of the level spacing distribution for some ensembles of random matrices*, Letts. Math. Phys. **25** (1992) 259–265.

[39] C. E. Porter, *Statistical Theory of Spectra: Fluctuations* (Academic, New York, 1965).

[40] D. Slepian, *Some asymptotic expansions for prolate spheroidal wave functions*, J. Math. Phys. **44** (1965) 99–140.

[41] A. D. Stone, P. A. Mello, K. A. Muttalib, and J.-L. Pichard, *Random matrix theory and maximum entropy models for disordered conductors*, in *Mesoscopic Phenomena in Solids*, eds. B. L. Altshuler, P. A. Lee, and R. A. Webb (North-Holland, Amsterdam, 1991), Chp. 9, 369–448.

[42] H. Weyl, *The Classical Groups* (Princeton, Princeton, 1946).

[43] H. Widom, *The asymptotics of a continuous analogue of orthogonal polynomials*, to appear in J. Approx. Th.

The Geometry of Elastic Waves Propagating in an Anisotropic Elastic Medium

Dirk-J. Smit[1] and Maarten V. de Hoop[2]

[1] Koninklijke/Shell Exploratie en Produktie Laboratorium
Volmerlaan 6, 2288 GD Rijswijk, The Netherlands
[2] Schlumberger Cambridge Research
High Cross, Madingley Road
Cambridge CB3 0EL, England

Abstract. We evaluate the fundamental solution of the hyperbolic system describing the generation and propagation of elastic waves in an anisotropic solid by studying the homology of the so-called slowness hypersurface defined by the characteristic equation. Our starting point is the Herglotz-Petrovsky-Leray integral representation of the fundamental solution. We find an explicit decomposition of the latter solution into integrals over vanishing cycles associated with the isolated singularities on the slowness surface. As is well known in the theory of isolated singularities, integrals over vanishing cycles satisfy a system of differential equations known as Picard-Fuchs equations. We discuss a method to obtain these equations explicitly. Subsequently, we use these to analyse the asymptotic behavior of the fundamental solution near wave front singularities in three dimensions. Our work sheds new light on how to compute the so-called Cagniard-De Hoop contour which is used in numerical integration schemes to obtain the full time behaviour of the fundamental solution for a given direction of propagation.

1 Introduction

Over the last few years, much attention has been paid to the evaluation of the fundamental solution (Green's tensor) of the hyperbolic system describing the generation and propagation of waves in generally anisotropic solids in $n =$ 3-dimensional space. One reason for this comes from the field of exploration geophysics; recently developed techniques in seismic surveys are powerful enough to reveal, in principle, anisotropic properties of rock in layers at great depth. Knowledge of this anisotropy is important to the oil and gas industries as some of the anisotropy is due to large joint systems (faults and fractures) in layered hydrocarbon reservoirs, which will affect the fluid and gas flows in production.

Wave propagation in anisotropic media is very different from propagation in isotropic media. Typical for anisotropic media are the phenomena of shear-wave splitting and conical refraction. In addition, several types of 'singularities' appear, for example, the self-intersections, cusps and swallow tails that a shear-wave front in a generic anisotropic medium develops. Figure 1 shows the situation for a typical hexagonal medium frequently observed in seismic experiments. The occurrence of a singularity affects the solution significantly and results in complicated propagation behavior; this cannot be described by standard 'asymptotic ray theory'. The latter approximation diverges in the vicinity of a singularity. As the physical solution at a wave front singularity is still regular, a more powerful theory is required to find this solution.

A geometrical description of waves leads to a generally valid solution of the hyperbolic system, and opens the way to rigorously check (future) numerical techniques for wave simulations in 'realistic' media. Further, the understanding of the time behavior near a singularity is crucial to interpret it properly on measured seismograms. Different approaches to solve a 3-dimensional anisotropic hyperbolic system, all of them based on (spectral-domain) plane-wave expansions of either the particle velocity or the Lamé potentials, have been explored to find closed-form integral representations of the solution. We mention the Sommerfeld-Weyl representation [42], Riesz's method [37], and the Herglotz-Petrovsky-Leray (HPL) representation [10, 36]. The reduction of the number of integrals was independently achieved, employing the time-Laplace transform domain, in the Cagniard-De Hoop method [12, 19, 22]; for particular symmetries the fundamental solution could be found explicitly (e.g., the work of Payton [34], Burridge, Chadwick and Norris [11]). In crystal acoustics (e.g., Musgrave [30]) and phonon focussing (see Every [16]) similar developments took place. However, most of these techniques lack a precise geometrical understanding of the features typical for the wave solution in anisotropic media.

The HPL formula gives the fundamental solution in the form of an (Abelian) integral of a rational closed $(n-1)$-form integrated over a complex $(n-1)$-dimensional algebraic hypersurface [8]. From the latter integral the full Green's tensor of the problem can be constructed. The algebraic surface, known as the 'slowness surface', is defined by the equation $H(\xi) = 0$ of degree $D = 2n$, the polynomial H being the complexified determinant of the symbol matrix of the hyperbolic operator. The integral is defined over a tube γ of properly oriented cycles $\partial\gamma$ on the real slowness surface; the generic form of the integral is

$$E = \int_\gamma \frac{\omega(\xi)}{\partial_{\xi_n} H}, \tag{1}$$

where $\omega(\xi)$ is the volume form on the slowness surface and ξ_n is the coordinate in slowness space along the direction of propagation of interest. The cycle $\partial\gamma$ is later on shown to be directly related to the so-called Cagniard-De Hoop contour. The theory of such integrals is extensive and well established in the mathematical literature. In particular, the Maslov theory of rapidly oscillating integrals [15], which is a frequency- rather than a time-domain approach, in conjunction with singularities defined through the critical points of so-called phase functions, see e.g. [8], deals with integrals of the type (1) near singularities.

This paper is a review and summary of our earlier work [41] in which we employ algebraic-geometrical techniques, known in complex singularity theory, to analyse integrals like (1) with the application to wave propagation in a 3-dimensional perfectly elastic anisotropic medium in mind. In particular, we are interested in the asymptotic behavior of the solution near the singularities on the wave fronts. In this context, any direction of propagation corresponds to a singular point on the slowness surface, but most of them are of a trivial type. Our angle of attack is different from the usual Maslov theory, in that we will derive this asymptotic behavior by analysing so-called Picard-Fuchs differential equations. These are differential equations associated with the singularities on the wave front. As these equations can have at most regular singular points, an asymptotic analysis is straightforward. The advantage of this approach is that, as the structure of the Picard-Fuchs equation is entirely fixed by monodromy around its regular singular points, it shows explicitly the geometrical content of the solution near the singularities on the wave front (which coincides with the high frequency limit; this relates to slowly varying medium properties).

To be able to derive the Picard-Fuchs equation, we first need to rewrite the integral in terms of a linear combination of integrals over certain $(n-2)$-dimensional cycles, called 'vanishing cycles' on the surface $H = 0$, which are directly associated with the singularities on the wave front. To be more precise: the cycle γ can be described in terms of the cohomology of its boundary $\partial\gamma$, which has a precise meaning in terms of the so-called Milnor fibration of a singularity in the Legendre transformation relating the slowness surface with the diffraction surface (wave front set). The fact that isolated singularities on wave fronts of hyperbolic operators correspond to singularities of Legendre transformations has been shown by Arnold [7]. The relation between a cycle $\partial\gamma$ and certain objects in singularity theory has been explored before by Vasiliev [43] and in fact been foreseen by Petrovsky, where the cycle was described in terms of a certain cohomology class, at present called the Petrovsky class, in relation with the recognition of lacunae in the fundamental solution of a hyperbolic operator. We will reestablish this relation, using the approach followed by Atiyah, Bott and Gårding [8]. We thus arrive at a canonical decomposition of the integral into integrals over vanishing cycles, which we will compute explicitly in the case of an arbitrarily anisotropic medium in three spatial dimensions. This decomposition is entirely topological and, hence, also valid *away* from the singularity. This is an interesting result in itself, as it reveals the geometry behind the so-called Cagniard-De Hoop contours used in numerical integration schemes to evaluate E in (1). In fact, we establish a classification of such contours near the singularities.

The Picard-Fuchs equation is satisfied by each of the integrals over a vanishing cycle. It arises from the fact that the homologies of the vanishing cycles, which are dependent on the deformation parameters of the singularity, are locally constant in these parameters. If we define a suitable residue we can differentiate under the integral with respect to these parameters and conclude that a sufficiently high-order derivative of the integrand must be a linear combination of the lower order ones. Applying standard results to the differential equation thus obtained, leads to the desired asymptotic behavior. As the coefficient functions in the equation are only depending on the monodromy of the regular singular points, the latter procedure encodes the topological content of the solution of the wave problem near a singularity. This then shows some of the more salient features of the contours used in previous approaches to evaluate (1), such as the Cagniard-De Hoop method [19].

The paper is organised as follows. In Section 2 we pose the seismic Cauchy problem and resume some of its basic properties. In Section 3 we derive explicitly the HPL representation of a general elastodynamic hyperbolic system from Gelfand's plane-wave expansion [17]. Along the way we discuss the geometrical properties of the slowness surface relevant to the later analysis. In Section 3 we also show how the HPL formula reduces to an integral over a $(n-2)$-dimensional cycle on the slowness surface, which can be associated with the Cagniard-De Hoop contour. In Section 4 the latter integral is decomposed into integrals over vanishing cycles associated with a singularity in the Legendre transformation. In Section 5, finally, we derive a Picard-Fuchs differential equation for the integrals over vanishing cycles and show how the asymptotic behavior in the high frequency limit follows from the monodromy properties around the regular singular points.

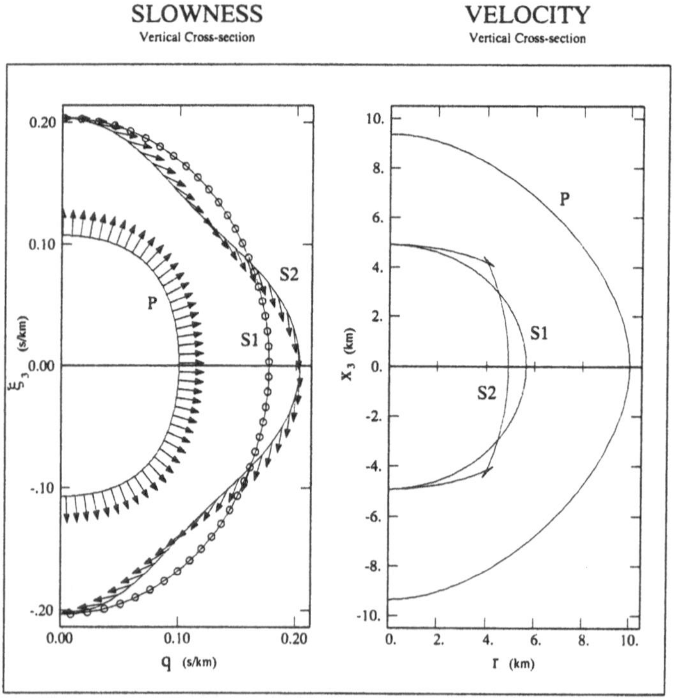

Fig. 1. Slowness surface and wave front for a typical hexagonal medium.

2 The Cauchy Problem

The tensorial Cauchy problem.
In this section we introduce the tensorial Cauchy problem arising in seismics. We define the physical quantities and transform the tensorial problem into an associated *scalar* Cauchy problem, from which the full Green's tensor can be constructed. The scalar problem is completely equivalent to the tensorial problem but more suitable for our analysis. Since the procedure to arrive at this scalar Cauchy problem is not so well known, though standard in functional analysis, we will present the relevant steps explicitly.

It is assumed that linear elasticity theory is applicable. Then the particle velocity v_m, $m = 1, \cdots, n$ satisfies the system of equations

$$[\delta_{km}\partial_t^2 - \rho^{-1}\partial_r c_{krms}\partial_s]\, v_m = \rho^{-1}\, [\partial_t f_k + \partial_r(c_{krms}h_{ms})], \tag{2}$$

where the summation convention applies and

$$\begin{aligned} \rho &= \text{volume density of mass,} \\ c_{krms} &= \text{stiffness,} \\ f_k &= \text{volume source density of force,} \\ h_{ms} = h_{sm} &= \text{volume source density of deformation rate.} \end{aligned}$$

Note that h_{ms} is related to the seismic moment density tensor, m_{kr} say, according to

$$(\partial_t m_{kr})\, \delta(n_j x_j) = c_{krms}h_{ms}, \tag{3}$$

where n_j is the unit normal to the plane of dislocation. The $(n \times n)$ principal part of the tensorial wave operator on the left-hand side of (2) is given by

$$\delta_{km}\partial_t^2 - C_{km}^{rs}\partial_r\partial_s, \tag{4}$$

with

$$C_{km}^{rs} = \rho^{-1}c_{krms}. \tag{5}$$

From the positive definite property of the strain-energy function it follows that the system is hyperbolic, that is, the system admits propagating waves as solutions; in general, the system is not strictly hyperbolic, which means that different modes may propagate with the same speed in a particular direction. The symmetries of the stiffness yield

$$C_{km}^{rs} = C_{ks}^{rm} = C_{rm}^{ks} = C_{mk}^{sr} = C_{rs}^{km}. \tag{6}$$

They follow from the condition that (angular) momentum is conserved and from the assumption that the processes of deformation are adiabatic. The macroscopic symmetry properties of the medium, collected in the 'point' group of transformations O, say, simply reflect themselves in the relations $O_{k'k}O_{m'm}O_{r'r}O_{s's}C_{m'k'}^{s'r'} = C_{km}^{sr}$. How to obtain the tensor C from microstructure of rocks can be found in several papers amongst which the ones by Hudson [21], Schoenberg and Muir [40] and Nichols, Muir and Schoenberg [32]. For the medium in Figure 1 elasticities are $C_{1111} = C_{2222} = 269.4$, $C_{3333} = 236.3$, $C_{2233} = C_{3311} = 66.1$, $C_{1122} = 96.1$, $C_{2323} = C_{1313} = 65.3$, $C_{1212} = 86.1$, in units of GPa; the density $\rho = 2700$ kg/m^3. For seismic applications the Green's tensor is introduced as the particle velocity due to a point body force (e.g., a vibrator) with signature $H(t)$ (the Heaviside function; this yields a δ behavior in time on the right-hand side of (2)). For mathematical convenience a linear combination of the causal and the anti-causal Green's tensor is taken to constitute the fundamental solution; the latter reduces to the causal Green's tensor on the positive real time axis. Thus, using Duhamel's principle [23], the tensorial Cauchy problem to construct the fundamental solution is introduced: determine the solution g_m of the homogeneous tensorial wave equation

$$[\delta_{km}\partial_t^2 - C_{km}^{rs}\partial_r\partial_s]\, g_m = 0 \ \text{ for } \ \{x,t\} \in \mathbb{R}^{n+1}, \tag{7}$$

satisfying the initial conditions

$$g_m\Big|_{t=0} = 0, \quad \partial_t g_m\Big|_{t=0} = a_m\delta(x_1,\cdots,x_n). \tag{8}$$

By taking for a_m the standard unit vectors, the columns of the fundamental solution are found. Later on, we will ignore the separation between the space and time coordinates and set $\{x,t\} \to \{x_1,\cdots,x_{n+1}\} \in \hat{Z}'$.

Through mutual elimination of the components of the system (7), it follows that any component g_m satisfies the $D = 2n$-degree equation in n dimensions

$$a(\partial_i,\partial_t)\, g_m = 0, \tag{9}$$

in which

$$a(\partial_i,\partial_t) = \det[\delta_{km}\partial_t^2 - C_{km}^{rs}\partial_r\partial_s], \tag{10}$$

with initial conditions

$$\partial_t^\ell g_m\Big|_{t=0} = \phi_{mk}^{(\ell)} a_k, \quad \ell = 0,\cdots,2n-1, \tag{11}$$

where, using our wave operator repeatedly, $\phi_{mk}^{(\text{even})} = 0$ and

$$\phi_{mk}^{(2\ell_0-1)} = ([C_{..}^{rs}\,\partial_r\partial_s]^{\ell_0})_{mk}\,\delta(x_1,\cdots,x_n), \quad \ell_0 = 1,\cdots,n \tag{12}$$

for $m = 1,\cdots,n$.

The scalar Cauchy problem.

Now, consider the particular family of scalar Cauchy problems

$$a(\partial_i,\partial_t)\, E^{(\ell)} = 0, \tag{13}$$

with

$$\partial_t^{\ell'} E^{(\ell)}\Big|_{t=0} = \delta_{\ell'\ell} U(x_1,\cdots,x_n), \quad \ell' = 0,\cdots,2n-1 \tag{14}$$

for $\ell = 0, \cdots, 2n-1$. The solution is denoted as $E^{(\ell)}[U]$ to indicate the initial value function explicitly. Note that for $\ell = 0, \cdots, 2n-2$,

$$E^{(\ell)}[U] = \partial_t^{2n-1-\ell} E^{(2n-1)}[U] - \sum_{\ell'=\ell+1}^{2n-1} E^{(\ell')}\left[\partial_t^{\ell'-\ell} \partial_t^{2n-1} E^{(2n-1)}[U]\right]. \quad (15)$$

This way, the $E^{(\ell)}$, $\ell = 0, \cdots, 2n-2$, follow from $E^{(2n-1)}$ by recursion. The further analysis will be focussed on the latter function. Thus, set

$$E = E^{(2n-1)}[\delta] \quad (16)$$

then

$$E^{(2n-1)}[U] = E *_{\mathbb{R}^n} U. \quad (17)$$

The $*_{\mathbb{R}^n}$ denotes convolution with U in $\mathbb{R}^n$. Note that the 'physical' scalar Cauchy problem is associated with $E^{(1)}$.

The function g_m can be constructed from the $E^{(\ell)}$ using (11):

$$g_m = a_k \left\{ \sum_{\ell_0=1}^{n} \Phi_{mk}^{(2\ell_0-1)}(E^{(2\ell_0-1)}[\delta]) \right\}, \quad (18)$$

where

$$\Phi_{mk}^{(2\ell_0-1)}(.) = \phi_{mk}^{(2\ell_0-1)} *_{\mathbb{R}^n} ., \quad \ell_0 = 1, \cdots, n. \quad (19)$$

In seismic applications the $n = 3$-dimensional problem is considered, although the 2-dimensional problem arises as well namely from propagation in a plane of symmetry in three dimensions. In that case a line source rather than a point source is considered.

Gel'fand's plane-wave expansion.

We will now employ Gel'fand's plane-wave expansion (the inversion formula for the Radon transformation) of the δ-source in the Cauchy problem to find a plane-wave expansion of the fundamental solution. We will employ explicit distributions on $\mathbb{R}$ which are boundary values of functions analytic in the lower or upper complex half-plane. According to Gel'fand [17], we have

$$\begin{aligned} \delta(x_1, \cdots, x_n) &= -\frac{i^{n-1}}{2^n \pi^{n-1}} \int_\Omega \Delta^{(n-1)}(\hat{\xi}_j x_j - i0)\, \mathrm{d}S(\hat{\xi}) \\ &= \frac{i^{n-1}}{2^n \pi^{n-1}} \int_\Omega \overline{\Delta^{(n-1)}(\hat{\xi}_j x_j + i0)}\, \mathrm{d}S(\hat{\xi}), \end{aligned} \quad (20)$$

where

$$\Delta(\tau) = \frac{i}{\pi \tau} \quad (21)$$

is the analytic Dirac distribution (odd in τ), Ω is the unit sphere and $\mathrm{d}S(\hat{\xi})$ is the volume form on this sphere. To employ this expansion, we introduce the family of analytic functions

$$\chi_z(\tau) = \Gamma(-z) \frac{\exp(-i\pi z)}{i\pi} \tau^z, \quad (22)$$

for $z \in \mathbf{C}\backslash\{0,1,2,\cdots\}$. (Note that Γ has simple poles at $0,-1,-2,\cdots$) This is a single-valued analytic function in the lower half-plane ($\text{Im}\{\tau\}<0$) but likewise it is a single-valued function in the upper half-plane ($\text{Im}\{\tau\}>0$). This family of functions satisfies the relation

$$\frac{\mathrm{d}\chi_z}{\mathrm{d}\tau} = \chi_{z-1}(\tau). \tag{23}$$

The analytic Dirac distribution is contained as a member in the family:

$$\chi_{-1}(\tau) = \Delta(\tau). \tag{24}$$

Using the Laurent series of (22) in z about the simple poles at $z=0,1,2,\cdots$, χ_z can be defined for positive integer values for z as follows. Let $m=0,1,2,\cdots$ then

$$\chi_m(\tau) \to \frac{\mathrm{d}}{\mathrm{d}\zeta}\zeta\chi_{m+\zeta}(\tau)|_{\zeta=0} = \frac{1}{m!}\left[1-\frac{i}{\pi}\left(\log(\tau^{-1})+c_m\right)\right]\tau^m, \tag{25}$$

with

$$c_0 = \Gamma'(1), \quad c_m = c_0 + \sum_{k=1}^{m}\frac{1}{k} \quad \text{for} \quad m=1,2,\cdots \tag{26}$$

The expression in (25) satisfies (23) with $z=m$. It is found that

$$\lim_{\text{Im}\{\tau\}\downarrow 0}\chi_m(\tau) = \frac{1}{m!}\left[\ \frac{1}{2}\text{sgn}(\tau)+\frac{1}{2}-\frac{i}{\pi}\left(\log(|\tau|^{-1})+c_m\right)\right]\tau^m, \tag{27}$$

$$\lim_{\text{Im}\{\tau\}\uparrow 0}\chi_m(\tau) = \frac{1}{m!}\left[-\frac{1}{2}\text{sgn}(\tau)+\frac{3}{2}-\frac{i}{\pi}\left(\log(|\tau|^{-1})+c_m\right)\right]\tau^m, \tag{28}$$

for $m=0,1,2,\cdots$. Now, using (23) for $m=0$, we find

$$\lim_{\text{Im}\{\tau\}\downarrow 0}\Delta(\tau) = \lim_{\text{Im}\{\tau\}\downarrow 0}\chi_{-1}(\tau) = \delta(\tau)+i(\mathcal{H}\delta)(\tau), \tag{29}$$

$$\lim_{\text{Im}\{\tau\}\uparrow 0}\Delta(\tau) = \lim_{\text{Im}\{\tau\}\uparrow 0}\chi_{-1}(\tau) = -\delta(\tau)+i(\mathcal{H}\delta)(\tau), \tag{30}$$

where $\mathcal{H}$ denotes the Hilbert transform.

Explicit evaluation reveals that

$$\int_\Omega \text{Re}\left\{\chi_{-n}(\hat{\xi}_j x_j - i0)\right\}\mathrm{d}S(\hat{\xi}) = 0 \quad \text{if } n \text{ is even} \tag{31}$$

$$\int_\Omega \text{Im}\left\{\chi_{-n}(\hat{\xi}_j x_j - i0)\right\}\mathrm{d}S(\hat{\xi}) = 0 \quad \text{if } n \text{ is odd.} \tag{32}$$

Thus, in fact (20) can be written as

$$\delta(x_1,\cdots,x_n) = -\frac{i^{n-1}}{2^n\pi^{n-1}}\begin{cases}\text{Re}\left\{\int_\Omega \chi_{-n}(\hat{\xi}_j x_j - i0)\,\mathrm{d}S(\hat{\xi})\right\} & \text{if } n \text{ is odd}\\ i\,\text{Im}\left\{\int_\Omega \chi_{-n}(\hat{\xi}_j x_j - i0)\,\mathrm{d}S(\hat{\xi})\right\} & \text{if } n \text{ is even}\end{cases} \tag{33}$$

or

$$\delta(x_1,\cdots,x_n) = -\frac{1}{2^n\pi^{n-1}}\operatorname{Re}\left\{\int_\Omega i^{n-1}\chi_{-n}(\hat{\xi}_j x_j - i0)\,\mathrm{d}S(\hat{\xi})\right\}, \tag{34}$$

but everywhere $\chi_{-n}(\hat{\xi}_j x_j - i0)$ can be replaced by

$$(1/2)\,[\chi_{-n}(\hat{\xi}_j x_j - i0) - \overline{\chi_{-n}(\hat{\xi}_j x_j + i0)}].$$

Given Gel'fand's plane-wave expansion of the δ-function, it is now straightforward to find the plane-wave expansion both of g_m in case of the tensorial problem and of E in case of the scalar problem. We will suppress the term $-i0$ in all the arguments to simplify our notation.

The plane-wave expansion of g_m yields:

$$g_m = -\frac{i^{n-1}}{2^n\pi^{n-1}}\sum_{p<0,p>0}\int_\Omega \alpha^{(p)}(\hat{\xi})e_m^{(p)}(\hat{\xi})\,\chi_{-n+1}(\hat{\xi}_j x_j - \lambda_p(\hat{\xi})t)\,\mathrm{d}S(\hat{\xi}). \tag{35}$$

Here, the λ_p, $p \in \{\pm 1, \pm 2, \cdots, \pm n\}$ are the phase velocities that must satisfy the dispersion relation

$$a(\hat{\xi}_i, -\lambda_p) = 0, \tag{36}$$

while the $e^{(p)}$ are the polarization vectors corresponding with the particle velocity; they satisfy the Christoffel equation

$$\mathcal{K}_{km}\, e_m^{(p)} = \lambda_p^2 e_k^{(p)}, \tag{37}$$

where

$$\mathcal{K}_{km} = C_{km}^{rs}\hat{\xi}_r\hat{\xi}_s \tag{38}$$

is the Christoffel matrix. Since this matrix is symmetric, the basis $\{e^{(p)}\}_{p>0}$ can be chosen to be orthonormal at every $\hat{\xi}$. Later on, we will group together the spectral-domain coordinates: $\{\hat{\xi}_1, \cdots, \hat{\xi}_n, -\lambda\} \to \{\hat{\xi}_1, \cdots, \hat{\xi}_{n+1}\} \in \hat{Z}$, say. In view of the symmetry under time reversal of the tensorial wave operator, we have

$$\lambda_{-p} = -\lambda_p, \tag{39}$$

and we can choose

$$e^{(-p)} = e^{(p)}. \tag{40}$$

Hence, (35) must be cubic in λ_p^2 and the values for λ_p can be found with the aid of Cardano's formula. Note the homogeneity of λ_p: $\lambda_p(\tau\hat{\xi}) = \tau\lambda_p(\hat{\xi})$, which also implies that $-\lambda_p(-\hat{\xi}) = \lambda_p(\hat{\xi})$.

The $\alpha^{(p)}$ follow from the initial conditions (cf. (11)),

$$\sum_{p<0,p>0}(-\lambda_p)^\ell\,\alpha^{(p)}e_m^{(p)} = \tilde{\phi}_{mk}^{(\ell)}a_k, \quad \ell = 0, \cdots, 2n-1, \tag{41}$$

where $\tilde{\phi}^{(\ell)}_{mk}$ follows from $\phi^{(\ell)}_{mk}$ by replacing ∂_r with $\hat{\xi}_r$, employing Gel'fand's formula. The resulting equations reduce to

$$\sum_{p<0,p>0} \alpha^{(p)} e^{(p)}_m = 0, \tag{42}$$

$$-\sum_{p<0,p>0} (e^{(p)}_m \lambda_p)\, \alpha^{(p)} = a_m. \tag{43}$$

In view of the $\pm p$ symmetry, the first equation implies

$$\alpha^{(-p)} = -\alpha^{(p)}; \tag{44}$$

then the second equation implies

$$\alpha^{(p)} = -\frac{1}{2\lambda_p}\,(e^{(p)}_k a_k). \tag{45}$$

The weighting functions $\alpha^{(p)} e^{(p)}_m = -a_k\,(2\lambda_p)^{-1}(e^{(p)}_k e^{(p)}_m)$ lead to the well-known dyadic form of the Green's tensor. Using that the distribution $\mathrm{Re}\{i^{n-1}\chi_{-n+1}(\hat{\xi}_j x_j \pm \lambda_p(\hat{\xi})t - i0)\}$ is even in $\hat{\xi}_j x_j \pm \lambda_p(\hat{\xi})t$, the expression in (35) can be reduced to a sum over positive values of p's and integrals over the hemisphere of Ω.

Using the symmetry in time again, the plane-wave expansion of the scalar function E is found to be

$$E = -\frac{i^{n-1}}{2^n \pi^{n-1}} \sum_{p>0} \int_{\Omega} A^{(p)}(\hat{\xi})\,[\chi_{n-1}(\hat{\xi}_j x_j - \lambda_p(\hat{\xi})t) - \chi_{n-1}(\hat{\xi}_j x_j + \lambda_p(\hat{\xi})t)]\,\mathrm{d}S(\hat{\xi}). \tag{46}$$

The initial conditions for E now lead to

$$2\sum_{p>0}(-\lambda_p)^{2\ell_0-1} A^{(p)} = \delta_{2\ell_0-1,2n-1}, \quad \ell_0 = 1,\cdots,n. \tag{47}$$

Since (39) implies

$$a(\hat{\xi},-\lambda) = \prod_{p>0}(\lambda^2 - \lambda_p^2(\hat{\xi})), \tag{48}$$

the solution of (47) is given by

$$A^{(p)} = -\frac{1}{(\partial_\lambda a)|_{\lambda=\lambda_p}}. \tag{49}$$

Using the homogeneity of a, viz., $a(\tau\hat{\xi},-\tau\lambda) = \tau^{2n} a(\hat{\xi},-\lambda)$, it follows that

$$(-\lambda\partial_\lambda a)|_{\lambda=\lambda_p} = (\hat{\xi}_i \partial_{\hat{\xi}_i} a)|_{\lambda=\lambda_p},$$

and (49) can be written as

$$A^{(p)} = \frac{\lambda_p}{(\hat{\xi}_i \partial_{\hat{\xi}_i} a)|_{\lambda=\lambda_p}}. \tag{50}$$

3 The Herglotz-Petrovsky-Leray Formulae

The slowness hypersurface.

We begin with summarizing the basic properties of the varieties which will play a rôle in the further analysis. For a detailed discussion we refer the reader to Duff [14], Musgrave [30], and Payton [34, 35].

Let the *slowness cone* $\hat{A} \subset \hat{Z}$ be defined through

$$\hat{A}: \ a = 0. \tag{51}$$

The slowness or ray vector ξ is introduced as

$$\hat{\xi}_j = \lambda_p \xi_j, \quad p = 1, \cdots, n. \tag{52}$$

If $\hat{\xi} \in \Omega$ then $|\xi| = 1/|\lambda_p|$ equals the phase slowness. In fact, (52) represents the transition from $\hat{Z} \simeq \mathbf{C}^{n+1}$ to the projective space[3] $Z \simeq \mathbf{CP}^n$. Let the function H be given by

$$H(\xi_1, \cdots, \xi_n) = a(\xi_1, \cdots, \xi_n, -1). \tag{53}$$

Then the *slowness hypersurface* is defined by

$$A: \ H = 0 \tag{54}$$

(this equals the intersection of the slowness cone $\hat{A}$ with the plane $\lambda = 1$). In any local cone (with its vertex at the origin) the entirely real solution of the latter equation, Re$\{A\}$, consists of n *sheets*. Every sheet corresponds, in any local cone, with a (double) mode $(\pm)p$ and can be covered with (two) almost everywhere holomorphic coordinate patches.

In general, the sheets may have a finite number of isolated multiple points, where two ('kiss' singularity) sheets are tangent [13], or for points where two sheets intersect. Sheets can also have curves or higher-dimensional surfaces of multiple points, in which case the space-time singularities are of a different nature: only a hypersurface of multiple points of codimension 2 leads to an additional arrival. There the coordinate patches cannot be holomorphic (for an analysis of the local slowness surface parametrization near a conical point in three dimensions, see Musgrave [31]). If points of tangency occur, the slowness surface as a whole is called singular; if the sheets are entirely disconnected, the surface is said to be regular. Each sheet is smooth and either locally convex or concave (elliptic points) or locally saddle shaped (hyperbolic points, Morse saddles) except at points or curves where (one of) the principal curvatures vanish (parabolic points, e.g., inflection points in a plane through the origin of the slowness space). In this paper we will focus exclusively on the latter singularities and postpone the treatment of conical refraction and kiss singularities to a future paper.

Integration over the real slowness surface.

The sum over the modes combined with the integral over the unit sphere can now be expressed as an integral over the real slowness surface (using local coordinates

[3] un-hatted variables are projective variables throughout.

in a small cone and the outward normal or group direction $\partial_\xi H/|\partial_\xi H|$). For a discussion see also Auld [9]. Let the volume form on $\mathrm{Re}\{A\}$ be denoted as $\mathrm{d}S(\xi)$. Then (see Figure 2)

$$\mathrm{d}S(\xi) = \frac{|\xi|^{n-1}}{|\cos\theta|}\,\mathrm{d}S(\hat{\xi}), \tag{55}$$

where θ is the angle between the group velocity and ray vectors. Hence

$$\mathrm{d}S(\hat{\xi}) = \frac{|\xi_j \partial_{\xi_j} H|}{|\xi|^n |\partial_\xi H|}\,\mathrm{d}S(\xi). \tag{56}$$

Further, the weighting function $A^{(p)}$ transforms as (cf. (50) and we extracted the positive phase velocities $p > 0$)

$$A^{(p)} \to \frac{|\xi|^{2n-1}}{\xi_j \partial_{\xi_j} H}, \tag{57}$$

while (cf. (22))

$$\chi_{z-1}(\hat{\xi}_j x_j - \lambda_p(\hat{\xi})t) \to |\xi|^{-z+1}\,\chi_{z-1}(\xi_j x_j - t). \tag{58}$$

Set

$$\mathrm{d}\sigma(\xi) = \frac{\mathrm{d}S(\xi)}{|\partial_\xi H|}\,\mathrm{sgn}(\xi_j \partial_{\xi_j} H); \tag{59}$$

upon substituting the latter results in (46) and taking the limit $z \to n$, we obtain

$$E = -\frac{i^{n-1}}{2^n \pi^{n-1}} \begin{cases} \mathrm{Re}\left\{\int_{\mathrm{Re}\{A\}} [\chi_{n-1}(\xi_j x_j - t) - \chi_{n-1}(\xi_j x_j + t)]\,\mathrm{d}\sigma(\xi)\right\} & n \text{ odd} \\ i\,\mathrm{Im}\left\{\int_{\mathrm{Re}\{A\}} [\chi_{n-1}(\xi_j x_j - t) - \chi_{n-1}(\xi_j x_j + t)]\,\mathrm{d}\sigma(\xi)\right\} & n \text{ even.} \end{cases} \tag{60}$$

This formula, known as the HPL formula, implies that the fundamental solution can be expressed in terms of so-called *non-evanescent* constituents only. Here, we made use of the fact that the system is non-dispersive. Note that the imaginary part involves a logarithm, while the real part does not. Also, note that $\mathrm{Re}\{i^{n-1}\,\chi_{n-1}(\xi_j x_j \pm t - i0)\}$ is even in $\xi_j x_j \pm t$. Hence, (60) may be written as

$$E = -\frac{i^{n-1}}{2^n \pi^{n-1}} \begin{cases} \mathrm{Re}\left\{\int_{\mathrm{Re}\{A\}} [\chi_{n-1}(t - \xi_j x_j) - \chi_{n-1}(t + \xi_j x_j)]\,\mathrm{d}\sigma(\xi)\right\} & n \text{ odd} \\ i\,\mathrm{Im}\left\{\int_{\mathrm{Re}\{A\}} [\chi_{n-1}(t - \xi_j x_j) - \chi_{n-1}(t + \xi_j x_j)]\,\mathrm{d}\sigma(\xi)\right\} & n \text{ even.} \end{cases} \tag{61}$$

We exploit the fact that the integrand is even in $t \pm \xi_j x_j$ further. It has been observed that $\mathrm{Re}\{A\}$ is invariant under the point reflection in the origin in ξ-space. Now, choose a direction of preference, x_n say, relative to the principal axes of symmetry of the medium. Set $\mathrm{Re}\{A\} = A_+ \cup A_-$, such that A_+ corresponds with the upgoing waves (identified by the wave front $\Sigma_w = \Sigma_+ \cup \Sigma_-$ through the polar reciprocal of $\mathrm{Re}\{A\}$) and A_- with the downgoing waves. The point reflection $\xi \to -\xi$ maps A_+ onto A_-, hence (still omitting the argument $-i0$)

$$\int_{A_\pm} \chi_{n-1}(t - \xi_j x_j)\,\mathrm{d}\sigma(\xi) = \int_{A_\mp} \chi_{n-1}(t + \xi_j x_j)\,\mathrm{d}\sigma(\xi). \tag{62}$$

Thus

$$E = -\frac{i^{n-1}}{2^{n-1}\pi^{n-1}} \int_{A_+} [\chi_{n-1}(t - \xi_j x_j) - \chi_{n-1}(t + \xi_j x_j)]\, \mathrm{d}\sigma(\xi). \tag{63}$$

This way we have introduced an orientation for $\mathrm{Re}\{A\}$. Now choose coordinates on A_+. For this purpose consider the orthogonal projection $\mathcal{C}$ of A_+ on the plane $\{\xi_n = 0\}$ and let $(\xi_1, \cdots, \xi_{n-1})$ be the coordinates. The parametric representation of the slowness surface A_+ is then given by $\{\xi_1, \cdots, \xi_{n-1}, f(\xi_1, \cdots, \xi_{n-1})\}$, where f represents a n-plet. Later on, the direction of preference will become the direction of observation. At the singular points contributing to this direction of propagation, a principal curvature or its derivative may vanish. Thus, in the preferred coordinate system, the occurrence of a singular point on a particular sheet corresponds with multiple ($\mu \geq 1$) roots of the equation

$$\partial_{\xi_1, \cdots, \xi_{n-1}} f = 0. \tag{64}$$

At these roots the normal to the slowness surface, i.e., the group direction, must be parallel to the ξ_n-axis, which in turn translates into the condition

$$\partial_{\xi_1, \cdots, \xi_{n-1}} H = 0 \quad \text{with} \quad H = 0. \tag{65}$$

We have (see Figure 2)

$$\frac{\mathrm{d}S(\xi)}{|\partial_\xi H|} = \frac{\mathrm{d}\xi_1 \cdots \mathrm{d}\xi_{n-1}}{|\partial_{\xi_n} H|}, \tag{66}$$

and hence, restricting to A_+,

$$\mathrm{d}\sigma(\xi) \to \frac{\mathrm{d}\xi_1 \cdots \mathrm{d}\xi_{n-1}}{\partial_{\xi_n} H}. \tag{67}$$

Note that this $(n-1)$-form has poles at the branch points of our parametrization f. *Cagniard Green's function.*

At this stage, to reduce the integral, E is written as

$$E(x,t) = -\frac{i^{n-1}}{2^{n-1}\pi^{n-1}(n-1)!} \int_{t'=0}^{t} (t-t')^{n-1} \begin{matrix} \mathrm{Re} \\ i\,\mathrm{Im} \end{matrix} \{E_{\mathcal{C}}(x,t')\}\, \mathrm{d}t' \quad \begin{matrix} \text{if } n \text{ is odd} \\ \text{if } n \text{ is even} \end{matrix} \tag{68}$$

which means that

$$\partial_t^{n-1} E(x,t) = -\frac{i^{n-1}}{2^{n-1}\pi^{n-1}} \int_{t'=0}^{t} E_{\mathcal{C}}(x,t')\, \mathrm{d}t'. \tag{69}$$

Here, $E_{\mathcal{C}}$ is denoted as the Cagniard Green's function and is given by

$$E_{\mathcal{C}}(x,t') = \int_{A_+} [\Delta(t' - \xi_j x_j - i0) - \Delta(t' + \xi_j x_j - i0)]\, \mathrm{d}\sigma(\xi). \tag{70}$$

We will redefine E, which will be the only form of the solution we will consider in the further analysis:

$$E(x,t) = \int_{t'=0}^{t} E_{\mathcal{C}}(x,t')\, \mathrm{d}t'. \tag{71}$$

In E_C we may replace the integrand by

$$\Delta(t' - \xi_j x_j - i0) - \Delta(t' + \overline{\xi_j x_j} - i0). \tag{72}$$

Integration over the complex slowness surface.
The surface A is invariant under complex conjugation in ξ-space (enabling the use of Schwartz' reflection principle in the holomorphic extension of the slowness integral). Thus, a complexification equivalent with the so-called causality 'trick' (Hubral and Tygel [20]) applies.

The n-plet f and the form $d\sigma(\xi)$ have holomorphic extensions with preservation of orientation so that the integration over the compact regions $C \subset \mathcal{R}^{n-1}$ can be extended over the full (level) plane

$$X_0 \; : \; \xi_n x_n = 0, \quad x_n > 0. \tag{73}$$

The holomorphic extension is carried out according to the condition $\mathrm{Im}\{f\} > 0$. In fact, we only need analytic function theory in one variable. For example, introduce polar coordinates

$$\begin{aligned} \xi_{n-1} &= q\sin(\psi_{n-2}), \; \xi_{n-2} = q\cos(\psi_{n-2})\sin(\psi_{n-3}), \\ \xi_{n-3} &= q\cos(\psi_{n-2})\cos(\psi_{n-3})\sin(\psi_{n-4}), \; \cdots, \\ \xi_2 &= q\cos(\psi_{n-2})\cdots\cos(\psi_2)\sin(\psi_1), \; \xi_1 = q\cos(\psi_{n-2})\cdots\cos(\psi_2)\cos(\psi_1), \end{aligned} \tag{74}$$

with $\psi_1 \in [0, 2\pi)$ and $\psi_2, \cdots, \psi_{n-2} \in [0, \pi)$. We will use the shorthand notation $\psi = (\psi_1, \cdots, \psi_{n-2})$ and $d\sigma(\xi) \to d\sigma(q, \psi)$. If we also introduce polar coordinates (r, ϕ) for $(x_1, \cdots, x_{n-1})$-plane, we can set $\xi_j x_j = qr\cos(\eta) + \xi_n x_n$ where η can be expressed in the angles ψ and ϕ. Then $C = \cup_\psi I_\psi$, where I_ψ denotes the collection of intervals on the positive real axis in the complex q-plane (the endpoints of which are branch points, where $\partial_{\xi_n} H = 0$) at the angle (azimuth) ψ. Note that there may be branch points in the q-plane off the real axis. (They follow directly from the f^2 vs. q^2 relations.) Upon integrating over the full slowness hypersurface, however, it follows that contributions from the associated branch cuts cancel.

Consider the n-plet $f(q, \psi)$ with $\mathrm{Re}\{f(0, \psi)\} > 0$. Set $\bar{I}_\psi = \mathcal{R}_{\geq 0} - I_\psi$ and $\bar{C} = \cup_\psi \bar{I}_\psi$. There exists a holomorphic extension of $f(q, \psi)$ on $\bar{I}_\psi + i0$ in the complex q-plane. This defines $\bar{A}_+ = f(\bar{I}_\psi, \psi) \subset A$. In a similar way $\bar{A}_-$ is defined. Under the transformation $\xi \to \bar{\xi}$, $\bar{A}_+$ maps onto $\bar{A}_-$. Thus, E is also given by

$$E = \int_{\tau=0}^{t} \int_{A_+ \cup \bar{A}_+} [\Delta(\tau - \xi_j x_j) - \Delta(\tau + \overline{\xi_j x_j})] \, d\sigma(\xi) \, d\tau. \tag{75}$$

(This expression follows more directly from the Sommerfeld-Weyl representation of the fundamental solution.) The first term on the right-hand side generates the causal component of the fundamental solution, whereas the second term generates the anti-causal component. We will denote $A_+ \cup \bar{A}_+$ by A_+^*. In view of (18)-(19) we also have to consider partial derivatives with respect to space of E. In general they yield

$$\partial_i^\nu E = \int_{A_+^*} [(-\xi_i)^\nu \chi_{-1-\nu}(t - \xi_j x_j) - \overline{\xi}_i^\nu \chi_{-1-\nu}(t + \overline{\xi_j x_j})] \, d\sigma(\xi). \tag{76}$$

It is observed that the integrals of the type (76) have integrands which are strictly rational in the slowness variables (in this respect note that $\partial_{\xi_n} H$ is polynomial).

Cycles.

At this stage the relation between cycles on the complex slowness hypersurface and the modified Cagniard contours will be established. Introduce the hyperplanes

$$X_\tau: \; \xi_j x_j - \tau = \sum_{j=1}^{n-1} \xi_j x_j + f x_n - \tau = 0. \tag{77}$$

Later on, we will employ the notation

$$\xi_j x_j = \sum_{j=1}^{n-1} \xi_j x_j + f x_n = x_n F(\xi_1, \cdots, \xi_{n-1}, x_1, \cdots, x_{n-1}). \tag{78}$$

Now, following Petrovsky, introduce the n-plet of cycles $\partial\gamma = \partial\gamma_+ \cup \partial\gamma_-$ on A through

$$\partial\gamma_\pm(x, \tau) = A_\pm^* \cap X_\tau. \tag{79}$$

On $\partial\gamma$ the argument of the Δ-function, denoted as the *phase function*, is real $(-i0)$.

The integral representation for E can be written as

$$E = \int_{\tau=0}^{t} \int_{X_0} \left[\Delta\left(\tau - \sum_{j=1}^{n-1} \xi_j x_j - f x_n\right) - \Delta\left(\tau + \sum_{j=1}^{n-1} \overline{\xi_j} x_j + \overline{f} x_n\right)\right] d\sigma(\xi) d\tau. \tag{80}$$

Integration of the resulting Δ-function and restricting to the time interval $[0, \infty)$ leads to

$$\begin{aligned} E &= (1 + i\mathcal{H}) \int_{\tau=0}^{t} \int_{\partial\gamma_+(x,\tau)} \frac{d\nu(\xi)}{\partial_{\xi_n} H} \, d\tau \\ &= (1 + i\mathcal{H}) \int_{\gamma_+(x,t)} \frac{\omega(\xi)}{\partial_{\xi_n} H}, \end{aligned} \tag{81}$$

where ω denotes the Leray form, and $\gamma_+(x,t)$ denotes the tube of cycles $\partial\gamma_+(x,\tau)$ for the interval $\tau \in [0, t)$. In this representation the time behaviour ('tail') of E is hidden as a parameter in the cycles at a given direction of propagation.

In polar coordinates the cycle $\partial\gamma_+$ can be parametrized as follows:

$$\begin{aligned} \partial\gamma_+(r, \phi, x_n, \tau) = \{ & q_+(r, \eta(\phi, \psi), x_n, \tau), \psi, f(q_+(r, \eta(\phi, \psi), x_n, \tau), \psi) \mid \\ & \psi \in ([0, 2\pi), [0, \pi), \cdots, [0, \pi))\}, \end{aligned} \tag{82}$$

where $q_+(r, \eta(\phi, \psi), x_3, \tau)$ denotes the n-plet of modified Cagniard contours. The latter contours, with parameter τ, are obtained from the cycles through intersection with the (q, ξ_n)-plane (fixed ψ) followed by an orthogonal projection on the complex q-plane.

Without restriction, we can assume that the direction of observation coincides with the x_n-axis. This implies that everywhere $\xi_i x_i$ can be replaced by $\xi_n x_n$, whereas (77) can be replaced by

$$X_\tau : \; f x_n - \tau = 0, \tag{83}$$

where the new coordinates are denoted as the old ones (see Figure 2). But then the time axis coincides with the ξ_n-axis and the tube $\gamma_+(x, t)$ can be defined through an interval along the ξ_n-axis.

At this point, it is observed that the construction of the fundamental solution reduces to the construction of the cycles on A, followed by an integration over these cycles. For isotropic media it is simple to find these cycles; for anistropic media, however, the construction is far from trivial. In the further analysis, a basis for these cycles in the homology group of the associated hypersurface will be found and a differential equation for the integral representation over any basis element will be derived.

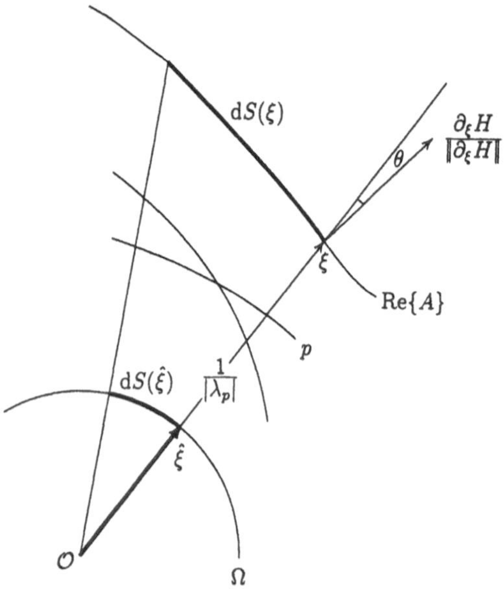

Fig. 2. The projection of the unit ξ-sphere onto the slownes surface

4 The Fundamental Solution Expressed as an Integral over Vanishing Cycles.

In this section we will show how the geometry of the slowness surface leads to a natural decomposition of the fundamental solution E into integrals over so-called vanishing cycles. As we shall see these cycles encode the topological information of the wave operator: they depend only on the topology of degenrate critical points on the slowness surface, which are defined by the polynomial equations describing the slowness surface around such points. To achieve this decomposition, we use the geometrical objects introduced in the two preceding sections. The decomposition of the fundamental solution into integrals over vanishing cycles, has in fact been foreseen already by Petrovsky [36], and used also in [43, 8].

Decomposition into vanishing cycles.

Let us begin with putting the result of the previous section (see also [8, 25, 36]) in a more general context. Expression (81) is of the generic form

$$E = \int_{\gamma} \frac{P(\xi_1, \cdots, \xi_n)}{\partial_{\xi_n} H(\xi_1, \cdots, \xi_n)} d\xi_1 \cdots d\xi_{n-1} \tag{84}$$

where P is a degree $k \leq \deg H - n - 1$ polynomial in ξ. The integral is over a complex $(n-1)$-dimensional cycle denoted by γ in the $(n-1)$-dimensional algebraic hypersurface $A : H(\xi_1, \cdots, \xi_n) = 0$, i.e., the cycle is a certain (connected) piece of the slowness surface, oriented such that the coordinate ξ_n corresponds to the 'vertical' direction. As was shown in the previous section the contour $\partial\gamma$ arises through intersection $A \cap X$. The integrand is in fact a closed rational form of degree $(n-1)$ with poles along $\partial_{\xi_n} H = 0$, hence the integral depends only on the homology class of γ. The quantity $\partial_{\xi_n} H$ corresponds to the group velocity in the n-th direction, i.e., the vertical group velocity. Since the cycle depends continuously on the coefficients of H, we may use these as deformation parameters of the cycle, without changing the value of the integral. The fact that the integral only depends on the homology class represented by γ has an important consequence which will play a central rôle in this section.

Define coordinates x_i on X_t such that its local equation is simply $\xi_n x_n = t$, where t has the interpretation of time. We are interested in the solutions of the following set of equations

$$H = 0, \quad \partial_{\xi_1, \cdots, \xi_{n-1}} H = 0, \tag{85}$$

where it is understood that the slowness hypersurface is oriented such that the group velocity is parallel with the ξ_n-axis. In general the roots of these equations will all be different, in which case there are $\mu = D(D-1)^{n-1}$ of them. Label them according to $\xi_n^{(k)}$, $k = 1, \cdots, \mu$. These points define singular points (for vertical propagation) on the slowness surface; we will study the topology of the slowness surface around those points in detail below.

The level surfaces $H(\xi_1, \cdots, \xi_{n-1}, \tau) = 0$ for generic values of τ, considered as a parameter, contain $(n-2)$-cycles $\Delta_k(\tau)$ which have the defining property that they 'vanish' when taken along a path u_k in the τ-plane (i.e. the target plane of the function H), connecting a generic point $\xi_n^{(0)}$ with the point $\xi_n^{(k)}$: upon approaching the point $\xi_n^{(k)}$, the cycle Δ_k shrinks to the point $\xi_n^{(k)}$. The system of paths $u_i, i = 1, \cdots, \mu$ is non-intersecting. The cycles $\Delta_i, i = 1, \cdots, \mu$ are referred to as *vanishing cycles*. (An explicit construction of these cycles will be given later.) The vanishing cycles define $(n-1)$-cycles δ_i on $H = 0$ by taking the union of all cycles $\Delta_k(\tau)$ as we move along u_k from $\xi_n^{(0)}$ to $\xi_n^{(k)}$. It is a standard fact on integrals of the type (84) that the cycles thus constructed are homologous to the cycle γ in the sense that there exists the following decomposition:

$$\gamma = \sum_{i=1}^{\mu} c_i \delta_i + C_{n-1}(\xi_n^{(0)}), \tag{86}$$

where the $(n-1)$-cycle C_{n-1} is homologous to zero on $H(\xi_1, \cdots, \xi_{n-1}, \xi_n^{(0)}) = 0$. This then results in the following decomposition of E over vanishing cycles:

$$\begin{aligned} E &= \int_\gamma \frac{P(\xi_1, \cdots, \xi_n)}{\partial_{\xi_n} H(\xi_1, \cdots, \xi_n)} d\xi_1 \cdots d\xi_{n-1} \\ &= \sum_{i=1}^{\mu} c_i \int_{\delta_i} \frac{P(\xi_1, \cdots, \xi_n)}{\partial_{\xi_n} H(\xi_1, \cdots, \xi_n)} d\xi_1 \cdots d\xi_{n-1} \\ &= \sum_{i=1}^{\mu} c_i \int_{u_i} d\tau \int_{\Delta_i(\tau)} \frac{P(\xi_1, \cdots, \xi_{n-1}, \tau)}{\partial_{\xi_n} H(\xi_1, \cdots, \xi_{n-1}, \tau)} d\xi_1 \cdots d\xi_{n-2}. \end{aligned} \tag{87}$$

This is just a rewriting of the formulae at the end of the previous section. The numbers c_i, describing the decomposition of the original integral, are referred to as intersection numbers and are of topological nature. The last integral in (87) is over the vanishing cycles; its detailed properties will be discussed in the next section. In the rest of this section we will apply elements of singularity theory to make the above decomposition explicit, that is we will show how to compute the intersection numbers c_i.

In order to do so, we will make more precise the duality between the slowness surface and the wave front. This will lead naturally to the study of singularities of specific families of hypersurfaces, which are deformations of the original slowness surface. The singularities on the slowness surface correspond to critical points of second (i.e., non-degenerate) or higher order (degenerate). Such points correspond to certain singularities on the wave front. We will show, following [4, 5, 7] that such singularities can be conveniently studied in terms of critical points in the Legendre transformation relating the slowness and wave front set (including the diffraction surface and wave front surface). As we shall discuss, this also will lead to a complete classification of wave front singularities that can arise in anisotropic elastic media.

A singularity on the wave front is the singularity of a Legendre transformation. The duality between the projective wave front Σ_d and $\mathrm{Re}\{A\}$ is described by a Legendre transformation and has extensively been discussed in [7]. The characteristic feature of a Legendre transformation is that it transforms functions on a vector space to functions on the dual of this vector space. To explain the concept, consider the following example. Let $y = f(\xi), \xi \in \mathbb{R}$ be a smooth function, with $f''(\xi) > 0$. A Legendre transformation of f is a new function g of a new variable x constructed in the following way. Let x be a given number and consider the line $y = x\xi$. Take the point $\xi(x)$ on the ξ-axis such that the vertical distance between the line and f is maximal. That is, for each x the function $F(\xi, x) \equiv \xi x + f(\xi)$ has a maximum at $\xi = \xi(x)$, following from $\partial_\xi F = 0$, i.e., from $f'(\xi) = x$. The function g is now defined as $g(x) = F(\xi(x), x)$. Note that since the second derivative of f does not change sign, the point $\xi(x)$ is unique. As an example take the cubic function $f(\xi) = -\frac{1}{3}\xi^3$. Then $F(\xi, x) = \xi x - \xi^3/3$, $\xi(x)^2 = x$, and hence $g(x) = 2x^{3/2}/3$.

An important property of a Legendre transformation is that its square is the identity, in other words, one can view it as a projection, which point of view will be adopted below. The extension to higher dimensions is straightforward: let $f = f(\xi_1, \cdots, \xi_{n-1})$ such that the Hessian $\det(\partial_i \partial_j f) > 0$, $i, j \in \{1, \cdots, n-1\}$, i.e., the function f is strictly convex. Then the Legendre transformation of f is a function

$$g(x) = F(\xi(x), x), \tag{88}$$

where $x(\xi)$ is the solution of $\partial_{\xi_1, \cdots, \xi_{n-1}} F = 0$ and $F = x_j \xi_j + f(\xi)$. The functions f and g are each other Legendre dual. An all too well known example of this duality is the duality between the Hamiltonian and Lagrangian functions of classical mechanics. The generic situation is summarized as follows [7]: for the function $F(\xi_i, x_j)$ of $2(n-1)$ variables, the formulae

$$\begin{aligned} \xi_i = \frac{\partial F}{\partial x_i}, \qquad x_j = -\frac{\partial F}{\partial \xi_j}, \\ \tau = f - \xi_i \frac{\partial F}{\partial \xi_i} = F \end{aligned} \tag{89}$$

define a Legendre variety in the 'big' $(2\,(n-1)+1)$-dimensional space coordinated by (x, ξ, τ), through the condition

$$\exists_{\xi(x)} : \partial_\xi F = 0, \; F = \tau.$$

The projection $(\xi, x, \tau) \to (x, \tau)$ is a Legendre transformation.

Apply this to the slowness surface defined by the equation $H(\xi_1, \cdots, \xi_n) = 0$. This surface has already been parametrized by $\xi_n = f(\xi_1, \cdots, \xi_{n-1})$. We defined the function

$$F(\xi, x) = \sum_{i=1}^{n-1} x_i \xi_i + x_n f(\xi_1, \cdots, \xi_{n-1}). \tag{90}$$

Upon dividing by $x_n > 0$, we may redefine the function F as

$$F(\xi, x) = f(\xi_1, \cdots, \xi_{n-1}) + \sum_{i=1}^{n-1} \frac{x_i \xi_j}{x_n}. \tag{91}$$

Henceforth, we will work exclusively with projective coordinates x_i/x_n, $i = 1,\cdots,n-1$, $x_n \neq 0$, denoted by x_i as well. Thus

$$\Sigma_d = \left\{(x_1,\cdots,x_n)\,|\,\exists\xi(x):\ \partial_\xi F = 0,\ F = \frac{\tau}{x_n}\right\} \tag{92}$$

is the Legendre dual of the slowness surface parametrized by f. Note that the function F is the Hamiltonian for the 'ray-system' and *not* the Hamiltonian for the full wave equation. That is, F is the linearized form of the full Hamiltonian, defined in terms of classical fields.

It is now evident what the origin of singularities on the wave front is: they are precisely the Legendre transformation of the critical points $\partial_\xi F = 0$ in the 'big' $2(n-1)+1$ space defined above. At these points the function f and its dual g fail to be convex, i.e., at these points at least one of the principal Gaussian curvatures on the slowness surface vanishes. Quite generally, the singular points on wave surfaces correspond to Legendre transformations of the critical points in ξ-space of the function F. The function F is called the generating function of the Legendre transformation. It can be considered as defining a family of slowness surfaces defined locally by f and 'parametrized' by $x_1,\cdots,x_{n-1}$. This is a useful interpretation in the context of singularities. In fact, the family F defines a *deformation* of the isolated singularities of the function f, such that $F(\xi,0) = f(\xi)$ parametrizes the original singularity defined by a critical point of f.

Singularity theory.

Smooth functions, like f, that define isolated singularities have been thoroughly studied, for example in [2, 7]. For us, the most important class of functions is the one formed by functions that define isolated, stable singularities, i.e., the ones that are invariant under a local reparametrization of the surfaces. In [7] a complete classification is given of all types of singularities that can occur in Legendre transformations of arbitrary slowness surfaces in low dimensions. In particular, it follows that possible singularities of fronts in three spatial dimensions fall into three classes. Correspondingly the critical points on the slowness surface define either a local extremum, an inflection point (i.e., vanishing Gaussian curvature) or a point at which a derivative of the curvature vanishes. Around such points on the slowness surface there exist *local* (curve-linear) coordinates $s(\xi,x), y(x)$, found upon applying a canonical transformation of the original ('horizontal') coordinates ξ, x, such that the function F at such points satisfies $F = \partial_s F = 0$, and close to the singularity is given by an expression of Table 1. Note that in projective space coordinates one of the y coordinates is set to 1. In general, the evaluation of the latter canonical transformation involves numerical computations. We postpone those computations to a future paper. The functions s have the physical interpretation of 'horizontal' slownesses, while the functions y_i are space(-time) coordinates.

These singularities define a point of transversal self intersection in the first case; in the second case a *cuspidal edge* in the (y_1, y_2)-plane, dividing the plane into regions of either three real roots of $F = 0$ or a real one and two complex conjugated ones. This singularity corresponds to a vanishing of Gaussian curvature

Table 1. Singularities on Wave fronts in 3-dimensional space.

Type	Normal form	μ
A_1	$F = s^2$	1
A_2	$F = \frac{1}{3}s^3 + y_1 s + y_2$	2
A_3	$F = \frac{1}{4}s^4 + y_1 s^3 + y_2 s + y_3$	3

on the slowness surface. In the third case the wave surface singularity is known as a *swallowtail* in (y_1, y_2, y_3)-space. On the slowness surface this corresponds to a point where the derivative of a Gaussian curvature vanishes. See Figure 3 for the possibile wave front singularities.

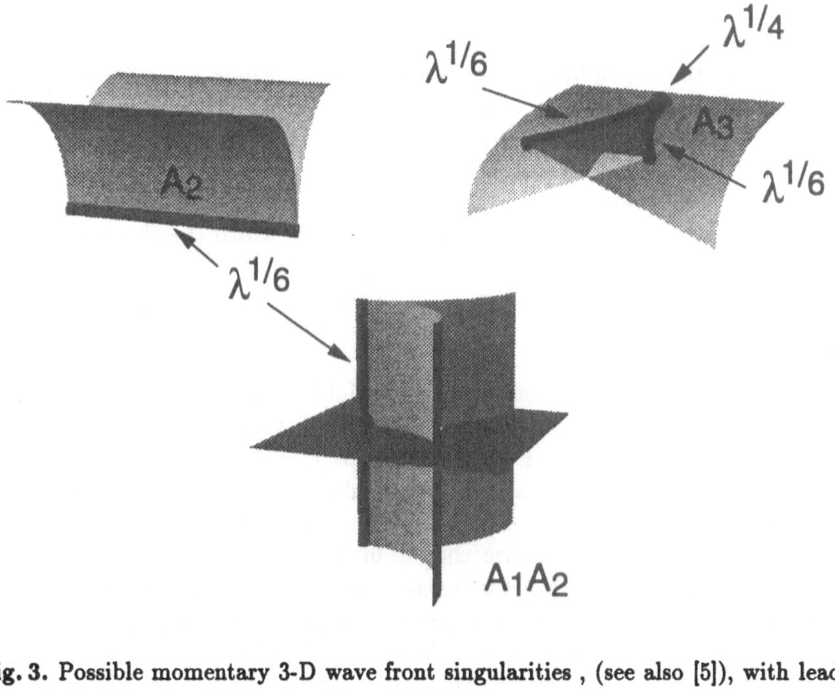

Fig. 3. Possible momentary 3-D wave front singularities , (see also [5]), with leading asymptotic behavior, (determined in section 5).

We will give a simple heuristic argument why this is indeed a complete classification. Possible points of inflection on the real slowness surface can be detected by considering the condition that a real valued *bitangent* exists. A bitangent line is a line tangent to a slowness sheet at *two* points. Obviously, this can only happen when the slowness sheet is locally not convex. For every bitangent line there will be automatically two inflection points. Note that the bitangent must be a real line as follows from the hyperbolicity condition. A limiting case thus occurs when the two inflection points coincide, as then the bitangent will become complex. This situation occurs at a point $(\zeta_1, \zeta_2, \zeta_3)$ for which up to the third derivative with respect to ξ vanish, implying that F is a fourth order polynomial proportional to $(\xi - \zeta)^4$.

The presence of cups and cuspidal edges is quite common in practical situations. For example, they occur in a generic 3-dimensional finely layered medium, consisting of different isotropic elastic material. The system of layers has a resulting hexagonal symmetry, but is not isotropic. One of the (quasi-)shear waves may exhibit two or four cusps. It is important to realize that the above classification can be recast in a classification of relations among the components of the stiffness tensor c_{ijkl} for which the above singularities occur.

From now on, we will assume that the function F is brought into normal form, and we will denote the normal coordinates s by ξ again. The labelling of the three cases is according to the types of reflection groups that act on the critical points. They encode the topological structure of the singularity as we shall now discuss.

In general, the multiplicity of the isolated critical point of $f(\xi)$ gets resolved into μ different critical points of the family $F(\xi, x)$, for a generic choice of parameters of x. In fact one can show that the set of all parameters x for which the F has μ distinct critical values is dense, i.e., F can be seen as a *non-singular Morsification of the function* f. This fact allows us to introduce $\mu \leq n-1$ parameters λ in place of the $n-1$ parameters x for which the the same is true:

$$F(\xi, \lambda) \equiv f(\xi) + \sum_{i=1}^{\mu} \lambda_i g_i(\xi). \tag{93}$$

The functions $g_i(\xi)$ are polynomials in all variables $\xi_1, \cdots, \xi_{n-1}$ forming a basis in the μ-dimensional vector space of polynomials in ξ modulo those polynomials generated by the the first partial derivatives of f. The fact that this is a finite μ-dimensional vector space is a standard result. This vector space, spanned by $g_i(\xi)$, is sometimes referred to as *the local ring of* f. The important point that we stress here, is that for generic choices of λ the critical points of F are all distinct and the set of those λ's is dense as well. In other words, in a suitably small neighborhood of the critical point of f every analytic function near f can be obtained by analytic changes of variables in $F(\xi, \lambda)$ for some value of λ. The form (93) of F is called a *versal* deformation of F.

The question now arises as to whether one requires a continuum or discrete set of values of λ's to unravel the singular points in a judiciously chosen neighborhood of f. Obviously, this depends on the nature of the singularity of f. Without discussing this (see e.g. [6] for a complete discussion) we state the result here: for those singularities appearing in the list above, there is always only a finite set of values of λ required. (The singularities in that list are of modality zero.)

The equation $F(\xi, \lambda) = 0$ defines for generic values of λ the nonsingular level surfaces of f. Introduce the set S of all λ_i, $i = 1, \cdots \mu$, and consider for each λ

the level surface

$$V_\lambda \equiv \{F(\xi,\lambda) = 0, |\xi| \leq \rho\}, \tag{94}$$

i.e., we intersect $F = 0$ with a small $(n-2)$-ball B of radius $\rho > 0$ centered at the critical point of f. (This is the region of validity of the normal form representing F.) The surfaces V_λ are for generic values of λ nonsingular and for those values they are all diffeomorphic to the non-singular level surfaces of f. The set of λ's for which V_λ is singular forms a set of codimension 1 in S. This set has an obvious geometrical interpretation: it defines a singularity on a wave front, now parametrized by λ instead of x. For example, for A_2 we find that the singular set in S (corresponding to the cusp) is constrained by the equation

$$\lambda_1^3 + \frac{27}{4}\lambda_2^3 = 0. \tag{95}$$

Loosely speaking, the projection $\pi : V \longrightarrow S$, of which the fibers are V_λ, replaces the rôle of the Legendre transformation. It should be realized that trading the parameters x_i for the parameters λ_i effectively reduces the dimensionality: the dimension of the space S of relevant parameters is usually *less* (at most equal) than the dimension of the original physical space. This 'collapse' of dimensions is typical for a singularity defined by a degenerate critical point and occurs in various physical problems (for example, in phase transitions).

Time evolution of fronts in three dimensions.

Having discussed the possible topologies of the wave fronts in a 3-dimensional (homogeneous) elastic medium, we now turn to the problem on how the fronts evolve in time. Quite generally, a moving front may change its shape in time. To study the possible time evolution of moving fronts, we consider the union of all momentary fronts classified in Table 1. This union defines a hypersurface in 4-dimensional space-time. In fact, it follows rather easily that this hypersurface is itself a front of a Legendre transformation acting on a Legendre variety of one dimension higher than the Legendre variety of the momentary front. One considers the time as an additional parameter; the Legendre transformation in the higher dimensional space projects on space-time parametrized by (x,t).

The singularities on the momentary fronts sweep out a subvariety (a lower dimensional surface) in the hypersurface swept out by the moving front. These singular surfaces are called *caustics*. In principle, caustics can be classified using similar techniques as used in the classification of singularities in the Legendre transformations defining the momentary fronts. In case the dimension of the initial space is three, the possible singularities in addition to the ones (intersections) given in Table 1 are given in Table 2 (see [5]).

The A_4 type singularity in 4-dimensional space-time corresponds again to a swallow tail, which we will discuss in detail below. The D_4-type singularities are qualitatively different from the A_k-type singularities in that they are *2*-dimensional rather than 1-dimensional (which is the case for the A_k-type singularities). They are usually referred to as the 'umbilical' singularities [15]. In Figure 4 we have depicted the typical singularities of the caustics in space-time swept by the singularities of momentary wave fronts in 3-dimensional space.

Table 2. Caustics swept in space-time by momentary fronts in 3-dimensional space (y_0 represents time).

Type	Normal form
A_4	$F = s^5 + y_3 s^3 + y_2 s^2 + y_1 s + y_0$
$D_4^{\pm}$	$F = s_2^2 s_1 \pm s_1^3 + y_2 s_1^2 + y_1 s_1 + y_0 s_2$

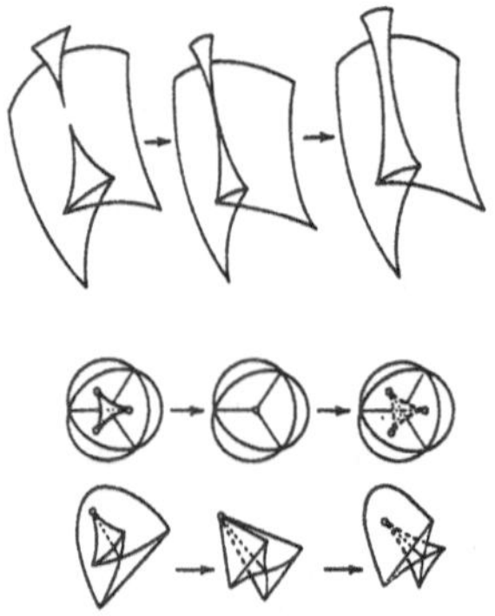

Fig. 4. Evolution of an A_4 and $D_{\pm}$ wave front singularity in 4-D space-time, (see also [5])

The swallowtail singularity A_4 in 4-dimensional space-time is most useful to illustrate the time evolution, as it is related to a common physical phenomenon. Consider a momentary wave front in 3-dimensional space and assume that it has cusp singularities. The cuspidal edges sweep out the caustic. Assume furthermore, that the caustic has a 3-dimensional swallowtail singularity. The caustic is locally diffeomorphic to the polynomial of type A_4 having a multiple root. The cuspidal edges sweep out a 2-dimensional surface on this hypersurface corresponding to a variety diffeomorphic to the A_4 polynomial having roots of multiplicity of at least 3. This variety is called an *open* swallowtail in 4-dimensional space-time. An important result by Arnold [3] states that the time evolution of any cuspidal edge can always be put upon a canonical transformation into the normal form of an open swallowtail, i.e., into a variety given by the zero locus of the A_4 polynomial with at least coinciding roots. This greatly facilitates the way the projection on to the physical 3-dimensional space has to be made: it corresponds to differentiation with respect to ξ. Note that differentiation lowers by one both the degree and the multiplicity of the roots. Thus the problem separates into two parts: first one studies the open swallowtail on the full hypersurface defined by the moving front, and secondly one projects onto the space of degree four polynomials having at least roots of multiplicity two by differentiation, i.e., on the swallowtail in three space dimensions.

As time proceeds, the edges disappear into a singularity of type A_4 in space-time, that, when projected onto 3-dimensional space, corresponds to an A_3-type singularity. This type of evolution is easily identified in a physical situation. Namely, suppose a cusp is generated (in a hexagonal medium, say) and moves into an isotropic medium. Then the cusp will eventually evolve into a singular point on the spherical wave front after which it disappears entirely. The singular point is necessarily of type A_3 in 3-dimensional space. It is interesting that the result in [3] implies that in general the cuspidal edge of a moving front passes the line of self intersection of the swallowtail in 3-dimensional space at two *different* times. Then it follows that if the time difference between the arrivals of the two edges is T, the time towards the vertex of the swallowtail, i.e., the point after which the cusp has disappeared, is $T^{2/5}$.

The decomposition of the cycle $\partial\gamma$.

We next describe the cycles in (87) in terms of the level surfaces $F(\xi, \lambda) = 0$ of $f(\xi)$. First, recall that at the end of the previous section we introduced the cycles as the intersection of $A \cap X_\tau$. From (77) we draw the conclusion that $A \cap X_0$ corresponds with $F(\xi, x) = 0$. To construct the contours $\partial\gamma$ near the arrivals, we intersect the level surface with a small ball of radius ρ centered at a critical point (which we have taken to be the origin of the affine coordinates ξ). Thus, we are looking 'closely' at the front, i.e., mathematically we consider the high frequency limit. Close to the critical point the function F can be brought into its normal form, i.e., one of the three representations given in Table 1. The (partial) contours $\partial\gamma$ are thus contained in the set

$$A \cap X_0 \cap B \sim \{F(\xi, x) = 0\} \cap B. \tag{96}$$

We set $X = X_0$. Upon constraining the full set of parameters x to the 'relevant' parameters λ, we conclude that the cycles $C_{n-2}(\xi_n) \in B$ are precisely cycles on the surfaces $F(\xi, \lambda) = 0$, which we denoted earlier as V_λ. This space is much more convenient, since we can now invoke results by Milnor on properties of the homology group formed by these cycles.

According to Milnor [29] the space of cycles C_{n-2} is of finite dimension μ, and any $(n-2)$-cycle can be expanded into a suitably defined basis of cycles, such that the coefficients of this expansion are all *integers*. More precisely, the μ-dimensional homology of the fibers V_λ for *generic* values of the parameters λ form a μ-dimensional lattice:

$$H_{n-2}(A \cap X \cap B) = H_{n-2}(V_\lambda) = \mathbb{Z}^\mu. \tag{97}$$

A basis in this homology is given by a system of μ cycles Δ_i which vanish along suitable chosen paths connecting a given non-critical value of F with a critical one. Milnor's theorem thus states that any cycle $\partial\gamma$ on a level surface of f can be decomposed into certain basis cycles drawn on the level surfaces defined by $F = 0$, such that the coefficients are all integers. In other words: there exists a basis of cycles such that the intersection numbers c_i in (87) are all integers. Such a basis is referred to as a basis of vanishing cycles.

We are now in a position to actually construct such systems of paths and the associated cycles. First we divide the critical values of $F(\xi, \lambda)$ into two sets: the real critical values, and the ones that come in complex conjugated pairs. (Recall that since we require that the function F is real on the real axis, and analytically continued in the complex plane according to Schwarz's reflection principle there are no arbitrary complex critical points.) In order to make contact with the original integral E it is obvious that we have to decide on a particular orientation of those cycles that correspond to the real critical values: they have to be in agreement with the chosen orientation of the slowness sheets. Assume that the origin is a non-critical value of F. Let U be that portion of the complex plane that contains all the critical values. Now connect the origin with any real critical point with a path in U such that it always consists of points whose imaginary parts are in absolute values less than the absolute values of the imaginary parts of any other complex critical point. Further, paths joining the origin with critical points that come in complex conjugated pairs, are conjugated and always contained in either the positive half plane or negative half plane according to the imaginary part of the critical value. We furthermore require that they intersect the real axis transversally (to be in agreement with Schwarz's reflection principle).

Let us denote a system of critical paths in the τ-plane obeying these conditions by $u_i(\lambda)$, $k = 1, \cdots, \mu$, with λ parametrizing the path. Next we apply a fundamental result in Morse theory, which ensures that close to a critical value Z_i of F, F has the expansion (considering the λ's as parameters)

$$F = Z_i + \sum_{j=1}^{k} z_j^2 - \sum_{j=k+1}^{n-1} z_j^2, \tag{98}$$

where the 'horizontal slowness' coordinates $z_j(\xi, \lambda)$ are *real*. The index k is the number of positive eigenvalues of the Hessian of F. (This is a coordinate independent quantity describing the number of directions at the critical point for which F is increasing.) Now introduce the vanishing sphere associated with the critical point Z_i as

$$S_i(\lambda) = \sqrt{u_i(\lambda) - Z_i}\, S^{n-2}, \tag{99}$$

with S^{n-2} the unit sphere in $n-2$ coordinates of which the tangent space is locally spanned by the vectors

$$(\partial/\partial z_1, \cdots, \partial/\partial z_k, i\partial/\partial z_{k+1}, \cdots, i\partial/\partial z_{n-1}).$$

The orientation of this sphere and hence of S_i is such that its volume form is positive with respect to the volume form orienting the (real) slowness surface. The homology class in $H_{n-2}(V_\lambda)$ represented by S_i is the vanishing cycle corresponding to the path u_i and is denoted by Δ_i. The system $\{\Delta_i\}_{i=1}^{\mu}$ forms a basis in this homology group. It is with respect to this basis that we will compute the numbers c_i in (87).

There is a lot of freedom in choosing a basis of vanishing cycles, which is refelcted in the invariance of this basis under automorphic transformations of the homology group H_{n-2}, i.e. the isometries of the Milnor lattice.These automorphisms are referred to as *monodromy* transformations. Rather than discussing this in detail we will simply list the results necessary for the computation of the intersection numbers c_i. It turns out that a monodromy transfromation can be described analytically by studying the properties of the vanishing cycles under the following transformations. Consider the combined transformation of going along a path u_i starting at a given non-critical value, going around anti-clockwise a given critical point Z_i and back along u_i. Under such transformations the *roots* of the equation $F = 0$ are permuted, but a vanishing cycle, i.e., a contour on a non-singular level surface which vanishes upon taking the parameters to zero, remains invariant. This implies the possibility of writing down a canonical representation of the vanishing cycles in terms of a suitably chosen parametrization of the level surface. This is reflected in the so-called *Picard-Lefschetz formula* , which gives an canonical form of the inner product on the vanishing cycles. Applying this formula yields the following representation of the monodromy operator h_i corresponding to a transformation encircling the i-th critical point:

$$h_i(\alpha) = \alpha + (-1)^{n(n-1)/2}(\alpha, \Delta_i)\Delta_i, \tag{100}$$

where α is an arbitrary cycle of dimension $n-2$. It is thus seen that the monodromy group acts as a (pseudo) reflection, $h_i(\Delta_i) = -\Delta_i$, and it is the identity on the hyperplane orthogonal to Δ_i. In general, the intersection product is hard to compute explicitly, and only for a few classes of singularities there exist explicit results. However, for simple singularities, among which are those of Table 1, it can be computed easily. In fact, for singularities defined by

$$f(\xi) = \xi^{k+1}, \tag{101}$$

it can be shown that one can bring a basis of vanishing cycles into the following form

$$(\Delta_i, \Delta_{i+1}) = -1, \qquad (\Delta_i, \Delta_j) = 0 \quad \text{for } |i-j| \geq 2. \tag{102}$$

Furthermore, we can construct the composition of all opertars h_i, denoted by h_*, which thus describes a loop based at a non-critical value, encircling all critical values at once. It turns out that such an operator in case of the singularity of A_k, has eigenvalues given by the roots of unity $\exp(2\pi ij/k)$, $j = 1, 2, \cdots k$. The operator h_* will be used in the next section.

For the simple singularities of type A_k the resulting reflection groups are just the permutation groups on respectively two or three elements, the Weyl groups of the Lie groups A_1,A_2 and A_3. For us the most important things are that the intersection indices appearing in (87) are first of all integers and, secondly, depend *only* on the monodromy group, i.e., on the topology of the singularity only.

Let us now finally compute explicitly these indices for real critical values, using a result by Vasiliev in [43]. In fact using the description of the vanishing cycles given above it follows that for three spatial dimensions the intersection of the contour $\partial\gamma$ with the vanishing cycle Δ_i corresponding to the i-th real critical point is given by

$$\begin{aligned} Z_i > 0 \; c_i &\equiv (\Delta_i, \partial\gamma) = -(1 + (-1)^{3-k}) - \textstyle\sum_j (\Delta_i, \Delta_j) \\ Z_i < 0 \; c_i &\equiv (\Delta_i, \partial\gamma) = -(1 + (-1)^{k}) + \textstyle\sum_j (\Delta_i, \Delta_j). \end{aligned} \tag{103}$$

set Σ_d at most at isolated points, such that locally $\lambda = 0$ at such a point. The integral over a vanishing cycle of f has the form

$$I = \int_{\Delta_i(\lambda)} \frac{P(\xi)d\xi}{\partial_{\xi_n} H(\xi, \lambda)} \tag{106}$$

associated with the $i-th$ critical value, and λ denoting the local coordinate. Upon varying λ the cycles will become dependent on λ, but as the integral is over a rational closed $(n-2)$ form they define locally constant classes in the homology $H_{n-2}(V_\lambda)$. Of course this is not globally so, as the cycles transform nontrivially under the monodromy group of the singularity. The local constancy of the classes in fact leads to the conclusion that the integrals over vanishing cycles satisfy a Picard-Fuchs differential equation in λ. We will now show how to compute these equations explicitly.

The integrand has poles precisely at the μ critical points of H. The differential equation which we are about to derive is actually a calculation in the co-homology (i.e., the Poincare dual of $H_{n-2}(V_\lambda)$). It is well known that the cohomology of projective hypersurfaces can be calculated from its rational differential forms having poles at infinity. To do the actual computation a bound on the order of the poles is required. Such bounds can be determined using the work of Griffiths [18].

The order of the pole of the rational integrand in (106), which *a priori* has a pole of 'arbitrary' order, can be put in some canonical form in which it has a pole of order one. This reduction-of-pole property is extensively discussed in [18]. Here, we will only briefly mention the idea, which is as follows. The middle cohomology of the complex hypersurface V_λ in $\mathbf{CP}^{n-2}$ is described by rational differentials in $\mathbf{CP}^{n-1}$ having poles of arbitrary order along V_λ. Each form $Pd\xi/\partial_{\xi_n}H$ defines a cohomology class by considering its residue on V_λ as follows. The $(n-2)$-cycle Δ_i on V_λ can be made into an $(n-1)$-cycle δ_i by considering the tube over Δ_i, i.e., δ_i is a small cylinder erected along the normal over Δ_i in $\mathbf{CP}^{n-1}$. Then we can define the *residue* as

$$\int_{\Delta_i} \mathrm{Res}_{V_\lambda} \left(\frac{Pd\xi}{\partial_{\xi_n} H} \right) = \frac{1}{2\pi i} \int_{\delta_i} \left(\frac{Pd\xi}{\partial_{\xi_n} H} \right). \tag{107}$$

We suppress the index i attached to the vanishing cycle from now.

Recall that the value of the integral on the r.h.s. does not change if we add to the integrand an exact differential form having poles on the hypersurface $H = \partial_{\xi_{1,\cdots,n-1}} H = 0$. This can be used to lower the order of the pole of the integrand (106) on V_λ as follows. Let l be the sum of the weights of the quasi-homogeneous polynomial defined by the principal symbol, i.e., $H = 0$ and let its degree be D. Consider quasi-homogeneous polynomials A_j, $j = 1, \cdots, n-1$, and the degree of A_j is $D + l_j - l$, with l_j the weight of the ξ_j in A_j. Define the $(n-3)$-form

$$\phi = \frac{1}{\partial_{\xi_n} H} \sum_{i<j} (l_i \xi_i A_j - l_j \xi_j A_i) d\xi_1 \wedge \cdots \wedge \widehat{d\xi_i} \wedge \cdots \wedge \widehat{d\xi_j} \wedge \cdots \wedge d\xi_{n-1}. \tag{108}$$

For complex conjugated critical values $Z_{\bar{i}} = \bar{Z}_i$ one has

$$(\Delta_i, \partial\gamma) = (\Delta_{\bar{i}}, \partial\gamma) = (\Delta_i, \Delta_{\bar{i}}). \tag{104}$$

The sum over j is over those vanishing cycles associated with real critical values that are in between the origin and Z_i. The original contour $\partial\gamma$ thus has the following decomposition in terms of vanishing cycles

$$\partial\gamma = \sum_{i=1}^{\mu} c_i \, \Delta_i, \tag{105}$$

with the intersection numbers c_i given in (103)-(104). This gives the decomposition of the original contour. As a side remark, note that in case all the intersection indices c_i vanish, the solution vanishes identically. In this case one speaks of a *lacuna*. In [8, 43] it is shown that the condition that all the c_i vanish (the so-called local Petrovsky condition) implies 'sharpness' of the wave-front in which case one speaks of a (local) lacuna. Note that since $\partial\gamma$ still depends on a coordinate x on the front, this condition is local, in particular, it depends from what side one approaches the front. We will briefly come back to this in the next section.

5 A Differential Equation for the Fundamental Solution

The main point of the previous section was that the geometry of the slowness surface, in particular its dependence on singular points, is conveniently studied in the homology of the level surfaces $H(\xi_1, \cdots, \xi_{n-1}, \lambda) = 0$ parametrized by $F(\xi, \lambda) = 0$, with λ playing the rôle of a parameter through the vertical slowness ξ_n. This has lead to an explicit decomposition of the fundamental solution into integrals over vanishing cycles.

In this section we will study the integral over a vanishing cycle in detail. We will show that it satisfies a linear differential equation in λ, of which the form only depends on the topology of the singularity through its monodromy group. This, among other things determines the high frequency asymptotics of the solution, which we will find for the three singularities A_1, A_2, A_3. The differential equation is well known among mathematicians, where it is called a *Picard-Fuchs* equation [18, 24]. The implications for the asymptotic behavior of the integrals in the high frequency limit are similar as obtained from steepest decent methods in Maslov theory on oscillatory integrals, where F plays the role of the phase function [15]. However, exploiting the geometrical content of the Picard-Fuchs equation, the topological origin of the asymptotics is immediately obvious.

Let $f(\xi_1, \cdots, \xi_{n-1})$ be the vertical slowness having an isolated singularity at the origin of criticality μ. Let $F(\xi, \lambda)$ be a versal deformation, with the property $F(\xi, 0) = f(\xi)$. Consider also the Milnor fibration $\pi : V \to S$ whose fibers V_λ over the space S of versal deformation parameters are the zero level surfaces $F(\xi, \lambda) = 0$ (which are diffeomorphic to the non-singular level surfaces of $f(\xi)$). Now, let $\lambda \in S$ be a local coordinate on a curve in S intersecting the singular

Its differential is defined as

$$d\phi = \frac{\left(\sum A_j \frac{\partial(\partial_{\xi_n} H)}{\partial \xi_j} - (\partial_{\xi_n} H) \sum \frac{\partial A_j}{\partial \xi_j}\right) d\xi}{(\partial_{\xi_n} H)^2} = \frac{\sum A_j \frac{\partial(\partial_{\xi_n} H)}{\partial \xi_j} d\xi}{(\partial_{\xi_n} H)^2} - \frac{\sum \frac{\partial A_j}{\partial \xi_j} d\xi}{(\partial_{\xi_n} H)}. \tag{109}$$

This shows that any form of which the numerator can be written as a linear combination of partial derivatives $\partial(\partial_{\xi_n} H)(\xi)$, is equivalent up to (rational) exact forms to a rational form with a smaller order pole. By this reduction technique, one can put the integrand in a canonical form, more precisely, one associates a specific cohomology class to such integrands.

Let us, before we proceed, illustrate this method by way of a simple example, where the integral (107) is simply a contour integral in the complex plane. The residue in (107) is the familiar Cauchy residue. Now assume that the integrand has a pole of order two at $\xi = 0$, say. Then it has a Laurent series expansion of the form $(b/\xi^2 + d/\xi + \cdots)d\xi$, b, d constants. Next let $\phi = b/\xi$. Then

$$Pd\xi/\partial_{\xi_n} H(\xi) + d\phi \simeq (d/\xi + \cdots)d\xi$$

has only a first order pole. Proceeding in this way, we can reduce the pole of the integrand to order one at any point on the hypersurface $H = \partial_{\xi_n} H = 0$. This brings the rational form in a canonical form.

It is obvious that in general the integral over a vanishing cycle as a function of λ will be multivalued due to the nontrivial monodromy. In particular, the μ-dimensional vector

$$I(\lambda) = \left(\int_{\Delta_1} \frac{Pd\xi}{\partial_{\xi_n} H}, \cdots, \int_{\Delta_\mu} \frac{Pd\xi}{\partial_{\xi_n} H}\right), \tag{110}$$

will be multivalued in λ. We will now show that this vector satisfies a differential equation in λ with ***unique*** holomorphic coefficient functions $p_i(\lambda)$, whose solutions are linear combinations of the integrals (110). For simplicity of notation, we consider here just one parameter λ, the multi prameter case is discussed in [41].

Recall that the cohomology class defined by the cycle α does not change upon varying λ locally, so we can differentiate with respect to λ under the integral:

$$\frac{d^l}{d\lambda^l} \int_\Delta \mathrm{Res}_{V_\lambda} \left(\frac{Pd\xi}{\partial_{\xi_n} H}\right) = \frac{1}{2\pi i} \int_\gamma \frac{d^l}{d\lambda^l} \left(\frac{Pd\xi}{\partial_{\xi_n} H}\right). \tag{111}$$

Now construct the vector $I^j(\lambda)$ by taking the j-th derivative w.r.t. λ, i.e.,

$$I^j(\lambda) \equiv \left(\frac{d^j}{d\lambda^j} \int_{\Delta_1} \frac{Pd\xi}{\partial_{\xi_n} H}, \cdots, \frac{d^j}{d\lambda^j} \int_{\Delta_\mu} \frac{Pd\xi}{\partial_{\xi_n} H}\right) \tag{112}$$

The vector spaces W_i formed by all vectors I^j, $j \leq i$, must have constant dimension as function of λ, as the integrands are all closed $(n-2)$-forms in λ. Also, the dimension $d_i = \dim W_i$ of these spaces cannot exceed μ. Thus there will be a smallest number of derivatives s, such that

$$I^s = -\sum_{j=0}^{s-1} p_j(\lambda) I^j. \tag{113}$$

That is, the vector I_i say, satisfies a differential equation of the type

$$\left(\frac{d^s}{d\lambda^s} + \sum_{i=0}^{s-1} p_i(\lambda)\frac{d^i}{d\lambda^i}\right)\left(\int_\Delta \frac{P d\xi}{\partial_{\xi_n} H}\right) = 0 \tag{114}$$

Such an equation is referred to as a Picard-Fuchs equation [6]. The unknown coefficient functions are determined by taking successive derivatives with respect to λ and repetitive use of the reduction technique discussed earlier. An important property of Picard-Fuchs equations is that they can have at worst regular-singular points, at which the coefficient functions may develop poles due to the monodromy around such points. By multiplying the the Picard-Fuchs operator

$$\frac{d^s}{d\lambda^s} + \sum_{i=0}^{s-1} p_i(\lambda)\frac{d^i}{d\lambda^i}$$

by λ^s, the equation takes the form

$$\left(\left(\lambda\frac{d}{d\lambda}\right)^s + \sum_{i=0}^{s-1} q_i(\lambda)\left(\lambda\frac{d}{d\lambda}\right)^i\right)\left(\int_\Delta \frac{P d\xi}{\partial_{\xi_n} H}\right) = 0 \tag{115}$$

whose coefficient functions are now holomorphic also at the regular singular points. (Upon changing λ the integral plus all its derivatives changes according to the monodromy operator, all in the same way.) We will see shortly that the order s of the equation is fixed by the monodromy group. (A more extensive analysis, valid for more general types of singularities, is given e.g. in [26].)

As a simple example consider $f = \xi^2$ (type A_1). Since $\mu = 1$, we only have one point representing the zeroth vanishing homology class. The differential equation in this case is thus of first order:

$$\frac{dI}{d\lambda} + \frac{1}{\lambda}\left(\frac{n}{2} - \frac{3}{2}\right)I = 0, \tag{116}$$

that is $I = \text{const.}\,\lambda^{(n-3)/2}$. This gives the asymptotic expansion near the critical point. The full solution E can in this case easily be obtained. From the above we conclude that it must be of the form

$$\lambda^{(n-3)/2} g(\lambda), \tag{117}$$

with $g(\lambda)$ a holomorphic function generally depending on λ. However, since F is quadratic, I is in fact an integral over an $(n-2)$-sphere of radius $\sqrt{\lambda}$, so that g is in this case independent of λ,

$$g(\lambda) = \frac{n-1}{2}\frac{\pi^{(n-1)/2}}{\Gamma((n-1)/2+1)}, \tag{118}$$

that is, half the surface of the unit $(n-2)$ sphere. (Compare this with the results in Section 3 for the HPL formula.) This immediately leads to the well known result for the full solution E. The quadratic case is generic around each local maximum on the slowness surface one can find coordinates ξ such that F is quadratic. The points corresponding to degenerate critical points, i.e., points where F is of type A_2 or A_3 are special. In those cases g *does* depend on λ and likewise the full solution is more involved, however, its asymptotics may be readily obtained.

For the A_2 type singularity the differential equation reads (after bringing it into its familiar form)

$$\left(\frac{d^2}{d\lambda^2} - \frac{1}{3}\lambda\right) I = 0 \tag{119}$$

of which the solution is the well known Airy function.

In fact by the techniques outlined above we find for any isolated singularity (not only the simple ones) a Picard-Fuchs equation whose solution has a *holomorphic* limit $\lambda \to 0$, and thus gives an exact solution for I at the singularity. The Airy-function is a well known example studied extensively in geometrical optics. Usually one arrives at the solution for the phase function using Maslov theory. Here we arrive at such solutions using a differential equation. As we will now discuss, this has the advantage that it illuminates the topological aspects of asymptotics of I near the singularity.

Asymptotic behavior of the integrals.

The fact that Picard-Fuchs equations are ordinary (matrix) differential equations with regular singular points allows us to apply classical techniques to find the asymptotic behavior of the integrals. One immediate conclusion is that the solution of the differential equation has an analytic extension at $\lambda = 0$. To find the exact behavior at the point $\lambda = 0$ we will rewrite (115) as a first-order matrix equation as follows. Define the $s \times s$ matrix

$$A(\lambda) = \begin{pmatrix} 0 & 1 & 0 \cdots\cdots & & 0 \\ 0 & 0 & 1 \;\; 0 \cdots & & 0 \\ \vdots & \vdots & \ddots \;\; \ddots & \vdots & \vdots \\ \vdots & \vdots & & \ddots \;\; \ddots & \vdots \\ \vdots & \vdots & & 0 & 1 \\ -q_0(\lambda) & -q_1(\lambda) & \cdots\cdots\cdots & & -q_{s-1}(\lambda) \end{pmatrix} \tag{120}$$

then the following matrix equation is equivalent to (115)

$$\lambda \frac{d}{d\lambda} d\xi(\lambda) = A(\lambda) d\xi(\lambda), \tag{121}$$

with

$$d\xi(\lambda) = \begin{pmatrix} \dfrac{Pd\xi}{\partial_{\xi_n} H} \\ \lambda \dfrac{d}{d\lambda} \dfrac{Pd\xi}{\partial_{\xi_n} H} \\ \vdots \\ \left(\lambda \dfrac{d}{d\lambda}\right)^{s-1} \dfrac{Pd\xi}{\partial_{\xi_n} H} \end{pmatrix} . \tag{122}$$

Standard theory now implies that around $\lambda = 0$ the solution can be written in terms of a matrix Ξ whose columns provide a basis in the s-dimensional solution space

$$\Xi(\lambda) = B(\lambda)\lambda^M \tag{123}$$

where $B(\lambda)$ is an $s \times s$-matrix regular at $\lambda = 0$ and M is a constant $s \times s$ matrix. The expression λ^M is defined as (for small λ)

$$\lambda^M = \mathbb{1} + M(\log \lambda) + \frac{(M \log \lambda)^2}{2!} + \cdots \tag{124}$$

The matrix B has a regular power series in λ with constant coefficient matrices, which is absolutely convergent near $\lambda = 0$.

The monodromy resulting from going around the critical point at $\lambda = 0$ is with respect to the basis defined by Ξ given as $\exp(2\pi i M)$. This is the monodromy operator $h_* = h_1 \cdot h_2$ defined in the previous section. It is easy to see that it is a unipotent operator, i.e., there is an index m such that $(\exp(2\pi i M) - \mathbb{1})^m = 0$, but $(\exp(2\pi i M) - \mathbb{1})^{m+1} \neq 0$. For the A_k-type singularities this index is equal to $k = \mu$, the multiplicity of the singularity.

As a result it follows (see [28] for a general proof) that close to the point $\lambda = 0$, i.e., close to the intersection point with the singular set Σ, $I(\lambda)$ has the following asymptotic expansion:

$$I_i = \sum_{\alpha,k} a^i_{k,\alpha} \lambda^\alpha (\log \lambda)^k. \tag{125}$$

This series is absolutely convergent if $|\lambda|$ is small. The coefficients of the series are vectors in $\mathbf{C}^\mu$. The numbers α are nonnegative rational numbers; all coefficients $a_{\alpha,k}$ vanish for $k > 0$. Furthermore, the numbers α have the property that $\exp(2\pi\alpha)$ is an eigenvalue of the monodromy operator h_*. This last property thus shows explicitly the relation of the asymptotic expansion and the monodromy group of the singularity.

As an example, consider the A_2 singularity discussed earlier. The eigenvalues of h_* are $\exp(\pm 2\pi i/6)$. The above result then implies that close to $\lambda = 0$

$$I = \sum_m a_m \lambda^{1/6+m} + \sum_m b_m \lambda^{5/6+m} \tag{126}$$

with all coefficients positive. This result agrees with the well known high frequency expansion of the fundamental solution near a cuspidal singularity described by the Airy-function. For the A_3- type singularity, we get

$$I \sim \lambda^{1/4}. \tag{127}$$

The first exponent in the expansion (125) is referred to as the *singularity index*. Its relation to the order of the monodromy group has been extensively discussed by Arnold [2], however, not from the differential point of view adopted in this paper. In fact, it is not hard to see that the general asymptotic formula for an isolated singularity defined by a quasi-homogeneous polynomial $F(s,y)$ in normal coordinates $s_1, \cdots, s_{n-1}$ is given by

$$I \sim \lambda^{n/2-\beta}, \qquad \beta = \frac{1}{2} - \sum d\xi_i. \tag{128}$$

For example, for an A_k-type singularity $\beta = \frac{1}{2} - \frac{1}{k+1}$.

Acknowledgement

We would like to thank J.J. Duistermaat for his interest in this work and his many valuable comments. We also thank W.S. Gast for his assistence in preparing Figures 1 and 2. This paper was published with the permission of Shell Research B.V., and Schlumberger Cambridge Research.

References

1. V.I. Al'shits, J. Lothe, Elastic waves in triclinic crystals. I. General theory and the degeneracy problem, Kristallografiya **24** (1979) 672-683;
   V.I. Al'shits, J. Lothe, Elastic waves in triclinic crystals. II. Topology of polarization fields and some general theorems, Kristallografiya **24** (1979) 683-693.
2. V.I. Arnold, Remarks on the stationary phase method and the Coxeter numbers, Russ. Math. Surv. **28**, No. 5 (1973) 19-48.
3. V.I. Arnold, Wave front evolution and equivariant Morse lemma, Comm. Pure and Appl. Math. **29**, 557-582, (1976).
4. V.I. Arnold, Surfaces defined by hyperbolic equations, Math. Zam. **44** (1988) 3-18.
5. V. I. Arnold, On the interior scattering of waves defined by hyperbolic variational principles, Journl. Geom. Phys. **5** (1988) 305-315.
6. V.I. Arnold, S.M. Gusein-Zade, A.N. Varchenko, *Singularities of Differentiable Maps* Vol. I and II, Birkhäuser, Boston, 1988.
7. V.I. Arnold, *Singularities of Caustics and Wave Fronts*, Kluwer Academic Publ., Dordrecht, 1990.
8. M.F. Atiyah, R. Bott, L. Gårding, Lacunae for hyperbolic differential operators with constant coefficients I and II, Acta. Math. **124** (1970) 109-189, and Acta Math. **131** (1973) 145-206;
   L. Gårding, Sharp fronts of paired oscillatory integrals, Publ. Res. Int. Math. Sci. **12** (1976).
9. B.A. Auld, *Acoustic fields and waves in solids*, John Wiley, New York, 1973.

10. R. Burridge, Lacunas in two dimensional wave propagation, Proc. Camb. Phil. Soc. **63** (1967) 819-825;
R. Burridge, The singularity on the plane lids of the wave surface of elastic media with cubic symmetry, Quart. Journ. Mech. Appl. Math. **XX** (1967) 41-56;
R. Burridge, Lamb's problem for an anisotropic half-space, Quart. Journ. Mech. Appl. Math. **XXIV** (1971) 81-98.
11. R. Burridge, P. Chadwick, A.N. Norris, Fundamental elastodynamic solutions for anistropic media with ellipsoidal slowness surfaces, submitted to the Proceedings of the Royal Society (1992).
12. L. Cagniard, *Réflexion et réfraction des ondes séismiques progressives*, Gauthes-Villars, Paris, 1939.
13. S. Crampin, M. Yedlin, Shear wave singularities of wave propagation in anisotropic media, J. Geophys. **49** (1981) 43-46.
14. G.F.D. Duff, The Cauchy problem for elastic waves in an anisotropic medium, Phil. Trans. R. Soc. A. **252** (1960) 249-273.
15. J. J. Duistermaat, Oscillatory integrals, Lagrange Immersions and Unfolding of singularities, Commun. Pure Appl. Math. **27**, No. 2 (1974) 207-281;
M.V. Fedoryuk, The stationary phase method and pseudodifferential operators, Russ. Math. Surv. **26**, No. 1 (1972) 65-115;
A.S. Mishchenko, V.E. Shatalov, B. Yu. Sternin, *Lagrangian Manifolds and the Maslov Operator*, Springer, New York, 1990;
J.-M. Kendall, C.J. Thomson, Maslov ray summation, pseudo-caustics, Lagrangian equivalence and transient seismic wave forms, submitted to Geophys. J. Int. (1992).
16. A.G. Every, General closed-form expressions for acoustic waves in elastically anistropic solids, Phys. Rev **B 22** (1980) 1746-1760.
17. I.M. Gel'fand, G.E. Shilov, *Generalized functions*, Vol. I, Academic Press, New York, 1964.
18. P. A. Griffiths, On the periods of certain rational integrals, I, Ann. of math. **90** (1969) 460-495.
19. A.T. de Hoop, A modification of Cagniard's method for solving seismic pulse problems, Appl. Sci. Res. **B8** (1960) 349-356.
20. P. Hubral, M. Tygel, Transient response from a planar acoustic interface by a new point-source decomposition into plane waves, Geoph. **50** (1985) 766-774.
21. J.A. Hudson, Wavespeed and attenuation of elastic waves in material containing cracks, G.J.R.A.S., **64** (1981) 133-150.
22. J.H.M.T. van der Hijden, *Propagation of transient elastic waves in stratified anisotropic media*, North Holland, Amsterdam, 1987.
23. F. John, *Partial differential equations*, Springer-Verlag, New York, 1982.
24. N. Katz, Differential equations for periods, Publ. Math. I.H.E.S. **35** (1968) 71.
25. J. Leray, *Hyperbolic differential equations*, The Institute for Advanced Study, Princeton N.J., 1952.
26. W. Lerche, D.-J. Smit, N.P. Warner, Differential equations for periods and flat coordinates in two-dimensional topological matter theories, Nucl. Phys. **B 372** (1991) 87;
D.R. Morrison, Picard- Fuchs equations and mirror maps for hypersurfaces, *in: Essays on Mirror Manifolds*, ed. S-T. Yau, International Press Honkong, 1992.
27. D. Ludwig, Exact and asymptotic solutions of the Cauchy problem, Comm. Pure Appl. Math. **13** (1960) 473-508.

28. B. Malgrange, Intégrale asymptotique et monodromie, Ann. Sci. École Normale Supp. **4**, No. 7 (1974) 405-430.
29. J. Milnor, *Singular Points of Complex Hypersurfaces*, Princeton University Press, Princeton N.J., 1968.
30. M.J.P. Musgrave, On the propagation of elastic waves in aelotropic media I, Proc. R. Soc. Lond. A **226** (1954) 339; *idem* II, Proc. R. Soc. Lond. A **226** (1954) 356; G.F. Miller, M.J.P. Musgrave, On the propagation of elastic waves in aelotropic media III, Proc. R. Soc. Lond. A **236** (1956) 352;
M.J.P. Musgrave, *Crystal acoustics*, Holden Day, San Francisco, 1970.
31. M.J.P. Musgrave, On an elastodynamic classification of orthorhombic media, Proc. R. Soc. Lond. A **374** (1981) 401-429;
M.J.P. Musgrave, Acoustic axes in orthorhombic media, Proc. R. Soc. Lond. A **401** (1985) 131-143.
32. D. Nichols, F. Muir, M. Schoenberg, Expanded Abstracts 59th Ann. Mtg. SEG (1989) 471.
33. A.N. Norris, A theory of pulse propagation in anisotropic elastic solids, Wave Motion **9** (1987) 509-532.
34. R.G. Payton, *Elastic wave propagation in transversely isotropic media*, Martinus Nijhoff Publishers, The Hague, 1983.
35. R.G. Payton, Int. J. Engng. Sci. **13**, 183 (1975).
R.G. Payton, Instituto Lombardo, (Rend. Sci), **A 108**, 684 (1974).
36. I. Petrovsky, On the diffusion of waves and the lacunas for hyperbolic equations, Math. Sbo. **17** (59) (1945) 289-370.
37. M. Riesz, L-intégrale de Riemann-Liouville et le problem de Cauchy, Acta Math. **81** (1949) 1-223.
38. G. Salmon, *Geometry of three dimensions*, Hodges, Foster and Figgis, Dublin, 1882.
39. P.M. Shearer, C.H. Chapman, Ray tracing in azimutally anisotropic media - I. Results for models of aligned cracks in the upper crust, G.J.R.A.S. **96** (1989) 51-64;
idem - II. Quasi-shear wave coupling, G.J.R.A.S. **96** (1989) 65-83.
40. M. Schoenberg, F. Muir, A calculus for finely layered anistropic media, Geoph. **54** (1989) 581-589;
J. Hood, M Schoenberg, NDE of fracture-induced anisotropy, Review of Progress in Quantitative Nondestructive Evaluation, 2101-2108, Plenum Press, New York, 1992.
41. D.-J. Smit, M.V. de Hoop, *The geometry of the hyperbolic system for an anisotropic perfectly elastic medium*, Shell Research preprint, Schlumberger Cambridge Research preprint (1993).
42. A. Sommerfeld, Über die Ausbreitung der Wellen in der drahtlosen Telegraphie, Ann. Physik **28** (1909) 665-737;
H. Weyl, Ausbreitung elektromagnetischer Wellen über einem ebenen Leiter, Ann. Physik **60** (1919) 481-500.
43. V.A. Vasil'ev, Sharpness and the local Petrovskii condition for strictly hyperbolic operators with constant coefficients, Math. USSR Izv. **28** (1987) 233-273.

MATHEMATICAL ADDENDA TO HOPPER'S MODEL OF PLANE STOKES FLOW DRIVEN BY CAPILLARITY ON A FREE SURFACE

J. de GRAAF

Eindhoven University of Technology, P.O. Box 513, 5600 MB Eindhoven, The Netherlands.

Introduction

In an interesting and stimulating series of papers [H1], [H2], [H3], Hopper presents some special exact solutions of the shape evolution of a piece of viscous matter driven by surface tension on the free boundary.
Hopper's paper [H1] is of a conceptual nature and consists of two parts. In his first part Hopper derives an evolution equation for the change of shape in time: The unknown function in this evolution equation is a Riemann mapping function from the unit disc onto the region occupied by the fluid at time t. Hopper's evolution equation is a partial differential equation of a very special nature, requiring 'compensation of analytic singularities'. In [H1] we find, what might be called, a pseudo Lagrangian description of the piece of matter and several other innovative concepts. However, a lot of important mathematical and physical details are missing in [H1]. In my view e.g. the kinematical aspects are completely neglected in [H1] (and also in [R]).
Chapter 1 in my paper might well be called: 'Mathematical addenda to Hopper's derivation of Hopper's equation'.
In the second part of [H1] and also in [H2], [H3], Hopper finds solutions of his equation which are of type $\Omega(z, \lambda(t))$. He makes a clever guess of a parametrized set of analytic functions $\Omega(z; \lambda)$, such that substitution of them in the evolution equation leads to one ordinary differential equation for $\lambda(t)$.
In Chapter 2 of this paper I study several mathematical aspects of Hopper's equation. On the 'state space', which is a part of an ellipsoid in Hilbert space, Hopper's evolution equation can be considered as an infinite system of ordinary differential equations. For this system there are 3 'exhausting' series of finite dimensional sub systems leading to solutions which are: 1. Complex polynomials with real coefficients, 2. Complex polynomials with complex coefficients, 3. Rational functions. Some local results on these finite dimensional sub systems are presented.
For numerical solutions to the same problem which use Lorentz– Ladyzhenskaja potentials I refer to work being done in Eindhoven [VM1], [VM2], [VM3].
I wish to thank Dr. H.K. Kuiken of Philips Research Laboratories for drawing my attention to these interesting problems.

1 A shape–evolution equation

1.1 Formulation of a Stokes problem with a free boundary

On a simply connected open domain $G_t \subset I\!R^2$ with a smooth boundary ∂G_t we consider the system of Navier–Stokes equations for the unknown velocity field $\underline{v}(\underline{x},t) = (v_1(\underline{x},t), v_2(\underline{x},t))$, $\underline{x} = (x,y)$ and the unknown pressure $p(\underline{x},t)$,

$$\left.\begin{array}{l} \rho\,\dfrac{D\underline{v}}{dt} = \rho\,\dfrac{\partial \underline{v}}{\partial t} + \rho(\underline{v}.\nabla)\underline{v} = -\nabla p + \eta\Delta\underline{v} + \rho\underline{g} \\ \\ \nabla.\underline{v} = 0\;, \end{array}\right\} \quad (x,y) \in G_t,\; t > 0\;,$$

with the boundary condition

$$T\underline{n} = -\gamma(\nabla.\underline{n})\underline{n} = -\gamma\kappa\underline{n} \quad \text{on } \partial G_t\;.$$

Here T is the stress–tensor (= stress–matrix)

$$T = -pI + \eta\Big[\Big(\frac{d\underline{v}}{d\underline{x}}\Big) + \Big(\frac{d\underline{v}}{d\underline{x}}\Big)^T\Big]\;.$$

Further, $\underline{n}(\underline{x})$ and $\kappa(\underline{x})$ are the outward normal and the curvature at points $\underline{x} \in \partial G_t$.
The relevant physical constants are: The density ρ $[ML^{-3}]$, the viscosity η $[ML^{-1}T^{-1}]$ and the surface tension γ $[MT^{-2}]$.
Note that with this boundary condition the surface is supposed to behave like a membrane.

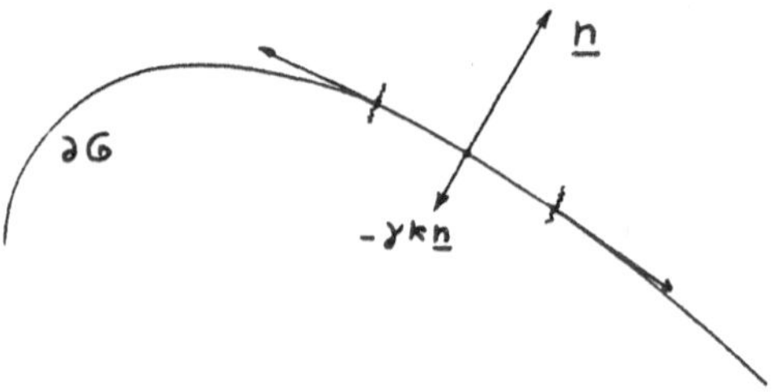

Next we introduce dimensionless quantities. Put

$$\underline{v} = \frac{\gamma}{\eta}\,\underline{\tilde{v}} \qquad \underline{x} = R\underline{\tilde{x}} \qquad t = \frac{\eta R}{\gamma}\,\tilde{t}$$

$$p = \frac{\gamma}{R}\,\tilde{p} \qquad \kappa = \frac{\tilde{\kappa}}{R} \qquad \iint\limits_G d\sigma = \pi R^2\;.$$

Then the Navier–Stokes system becomes

$$\left.\begin{array}{l} S\,\dfrac{D\underline{\tilde{v}}}{d\tilde{t}} = -\tilde{\nabla}\tilde{p} + \tilde{\Delta}\underline{\tilde{v}} + \dfrac{R^2\rho}{\gamma}\,\underline{g} \\ \\ \tilde{\nabla}.\underline{\tilde{v}} = 0 \end{array}\right\} \quad \underline{\tilde{x}} \in \tilde{G}_t$$

$$\tilde{T}\underline{n} = -\tilde{\kappa}\underline{n} \qquad \text{on } \partial\tilde{G}_t$$

with

$$\tilde{T}_{ij} = \frac{R}{\gamma} T_{ij} = -\tilde{p}\delta_{ij} + \left(\frac{\partial\tilde{v}_i}{\partial\tilde{x}_j} + \frac{\partial\tilde{v}_j}{\partial\tilde{x}_i}\right).$$

If the Suratmannumber $S = \frac{\rho\gamma}{\eta^2} gR$ and the Bondnumber $B = \frac{R^2\rho}{\gamma}$ are very small, e.g. if R is very small, it suffices to solve Stokes' equations on G_t. (We omit the tilde $\sim$)

$$\left.\begin{aligned} \Delta\underline{v} &= \nabla p \\ \nabla.\underline{v} &= 0 \end{aligned}\right\} \quad \text{in } G_t$$

$$T\underline{n} = -\kappa\underline{n} \qquad \text{on } \partial G_t\ .$$

An equivalent formulation is

$$\left.\begin{aligned} \nabla T = \partial_i T_{ij} &= 0 \\ \nabla.\underline{v} &= 0 \end{aligned}\right\} \quad \text{in } G_t$$

$$T\underline{n} = -\kappa\underline{n} \qquad \text{on } \partial G_t\ .$$

1.2 The general solution of Stokes' equations

In this section we want to describe the general solution of the Stokes system on a fixed, simply connected open domain $G \subset \mathbb{R}^2$

$$\frac{\partial^2 v_1}{\partial x^2} + \frac{\partial^2 v_1}{\partial y^2} = \frac{\partial p}{\partial x}$$

$$\frac{\partial^2 v_2}{\partial x^2} + \frac{\partial^2 v_2}{\partial y^2} = \frac{\partial p}{\partial y}$$

$$\frac{\partial v_1}{\partial x} + \frac{\partial v_2}{\partial y} = 0\ .$$

Suppose that the pair $(\underline{v}, p)$ solves this system on G and let T be stress tensor obtained from this solution.
Then because of the simple connectedness of G there exists a 'streamfunction' ψ and an 'Airy function' ϕ such that

$$\underline{v} = (v_1, v_2) = (\psi_y\ , -\psi_x)$$

$$T = \begin{pmatrix} T_{11} & T_{12} \\ T_{21} & T_{22} \end{pmatrix} = \begin{pmatrix} -\phi_{yy} & \phi_{xy} \\ \phi_{xy} & -\phi_{xx} \end{pmatrix}.$$

The latter can be argued as follows: Since $\nabla \cdot T = 0$ the stress tensor T must be of the form

$$T = \begin{pmatrix} f_y & g_y \\ -f_x & -g_x \end{pmatrix} .$$

The symmetry of T then requires $-f_x = g_y$ which says $\nabla \cdot (f, g) = 0$. Hence $(f, g) = (-\phi_y, \phi_x)$ for some function ϕ. Note that, if $\underline{v}$ is given, the streamfunction ψ is determined up to a constant C and the Airy function is determined up to a linear function $Ax + By + C_1$.
Taking the trace of T we find

$$p = \tfrac{1}{2}(\phi_{xx} + \phi_{yy}) = \tfrac{1}{2}\Delta\phi .$$

Combining this with the equation $\Delta\underline{v} = \nabla p$ we find the Stokes equations in Cauchy–Riemann form

$$\frac{\partial}{\partial x}\,(\tfrac{1}{2}\Delta\phi) - \frac{\partial}{\partial y}\,(\Delta\psi) = 0$$

$$\frac{\partial}{\partial y}\,(\tfrac{1}{2}\Delta\phi) + \frac{\partial}{\partial x}\,(\Delta\psi) = 0 .$$

So $(\tfrac{1}{2}\Delta\phi) + i(\Delta\psi)$ is an analytic function on G, therefore $\Delta\Delta\phi = 0$ and $\Delta\Delta\psi = 0$, so the functions ϕ and ψ are biharmonic.
Any biharmonic function ϕ on a simply connected domain G can be represented as

$$\phi = 2Re(\overline{z}f_1 + g_1) , \quad z = x + iy$$

with f_1 and g_1 analytic on G. Cf. [M], pp. 106-111.
Following the same reasoning we put

$$\psi = Im(\overline{z}f_2 + g_2)$$

with f_2 and g_2 analytic on G.
From the Cauchy–Riemann representation of Stokes' equations it follows that $f_1'' = f_2''$. Consistency in the stress tensor requires $g_1'' = g_2''$. So there are constants $A, B, D, E \in \mathbb{R}$ and $C, F \in \mathbb{C}$ such that

$$f_1 = f_2 + Az + iBz + C , \quad g_1 = g_2 + Dz + iEz + F .$$

Define

$$\varphi = f_2 + Az , \quad \chi = g_2 .$$

Then

$$\psi = Im(\overline{z}\varphi + \chi)$$

$$\phi = 2Re(\overline{z}\varphi + \overline{z}(iBz + C) + \chi + (D + iE)z + F)$$

$$= 2Re(\overline{z}\varphi + \chi) + 2Re(\overline{z}C + (D + iE)z + F) .$$

Omission of the second term leads to the same stress tensor. Summarizing: The state of the system is described by the analytic functions φ and χ and

$$\tfrac{1}{2}\,\phi + i\psi = \bar{z}\varphi + \chi\,.$$

Note that φ is uniquely determined by the state $(\underline{v}, p)$ and that addition of a complex constant to χ leads to the same state. Conversely, any pair of analytic functions φ and χ leads to a solution of Stokes' equations.

1.3 Kinematic and Dynamic Quantities expressed in φ and χ

In a straightforward way the velocity field $\underline{v}(\underline{x}) = (v_1(x,y), v_2(x,y))$ and the stress tensor field $T(\underline{x}) = [T_{ij}(x,y)]$ can be expressed in the analytic potentials φ and χ. Write $z = x + iy$.

- $$v_1 + iv_2 = \psi_y - i\psi_x = \frac{\partial}{\partial y}\, Im(\bar{z}\varphi + \chi) - i\,\frac{\partial}{\partial x}\, Im(\bar{z}\varphi + \chi)$$

$$= Im(-i\varphi + i\bar{z}\varphi' + i\chi') - i\ Im(\varphi + \bar{z}\varphi' + \chi')$$

$$= Re(-\varphi + \bar{z}\varphi' + \chi') - i\ Im(\varphi + \bar{z}\varphi' + \chi')$$

$$= -Re\ \varphi - i\ Im\ \varphi + \overline{z\varphi' + \chi'}$$

$$= -\varphi + z\overline{\varphi'} + \overline{\chi'}$$

- $$T_{11} + T_{22} = -2p = -\Delta\phi = -2\Delta Re(\bar{z}\varphi + \chi)$$

$$= -8\ Re\ \varphi' = -4(\varphi' + \overline{\varphi'})$$

- $$T_{22} - T_{11} + i2T_{12} = -\phi_{xx} + \phi_{yy} + 2i\phi_{xy}$$

$$= (-2\frac{\partial^2}{\partial x^2} + 2\ \frac{\partial^2}{\partial y^2})Re(\bar{z}\varphi + \chi) + 4i\ \frac{\partial^2}{\partial x \partial y}\ Re(\bar{z}\varphi + \chi)$$

$$= 2\ \frac{\partial v_2}{\partial y} - 2\ \frac{\partial v_1}{\partial x} + 2i\phi_{xy} = -4\psi_{xy} + 2i\phi_{xy}$$

$$= 4i\ \frac{\partial^2}{\partial x \partial y}\ [i\ Im(\bar{z}\varphi + \chi) + Re(\bar{z}\varphi + \chi)] = -4(\bar{z}\varphi'' + \chi'')\,.$$

Stress orthogonal to a given curve

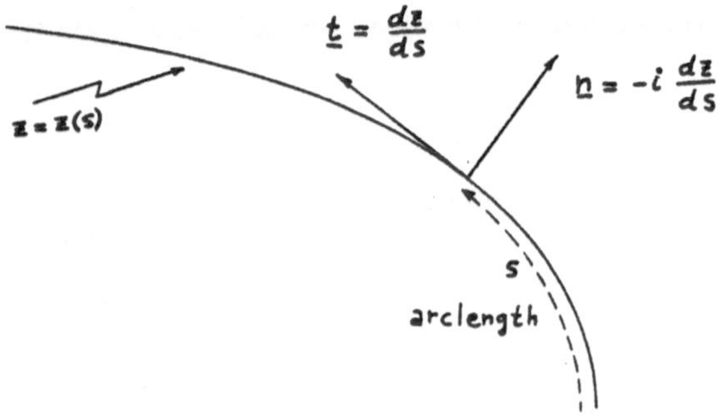

$$T\underline{n} = \begin{pmatrix} -\frac{1}{2}\Delta\phi + 2\psi_{yx} & \phi_{xy} \\ \phi_{xy} & -\frac{1}{2}\Delta\phi - 2\psi_{xy} \end{pmatrix} \begin{pmatrix} Im\ \dot{z} \\ -Re\ \dot{z} \end{pmatrix} =$$

$$= \tfrac{1}{2}\Delta\phi(-Im\ \dot{z} + i\ Re\ \dot{z}) + (2\psi_{yx} Im\ \dot{z} - \phi_{xy} Re\ \dot{z}) +$$

$$+ i(\phi_{xy} Im\ \dot{z} + 2\psi_{xy} Re\ \dot{z})$$

$$= \tfrac{1}{2}i(\Delta\phi)\dot{z} - 2\overline{(\tfrac{1}{2}\phi_{xy} + i\psi_{xy})\dot{z}}$$

$$= i\{\Delta\ Re(\overline{z}\varphi + \chi)\}\dot{z} - 2\{\frac{\partial^2}{\partial x \partial y}\ \overline{(\tfrac{1}{2}\phi + i\psi)\}\dot{z}}$$

$$= 2i\{(\varphi' + \overline{\varphi'}\dot{z}) + (z\overline{\varphi''} + \overline{\chi''})\dot{z}\}$$

$$= 2i\ \frac{d}{ds}\ (z\overline{\varphi'} + \varphi + \overline{\chi'})\ .$$

Note that if we replace φ by $\varphi_1 = \varphi + C + i\beta z$ with $C \in \mathbb{C}$, $\beta \in \mathbb{R}$ and keep the same χ than a rigid motion is added to the velocity field $v_1 + iv_2$. However this modification does not affect T and $T\underline{n}$.

1.4 A road to Hopper's equation

In continuum mechanics there are two conventional ways of describing the motion of matter. In the *Lagrangian* description each matter particle gets its own label, $\underline{X}$ say, and one wants to find the position $\underline{x}$ of each particle $\underline{X}$ as a function of time, i.e. one looks for the function $\underline{x} = \underline{F}(\underline{X}, t)$.
On the other hand, users of the *Eulerian* description are not so much interested in the position of each particle. In the Eulerian description one wants to calculate the velocity field

$$\underline{v}(\underline{x},t) = \underline{\dot{F}}(\underline{F}^{\leftarrow}(\underline{x},t),t)$$

with

$$\underline{\dot{F}}(\underline{x},t) = \frac{\partial}{\partial t}\,\underline{F}(\underline{x},t)\ .$$

In this paper we want to determine the evolution of the shape of a piece of matter and the positions $\underline{F}(\underline{x},t)$ of the particular and the velocity fields $\underline{v}(\underline{x},t)$ are not so relevant.
Instead of the Lagrangian or Eulerian approach we use what we call the "Pseudo–Lagrangian picture": At each time t a fixed domain D in $\underline{\xi}$-space is mapped by a function $\underline{x} = \underline{\Omega}(\underline{\xi},t)$ onto the actual configuration of the piece of matter. The function Ω is made 'more or less rigid' by requiring extreme smoothness of it. In our 2 dimensional case, following Hopper [H1], we require it to be analytic.

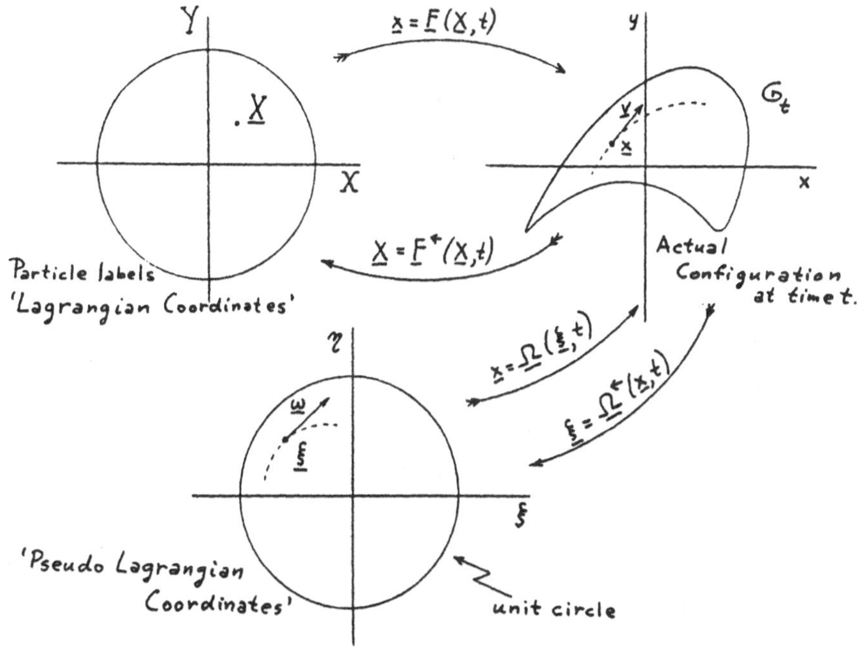

We now gather some convenient kinematical expressions.
The trajectory of particle $\underline{X}$ in configuration space is

$$t \mapsto \underline{x}(t) = \underline{F}(\underline{X},t)\ .$$

The trajectory of particle $\underline{X}$ in Pseudo–Lagrangian coordinates is

$$t \mapsto \underline{\xi}(\underline{X},t) = \Omega^{\leftarrow}(\underline{F}(\underline{X},t),t)\ .$$

Differentiating the identity

$$\underline{\Omega}^{\leftarrow}(\underline{\Omega}(\underline{\xi},t),t) = \underline{\xi}$$

according to t leads to

$$(D\underline{\Omega}^{\leftarrow})(\underline{\Omega}(\underline{\xi},t))\dot{\underline{\Omega}}(\underline{\xi},t) + (\underline{\Omega}^{\leftarrow})^{\bullet}(\underline{\Omega}(\underline{\xi},t),t) = \underline{0} \ .$$

Here the dot $\cdot$ denotes partial differentiation to t. So

$$(\Omega^{\leftarrow})^{\bullet}(\underline{x},t) = -(D\Omega)^{-1}(\Omega^{\leftarrow}(\underline{x},t))\dot{\Omega}(\underline{\Omega}^{\leftarrow}(\underline{x},t),t) \ .$$

For the velocity field in configuration space we find

$$\underline{v}(\underline{x},t) = \dot{\underline{F}}(\underline{F}^{\leftarrow}(\underline{x},t),t) \ .$$

Since

$$\frac{\partial \underline{\xi}(\underline{X},t)}{\partial t} = (D\underline{\Omega}^{\leftarrow})(\underline{F}(\underline{X},t),t)\dot{\underline{F}}(\underline{X},t) + (\underline{\Omega}^{\leftarrow})^{\bullet}(\underline{F}(\underline{X},t),t)$$

we find for the velocity field in Pseudo–Lagrangian coordinates

$$\underline{\omega}(\underline{\xi},t) = (D\Omega^{\leftarrow})(\underline{\Omega}(\underline{\xi},t),t)\underline{v}(\underline{\Omega}(\underline{\xi},t),t) + (\Omega^{\leftarrow})^{\bullet}(\underline{\Omega}(\underline{\xi},t),t)$$

$$= (D\underline{\Omega})^{-1}(\underline{\xi},t)[\underline{v}(\underline{\Omega}(\underline{\xi},t),t) - \dot{\underline{\Omega}}(\underline{\xi},t)] \ .$$

Our ultimate goal is to calculate

$$\underline{\Omega}(\underline{\xi},t) \quad \text{for } |\underline{\xi}| = 1$$

which represents the shape of our piece of matter.
Now suppose that at time t the fluid occupies a domain $G_t \subset \mathbb{R}^2$. Fix a point in G_t and choose $x+iy$ coordinates such that this point becomes the origin. Introduce a conformal mapping $\Omega: \ D \to G_t$, with $z = \Omega(\zeta,t)$, $\zeta = \xi + i\eta$. D is the unit disc in the complex ζ plane and $\Omega(0,t) = 0$. Note that Ω is uniquely determined if we require $\Omega'(0,t) > 0$. Suppose further that at time t the state $(\underline{v},p)$ of the Stokes system is described by the complex analytic 'potentials' φ and χ. If necessary we may add a uniform rotation velocity field to $\underline{v}(\underline{x},t)$ in order to arrange that $\varphi'(0) \in \mathbb{R}$. Cf. the remark at the end of section 1.3.

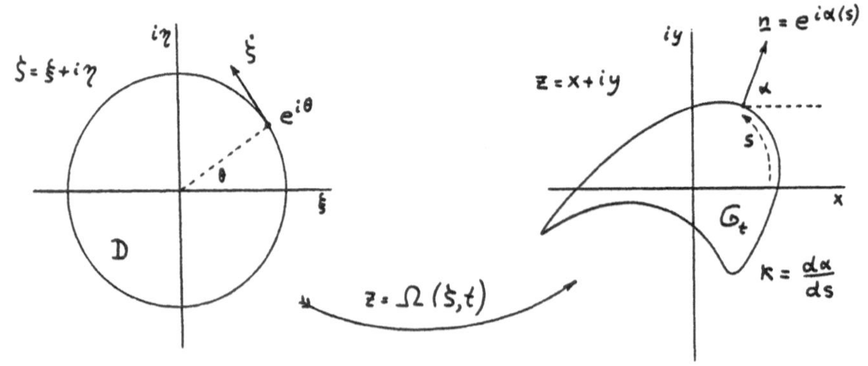

At the boundary ∂G_t of G_t, see picture, we have the boundary condition $T\underline{n} = -\kappa\underline{n}$.
With the potentials φ and χ this becomes, in complex notation,

$$2i\,\frac{d}{ds}(z\overline{\varphi'}+\varphi+\overline{\chi'})=i\,\frac{d}{ds}\,e^{i\alpha(s)}\;,$$

which leads to

$$2(z\overline{\varphi'}+\varphi+\chi')=e^{i\alpha(s)}+C_1\;,\qquad \text{at } \partial G_t\;.$$

The constant C_1 can be made zero by addition of a suitable constant velocity field to $\underline{v}(\underline{x},t)$.
Then combination at ∂G_t of the latter result with

$$v_1+iv_2=(-\varphi+z\overline{\varphi'}+\overline{\chi'})$$

yields

$$2(v_1+iv_2)=-4\varphi+e^{i\alpha(s)}\;,\qquad \text{at } \partial G_t\;.$$

With the parametrization $\Omega(e^{i\theta},t)$ for ∂G_t this becomes

$$2v_1(\Omega(e^{i\theta},t),t)+2iv_2(\Omega(e^{i\theta},t),t)=-4\varphi(\Omega(e^{i\theta},t),t)+\frac{e^{i\theta}\Omega'(e^{i\theta},t)}{|\Omega'(e^{i\theta},t)|}\;.$$

Rewrite

$$\underline{\omega}=(D\underline{\Omega})^{-1}\underline{v}-(D\underline{\Omega})^{-1}\dot{\underline{\Omega}}$$

in the complex ζ-plane

$$\dot{\zeta}(\zeta,t)=\frac{v_1(\Omega(\zeta,t),t)+iv_2(\Omega(\zeta,t),t)}{\Omega'(\zeta,t)}-\frac{\dot{\Omega}(\zeta,t)}{\Omega'(\zeta,t)}\;.$$

At ∂D we have $\zeta=e^{i\theta}$, $\theta\in\mathbb{R}$, and

$$Re\;e^{-i\theta}\dot{\zeta}(e^{i\theta},t)=0\;.$$

With $e^{i\theta}=\sigma$ and $\tilde{\varphi}(\zeta,t)=\varphi(\Omega(\zeta,t),t)$ we arrive at

$$\frac{2\tilde{\varphi}(\sigma,t)}{\sigma\Omega'(\sigma,t)}+\frac{\dot{\Omega}(\sigma,t)}{\sigma\Omega'(\sigma,t)}=\frac{1}{2|\Omega'(\sigma,t)|}-\frac{\dot{\zeta}(\sigma,t)}{\sigma}\;.$$

We now make the important observation that the two terms on the right hand side are the respective real and imaginary parts of an analytic function $\mathcal{F}(|\Omega'(\zeta,t)|)$ restricted to the boundary ∂D. This analytic function $\mathcal{F}$ is uniquely defined by

$$Re\;\mathcal{F}(|\Omega'(\zeta,t)|)=\frac{1}{2|\Omega'(\zeta,t)|}\;,\qquad \zeta\in\partial D$$

$$Im\;\mathcal{F}(|\Omega'(0,t)|)=0\;.$$

Summarizing, we find on D the relation

$$2\tilde{\varphi}(\zeta,t)=\zeta\Omega'(\zeta,t)\;\mathcal{F}(|\Omega'(\zeta,t)|)-\dot{\Omega}(\zeta,t)\;.$$

We now proceed to derive an evolution equation for Ω in which the unknown complex analytic potentials φ and χ play no role.
In $2(z\overline{\varphi'}+\varphi+\overline{\chi'})$ on G_t substitute $z=\Omega(\zeta,t)$ and put $\tilde{\chi}(\zeta,t)=\chi(\Omega(\zeta,t),t)$, $\tilde{\varphi}(\zeta,t)=\varphi(\Omega(\zeta,t),t)$. At the boundary ∂D this leads to

$$2\Big(\Omega\,\frac{\overline{\tilde{\varphi}'}}{\overline{\Omega}'}+\tilde{\varphi}+\frac{\overline{\tilde{\chi}'}}{\overline{\Omega}'}\Big)=\frac{\sigma\Omega')\sigma)}{|\Omega'(\sigma)|}\,,\qquad \sigma\in\partial D\,.$$

Suppress $\sim$ and substitute $2\varphi=\zeta\Omega'\mathcal{F}-\dot{\Omega}$ and its derivative. Then, at ∂D

$$-\zeta\Omega'\overline{\Omega'}(\mathcal{F}-\frac{1}{|\Omega'|})+\frac{d}{dt}(\Omega\overline{\Omega'})-\Omega\overline{(\zeta\Omega'\mathcal{F})'}=2\overline{\chi'}\,.$$

After complex conjugation and writing $\zeta=\sigma$, $|\sigma|=1$,

$$\overline{\sigma}\,\overline{\Omega'}\,\Omega'\,\mathcal{F}-\overline{\Omega}(\sigma\Omega'\mathcal{F})'+\frac{d}{dt}(\overline{\Omega}\,\Omega')=2\chi'\,.$$

With $\sigma=e^{i\theta}$ and $\dfrac{d}{dz}=-ie^{-i\theta}\,\dfrac{d}{d\theta}$ this becomes

$$ie^{-i\theta}\,\frac{d}{d\theta}(e^{i\theta}\,\overline{\Omega}\,\Omega'\,\mathcal{F})+\frac{d}{dt}(\overline{\Omega}\,\Omega')=2\chi'\qquad\text{on }\partial D\,.$$

Hopper [H1], writes $\frac{d}{d\sigma}=-ie^{-i\theta}\,\frac{d}{d\theta}$ for 'differentiation along the unit circle'. Then

$$\frac{d}{dt}\,(\overline{\Omega}\,\Omega')-\frac{d}{d\sigma}(\sigma\overline{\Omega}\,\Omega'\,\mathcal{F}(|\Omega'|))=2\chi'$$

which is Hopper's evolution equation for the shape of a piece of viscous matter driven by surface tension.

2 Some mathematical analysis on Hopper's equation

2.1 Mathematical generalities on Hopper's evolution equation

On the closed unit disk $\overline{D}\subset\mathbb{C}$ we look for solutions $\Omega(\zeta,t)$, $\zeta\in\overline{D}$, $t\geq 0$ of the evolution equation

$$\text{(H)}\qquad \frac{d}{dt}(\overline{\Omega}\,\Omega')-\frac{d}{d\zeta}(\zeta\overline{\Omega}\,\Omega'\,\mathcal{F}(|\Omega'|))=-2\chi'=\text{ analytic on }D\,.$$

Solutions Ω are required to be (at least) analytic on D and continuous on $\overline{D}$. Remind that, by definition,

$$\overline{\Omega}(\zeta,t)=\overline{\Omega(\frac{1}{\overline{\xi}}\,,t)}$$

and also that $\mathcal{F}(|\Omega'|;\zeta)$ is analytic on D and uniquely defined by

$$\text{(F)}\qquad \begin{cases} Re\,\mathcal{F}(|\Omega'(\zeta,t)|)=\dfrac{1}{2|\Omega'(\zeta,t)}|\,, & \zeta\in\partial D\\[2ex] Im\,\mathcal{F}(|\Omega'(0,t)|)=0 \end{cases}$$

The righthand side $-2\chi'$ of the evolution equation is an unknown analytic function. Therefore the question:
Does the cancellation of singularities inside D determine the shape evolution?

Note that if $\Omega(\zeta, t)$ solves (H) then also $e^{i\varphi(t)}\Omega(\zeta, t)$, $\varphi : \mathbb{R} \to \mathbb{R}$ arbitrary, is a solution. This type of nonuniqueness can be resolved by requiring $\Omega'(0,t) > 0$.

DEFINITION. (Set of states $\Sigma \subset \overline{\Sigma}$)

$\Sigma = \{\Omega\}$, with

- $$\Omega(\zeta) = \sum_{n=1}^{\infty} n\zeta^n \; ; \quad \overline{D} \to \mathbb{C} \, , \quad \text{analytic}$$

- $$\sum_{n=1}^{\infty} n|a_n|^2 = 1 \, , \quad \text{means : Area } \Omega(\overline{D}) = \pi$$

- $$\forall \zeta, \; |\zeta| \leq 1 \quad \Omega'(\zeta) \neq 0 \, .$$

$\Omega \in \overline{\Sigma}$ means: Ω is analytic on D, Ω is continuous on $\overline{D}$ and $\Omega'(\zeta) \neq 0$ for $|\zeta| < 1$.

Note that Σ is a part of an ellipsoid in the Hilbert space.

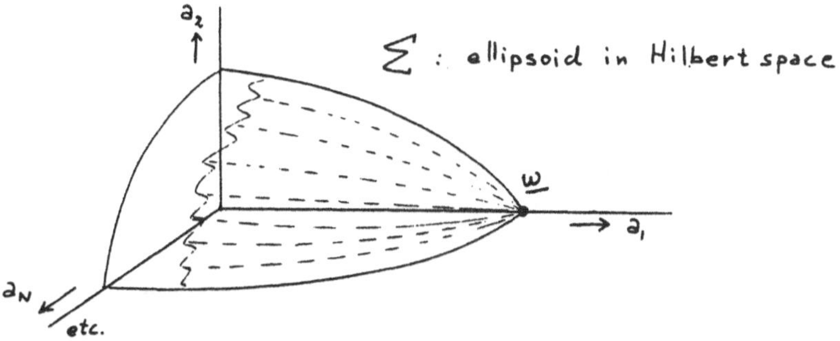

Calculate

$$\Omega'\overline{\Omega} = \sum_{k \geq 0, \, \ell \geq 0} k a_k \overline{a}_\ell \zeta^{k-\ell-1} = \, .$$

$$= \sum_{m=-\infty}^{\infty} \left\{ \sum_{\ell=\max(0,m)}^{\infty} (\ell - m + 1)\overline{a}_\ell a_{\ell-m+1} \right\} \zeta^{-m} =$$

$$= \sum_{m=-\infty}^{\infty} u_m \zeta^{-m} \, .$$

Note:

$$m \geq 1 \;\Rightarrow\; u_m = \sum_{\ell=m}^{\infty} (\ell - m + 1)\overline{a}_\ell a_{\ell-m+1} \;.$$

There are two matrix forms for this expression:

$$\begin{pmatrix} u_1 \\ u_2 \\ u_3 \\ u_4 \\ \vdots \end{pmatrix} = \begin{pmatrix} \overline{a}_1 & 2\overline{a}_2 & 3\overline{a}_3 & 4\overline{a}_4 & \cdots \\ \overline{a}_2 & 2\overline{a}_3 & 3\overline{a}_4 & \cdot & \cdots \\ \overline{a}_3 & 2\overline{a}_4 & \cdot & \cdot & \cdots \\ \overline{a}_4 & \cdot & \cdot & \cdot & \cdots \\ \vdots & \vdots & \vdots & \vdots & \cdots \end{pmatrix} \begin{pmatrix} a_1 \\ a_2 \\ a_3 \\ a_3 \\ \vdots \end{pmatrix} =$$

$$= \begin{pmatrix} a_1 & 2a_2 & 3a_3 & 4a_4 & \cdots \\ 0 & a_1 & 2a_2 & 3a_3 & \cdots \\ 0 & 0 & a_1 & 2a_2 & \cdots \\ 0 & 0 & 0 & a_1 & \cdots \\ \vdots & \vdots & \vdots & \vdots & \cdots \end{pmatrix} \begin{pmatrix} \overline{a}_1 \\ \overline{a}_2 \\ \overline{a}_3 \\ \overline{a}_4 \\ \vdots \end{pmatrix} .$$

In short $\underline{u} = N(\underline{\overline{a}})\underline{a} = M(\underline{a})\underline{\overline{a}}$.
Note that if $\Omega(\zeta, t) = \sum_{n=1}^{\infty} a_n(t)\zeta^n$ satisfies (H) then

$$\frac{d}{dt} \int\limits_{|z|=1} \overline{\Omega}(z)\Omega'(z)dz = \frac{d}{dt} \sum_{\ell=1}^{\infty} \ell |a_\ell(t)|^2 = \frac{d}{dt}\, u_1(t) = 0 \;.$$

Hence

$$\pi \sum_{\ell=1}^{\infty} \ell |a_\ell(t)^2 = \text{constant} = \pi \;,$$

which means 'conservation of area'.

In the next theorem we gather some results on a Taylor series representation of $\mathcal{F}(|\Omega'|)$.

THEOREM.

- For all $\Omega \in \Sigma$ the function

$$\tfrac{1}{2}[\Omega'(\zeta)\overline{\Omega'(\tfrac{1}{\overline{\zeta}})}]^{-\frac{1}{2}} = \sum_{n=-\infty}^{\infty} \alpha_n \zeta^n$$

is analytic and single valued on an annulus $\frac{1}{R} < |\zeta| < R,\; R > 1$.

- Define

$$\mathcal{F}(|\Omega'(\zeta, t)|) = \alpha_0 + 2\sum_{n=1}^{\infty} \alpha_n \zeta^n = \sum_{n=0}^{\infty} \beta_n \zeta^n$$

then $\mathcal{F}$ satisfies the conditions (F)

- $$\alpha_n = \frac{1}{4\pi}\int_0^{2\pi}\Big\{\sum_{m=1}^{\infty}\sum_{\ell=1}^{\infty} m\ell\; a_m\overline{a}_\ell e^{i(m-\ell)s}\Big\}^{-\frac{1}{2}} e^{-ins}ds$$

- $$\alpha_0 = \beta_0 > 0 \qquad \alpha_{-n} = \overline{\alpha_n}$$

- $$\{a_m\} \subset I\!R \quad \Rightarrow \quad \{\alpha_n\} \subset I\!R\,.$$ □

We now calculate the second term in (H)

$$\frac{d}{d\zeta}(\zeta\overline{\Omega(\frac{1}{\overline{\zeta}})}\;\Omega'(\zeta)\;\mathcal{F}(|\Omega'(\zeta)|) =$$

$$\frac{d}{d\zeta}\{\zeta\sum_{m=-\infty}^{\infty} u_m\zeta^{-m}\cdot\sum_{n=0}^{\infty}\beta_n\zeta^n\} =$$

$$\sum_{m=-\infty}^{\infty}\sum_{n=0}^{\infty}(n-m+1)\beta_n u_m\zeta^{n-m} = \sum_{k=-\infty}^{\infty}(-k+1)\Big[\sum_{n=0}^{\infty}\beta_n u_{n+k}\Big]\zeta^{-k}\,.$$

The singular part is

$$\sum_{k=1}^{\infty}(-k+1)\Big[\sum_{n=0}^{\infty}\beta_n u_{n+k}\Big]\zeta^{-k}\,.$$

Now the condition of cancellation of singularities leads to the following infinite system of ordinary differential equations

$$\frac{d}{dt}\begin{pmatrix} a_1 & 2a_2 & 3a_3 & 4a_4 & \cdots \\ 0 & a_1 & 2a_2 & 3a_3 & \cdots \\ 0 & 0 & a_1 & 2a_2 & \cdots \\ 0 & 0 & 0 & a_1 & \cdots \\ \vdots & \vdots & \vdots & \vdots & \cdots \end{pmatrix}\begin{pmatrix}\overline{a}_1\\ \overline{a}_2\\ \overline{a}_3\\ \overline{a}_3\\ \vdots\end{pmatrix} =$$

$$-\begin{pmatrix} 0 & 0 & 0 & 0 & \cdots \\ 0 & \beta_0 & \beta_1 & \beta_2 & \cdots \\ 0 & 0 & 2\beta_0 & 2\beta_1 & \cdots \\ 0 & 0 & 0 & 3\beta_0 & \cdots \\ \vdots & \vdots & \vdots & \vdots & \cdots \end{pmatrix}\begin{pmatrix} a_1 & 2a_2 & 3a_3 & 4a_4 & \cdots \\ 0 & a_1 & 2a_2 & 3a_3 & \cdots \\ 0 & 0 & a_1 & 2a_2 & \cdots \\ 0 & 0 & 0 & a_1 & \cdots \\ \vdots & \vdots & \vdots & \vdots & \cdots \end{pmatrix}\begin{pmatrix}\overline{a}_1\\ \overline{a}_2\\ \overline{a}_3\\ \overline{a}_4\\ \vdots\end{pmatrix}.$$

In short

$$= \frac{d}{dt}\{M(\underline{a})\underline{\overline{a}}\} = -B(\underline{a})\underline{\overline{a}}\}\,,\quad \underline{a}(0) = \underline{a}_0\,.$$

Let there be given an initial condition $\underline{a}(0) = \underline{a}_0$ and put $\underline{u}_0 = M(\underline{a}_0)\underline{\overline{a}}_0$. Now if from the infinite system of quadratic equations

$$M(\underline{a})\overline{\underline{a}} = \underline{u} .$$

$\underline{a}$ is locally solvable as a function of $\underline{u}$ around $\underline{u} = \underline{u}_0$ then the initial value problem is reduced to an initial value problem for the infinite system of quasi–linear differential equations

$$\frac{d}{dt}\,\underline{u} = -B(\underline{a}(\underline{u}))\underline{u} \;, \quad \underline{u}(0) = \underline{u}_0 \;.$$

Not much can be said about the solvability of this dynamical system at this moment. If every solution is a trajectory on the above mentioned ellipsoid in Hilbert space then the shape would remain simply connected if the initial domain is simply connected. Most probably such a deep result does not have a simple proof.

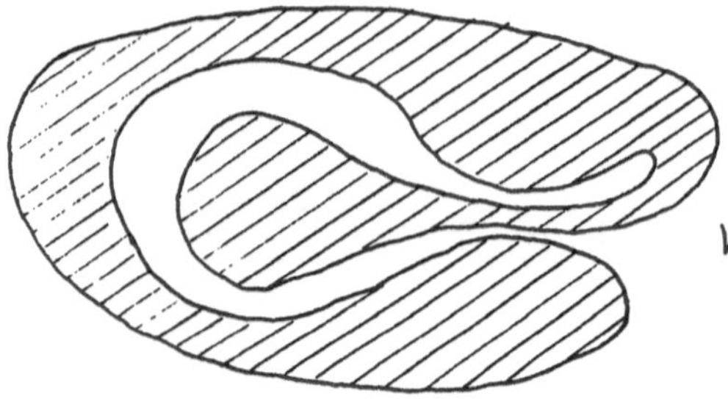

2.2 The real polynomial Hopper problem

If we substitute the Ansatz

$$\Omega(\zeta, t) = \sum_{n=1}^{N} a_n(t)\zeta^n \;, \quad n \in \mathbb{N},\; a_n :\; [0,\infty) \to \mathbb{R}$$

in Hopper's equation (H) we find, e.g. for $N = 4$, the following finite system of ordinary differential equations

$$\frac{d}{dt}\begin{pmatrix} u_1 \\ u_2 \\ u_3 \\ u_3 \end{pmatrix} = -\begin{pmatrix} 0 & 0 & 0 & 0 \\ - & \beta_0 & \beta_1 & \beta_2 \\ 0 & 0 & 2\beta_0 & 2\beta_1 \\ 0 & 0 & 0 & 3\beta_0 \end{pmatrix}\begin{pmatrix} u_1 \\ u_2 \\ u_3 \\ u_4 \end{pmatrix} ,$$

in short $\frac{du}{dt} = -B(\underline{a})\underline{u}$ with

$$\begin{pmatrix} u_1 \\ u_2 \\ u_3 \\ u_4 \end{pmatrix} = \begin{pmatrix} a_1 & 2a_2 & 3a_3 & 4a_4 \\ 0 & a_1 & 2a_2 & 3a_3 \\ 0 & 0 & a_1 & 2a_2 \\ 0 & 0 & 0 & a_1 \end{pmatrix}\begin{pmatrix} a_1 \\ a_2 \\ a_3 \\ a_4 \end{pmatrix}$$

$$= \begin{pmatrix} a_1 & 2a_2 & 3a_3 & 4a_4 \\ a_2 & 2a_3 & 3a_4 & 0 \\ a_3 & 2a_4 & 0 & 0 \\ a_4 & 0 & 0 & 0 \end{pmatrix} \begin{pmatrix} a_1 \\ a_2 \\ a_3 \\ a_4 \end{pmatrix}$$

in short $\underline{u} = M(\underline{a})\underline{a} = N(\underline{a})\underline{a}$ and

$$\beta_n(\underline{a}) = (2 - \delta_{n0}) \frac{1}{4\pi} \int_0^{2\pi} \Big\{ \sum_{m=1}^{N} \sum_{\ell=1}^{N} m\ell a_m a_\ell e^{i(m-\ell)s} \Big\}^{-\frac{1}{2}} e^{ins} ds \,.$$

Denote $\underline{a} = (1, 0, 0, \ldots, 0) = \omega$.

The following properties are straightforward

PROPERTIES of β_n, $1 \le n \le N$

- $\beta_0(\underline{a}) > \frac{1}{N}$, for all $\underline{a} \in \Sigma$
- $\beta_n(\underline{\omega}) = \frac{1}{2}\delta_{n0}$, $1 \le n \le N$
- $\beta_n(\underline{a}) \to 0$ als $\underline{a} \to \underline{\omega}$, $1 \le n \le N$
- $\beta_0(\underline{a}) \uparrow \infty$ if $\underline{a} \to \partial\Sigma$.

The derivative of $\underline{u}$, e.g. is $N = 4$, is found to be

$$\frac{d\underline{u}}{d\underline{a}} = \begin{pmatrix} a_1 & 2a_2 & 3a_3 & 4a_4 \\ a_2 & 2a_3 & 3a_4 & 0 \\ a_3 & 2a_4 & 0 & 0 \\ a_4 & 0 & 0 & 0 \end{pmatrix} + \begin{pmatrix} a_1 & 2a_2 & 3a_3 & 4a_4 \\ 0 & a_1 & 2a_2 & 3a_3 \\ 0 & 0 & a_1 & 2a_2 \\ 0 & 0 & 0 & a_1 \end{pmatrix} .$$

So

$$\frac{d\underline{u}}{d\underline{a}}(\underline{\omega}) = \begin{pmatrix} 2 & 0 & 0 & 0 \\ 0 & 1 & 0 & 0 \\ 0 & 0 & 1 & 0 \\ 0 & 0 & 0 & 1 \end{pmatrix} .$$

Applying the Inverse Function Theorem we find that $\underline{a}$ can be solved locally as a function of $\underline{u}$ in a neighbourhood of $\underline{u} = \underline{\omega}$.
Via the method of variation of constants we find for the components u_j, $2 \le j \le N$,

$$|u_j(t)| \le C_j e^{-\frac{(j-1)}{n}t} .$$

So $\underline{\omega}$ is a local attractor. Near $\underline{\omega}$ there is exponential decay: $|\underline{a} - \underline{\omega}| \le Ce^{-\frac{1}{2}t}$.

Note that there are special solutions $\underline{a}(t)$ with $a_1(t) \neq 0$, $a_N(t) \neq 0$ and $a_2(t) = \ldots = a_{N-1}(t) = 0$. These are the typical solutions in Hopper's work. He 'guesses' shapes with one parameter and then solves an ordinary differential equation for this parameter as a function of t, cf. [H1], [H2], [H3].

EXAMPLE (Hopper 1990).
Try to solve Hopper's equation by the function

$$\Omega(\zeta, t) = a(t)\zeta - \frac{b(t)}{N+1}\,\zeta^N$$

with $a(t) \geq b(t) > 0$ and $N \in I\!N$ fixed.
Calculate

- $$\Omega' = a - b\zeta^N$$

- $$\overline{\Omega} = a\,\frac{1}{\zeta} - \frac{b}{N+1}\,\zeta^{-(N+1)}$$

- $$\overline{\Omega'} = a - b\zeta^{-N}$$

- $$\overline{\Omega}\Omega' = (a^2 + \frac{b^2}{N+1})\zeta^{-1} - \frac{ab}{N+1}\,\zeta^{-(N+1)} - ab\zeta^{-(N-1)}$$

- $$\frac{1}{2}(\Omega'\overline{\Omega'})^{-\frac{1}{2}} = \frac{1}{2}(a^2 + b^2)^{-\frac{1}{2}}[1 - \frac{ab}{a^2+b^2}(\zeta^N + \zeta^{-N})]^{-\frac{1}{2}}$$

if $a > b$ then $\frac{ab}{a^2+b^2} < \frac{1}{2}$

- $$\mathcal{F} = \alpha_0 + 2\alpha_N\zeta^N + 2\alpha_{2N}\zeta^{2N} + \cdots\,,$$

$$\alpha_0(a,b) = \frac{1}{4\pi}(a^2+b^2)^{-\frac{1}{2}} \int_0^{2\pi} (1 - \frac{2ab}{a^2+b^2}\;\cos\theta)^{-\frac{1}{2}}\,d\theta$$

- $$\frac{d}{dt}(\overline{\Omega}\Omega') = 2(a\dot{a} + \frac{b\dot{b}}{N+1})\zeta^{-1} + (\frac{\dot{a}b + a\dot{b}}{N+1})\zeta^{-(N+1)} + \text{ Taylorseries}$$

- $$\frac{d}{d\zeta}[\zeta\overline{\Omega}\Omega'(\alpha_0 + 2\alpha_N\zeta^N)] = ab\alpha_0(a,b)\frac{N}{N+1}\,\zeta^{-(N+1)}\,.$$

The system of two ordinary differential equations

$$\begin{cases} a\dot{a} + \dfrac{b\dot{b}}{N+1} = 0 \\[2ex] \dot{a}b + a\dot{b} = -ab\alpha_0(a,b)N \end{cases}$$

can be written explicitely

$$\begin{cases} \dot{a} = \alpha_0(a,b)\dfrac{ab^2N}{(N+1)a^2-b^2} \\ \\ \dot{b} = -\alpha_0(a,b)\dfrac{a^2bN(N+1)}{(N+1)a^2-b^2} \end{cases}$$

This system is singular if $a = b$. At this point the decay of b is faster than exponential. For small b there is exponential decay

$$b(t) \approx b_0 e^{-\frac{1}{2}tN} \quad (e^{-\frac{1}{2}\frac{\gamma}{\eta}\frac{N}{R}t}) \, .$$

2.3 The Complex polynomial Hopper problem

If we substitute the Ansatz

$$\Omega(\zeta,t) = \sum_{n=1}^{N} a_n(t)\zeta^n \, , \quad N \in \mathbb{N} \, , \quad n : [0,\infty) \to \mathbb{C}$$

in Hopper's equation (H) we find, again, the finite system of ordinary differential equations

$$\frac{d}{dt}\underline{u} = -B(\underline{a})\underline{u} \, ,$$

but now with

$$\beta_n(\underline{a}) = (2-\delta_{n0})\frac{1}{4\pi}\int_0^{2\pi} \Big\{ \sum_{m=1}^{N}\sum_{\ell=1}^{N} m\ell a_m \overline{a}_\ell e^{i(m-\ell)s} \Big\}^{-\frac{1}{2}} e^{-ins} ds \, ,$$

and

$$\underline{u} = M(a)\underline{\overline{a}} = N(\underline{\overline{a}})\underline{a} \qquad (S) \, .$$

Note that $u_1 = \overline{a}_1 a_1 + 2\overline{a}_2 a_2 + \ldots + N\overline{a}_N a_N \in \mathbb{R}$. So in order to make quasi linearization possible, at least locally, we require $u_1 \in \mathbb{R}$. Then (unlike Hopper in [H1]) we find that the system (S) consists of $2n-1$ real equations with $(2n-1)$ unknowns.

Now define

$$H : \mathbb{R}^{2n-1} \to \mathbb{R}^{2n-1} \quad : \quad \underline{a} \mapsto \underline{u} = M(\underline{a})\underline{\overline{a}}$$

with $a_1 \in \mathbb{R}$, $u_1 \in \mathbb{R}$.
The *real* derivative DH of H at $\underline{a}$ is, with complex notation and $z = (z_1, \ldots, z_N)$, $z_1 = \overline{z}_1$

$$\underline{a} \mapsto DH(\underline{a})\underline{z} = \begin{bmatrix} a_1 & 2a_2 & 3a_3 & 4a_4 \\ 0 & a_1 & 2a_2 & 3a_3 \\ 0 & 0 & a_1 & 2a_2 \\ 0 & 0 & 0 & a_1 \end{bmatrix} \begin{bmatrix} \overline{z}_1 \\ \overline{z}_2 \\ \overline{z}_3 \\ \overline{z}_4 \end{bmatrix}$$

$$+\begin{bmatrix} \overline{a}_1 & 2\overline{a}_2 & 3\overline{a}_3 & 4\overline{a}_4 \\ \overline{a}_2 & 2\overline{a}_3 & 3\overline{a}_4 & 0 \\ \overline{a}_3 & 2\overline{a}_4 & 0 & 0 \\ \overline{a}_4 & 0 & 0 & 0 \end{bmatrix} \begin{bmatrix} z_1 \\ z_2 \\ z_3 \\ z_4 \end{bmatrix} .$$

The real linear mapping $DH(\underline{a})$ is invertible at $\underline{a} = \underline{\omega}$. So also in the complex case $\underline{\omega}$ turns out to be a local attractor. With the modifications mentioned in this section, the complex polynomial Hopper problem can be attacked with the same methods as the real polynomial Hopper problem in the preceding section.

2.4 The Rational Hopper Problem

The Hopper equation

$$\frac{d}{dt}(\overline{\Omega}\Omega') - \frac{d}{d\zeta}(\zeta\overline{\Omega}\Omega'\mathcal{F}) = -2\chi' = \text{ analytic on } \overline{D}$$

can be written

$$\overline{\Omega}[(1+\zeta\frac{\Omega''}{\Omega'})\mathcal{F} + \zeta\mathcal{F}' - \frac{\dot{\Omega}'}{\Omega'}] - \zeta\overline{\Omega}'\mathcal{F} - \dot{\overline{\Omega}} = -\frac{2\chi'}{\Omega'} = \text{ analytic on } \overline{D} .$$

Following Hopper we take the Ansatz

$$\Omega = \zeta \sum_{n=1}^{N} \frac{A_n}{1-\alpha_n\zeta} , \qquad \overline{\Omega} = \sum_{n=1}^{N} \frac{\overline{A}_n}{\zeta - \overline{\alpha}_n} ,$$

$$\overline{\Omega}' = - \sum_{n=1}^{N} \frac{\overline{A}_n}{(\zeta - \overline{\alpha}_n)^2} ,$$

$$\dot{\overline{\Omega}} = \sum_{n=1}^{N} \frac{\dot{\overline{A}}_n}{(\zeta - \overline{\alpha}_n)} + \sum_{n=1}^{N} \frac{\overline{A}_n \dot{\overline{\alpha}}_n}{(\zeta - \overline{\alpha}_n)^2} .$$

After substitution in Hopper's equation and rearranging

$$\Big(\sum_{n=1}^{N} \frac{\overline{A}_n}{\zeta - \overline{\alpha}_n}\Big) \Big[(1+\zeta\,\frac{\Omega''}{\Omega'})\mathcal{F} + \zeta\mathcal{F}' - \frac{\dot{\Omega}'}{\Omega'}\Big] +$$

$$-\Big[\sum_{n=1}^{N} \frac{\overline{A}_n}{(\zeta - \overline{\alpha}_n)^2}\Big]\zeta\mathcal{F} - \sum_{n=1}^{N} \frac{\dot{\overline{A}}_n}{(\zeta - \overline{\alpha}_n)} - \sum_{n=1}^{N} \frac{\overline{A}_n \dot{\overline{\alpha}}_n}{(\zeta - \overline{\alpha}_n)^2} = \text{ analytic on } D .$$

Since 2nd order and 1st order poles have to compensate each other on D we find the following two sets of ordinary differential equations

$$\begin{cases} \dfrac{\dot{\alpha}_n}{\alpha_n} = -\overline{\mathcal{F}(|\Omega'(\overline{\alpha}_n, t)|} \\[2ex] \dfrac{\dot{A}_n}{A_n} = -\Big[\alpha_n \overline{\dfrac{\Omega''(\overline{\alpha}_n)}{\Omega'(\overline{\alpha}_n)}}\, \dfrac{\dot{\alpha}_n}{\alpha_n} + \overline{\dfrac{\dot{\Omega}'(\overline{\alpha}_n)}{\Omega'(\overline{\alpha}_n)}}\Big] \end{cases}$$

Remind that, in this special case, $\mathcal{F}$ is a function of $\alpha_1, \ldots, \alpha_N, A_1, \ldots, A_N$.

So we have $2N$ complex, coupled, explicit ordinary differential equations which are locally solvable.

EXAMPLE (Hopper 1989).

Exact solution of the problem of coalescence of 2 equal cylinders.
Take

$$z = \Omega(\zeta, \nu(t)) = \frac{1-\nu^2}{(1+\nu^2)^{\frac{1}{2}}} \, \frac{\zeta}{1+\nu\zeta^2} \, .$$

The inverse of the 'parameter function' $\nu(t)$ is

$$t = \tfrac{1}{2}\nu \int\limits_{\nu}^{1} [k(1+k^2)^{\frac{1}{2}} \; K(k)]^{-1} \; dk$$

with K an elliptic integral. See [H1].

REFERENCES

[H1] HOPPER, R.W. 1990 Plane Stokes flow driven by capillarity on a free surface. *J. Fluid Mech.* **213**, 349-375.

[H2] HOPPER, R.W. 1991 Plane Stokes flow driven by capillarity on a free surface, 2: Further developments. *J. Fluid Mech.* **230**, 355-364.

[H3] HOPPER, R.W. 1992 Stokes flow of a cylinder and half–space driven by capillarity. *J. Fluid Mech.* **243**, 171-181.

[M] MUSKHELISHVILI, N.I., Some basic problems of the mathematical theory of elasticity. Groningen–Holland, 1953.

[R] RICHARDSON, S., Two dimensional flows with time–dependent free boundaries driven by surface tension. *Euro J. of Appl. Maths* (1992), **3**, 193-207.

[VM] VAN DE VORST, G.A.L., MATTHEIJ, R.M.M., KUIKEN, H.K. 1992a A boundary element solution for two–dimensional viscous sintering. *J. Comput. Phys.* **100**, 50-63.

[VM2] VAN DE VORST, G.A.L., MATTHEIJ, R.M.M. 1992b A BDF-BEM Scheme for modelling viscous sintering. In *Proc. Conf. on Boundary Element Technology VII,* (ed. C.A. Brebbia and M.S. Ingber), pp. 59-74. Computational Mechanics Publications, Southampton.

[VM3] VAN DE VORST, G.A.L., MATTHEIJ, R.M.M. 1992c Numerical analysis of a 2-D viscous sintering problem with non smooth boundaries. *Computing* **49**, 239-263 (1992).

QUANTIZATION OF INTEGRABLE MAPPINGS

F.W. Nijhoff[1] and H.W. Capel[2]

1 Department of Mathematics and Computer Science
and Institute for Nonlinear Studies,
Clarkson University, Potsdam NY 13699-5815, USA
2 Institute for Theoretical Physics, University of Amsterdam,
Valckenierstraat 65, 1018 XE Amsterdam, The Netherlands

Abstract We study a quantum Yang-Baxter structure associated with non-ultralocal lattice models. We discuss the canonical structure of a class of integrable quantum mappings, i.e. iterative canonical transformations that can be interpreted as the time-one step of a discrete-time evolution. As particular examples we consider quantum mappings associated with the lattice analogues of the KdV and MKdV equations, together with their exact quantum invariants.

1 Introduction

Discrete integrable models, in which the spatial dimension is discretized, but the time is continuous, have traditionally played an important role in mathematics and physics, both in the classical as well as in the quantum regime. On the quantum level the algebraic structure of integrable systems is discussed in terms of quantum groups [1]-[3]. The discretized version of such models has played a particular role in this respect, e.g. in the quantum inverse scattering method [4]. The models, in which also the time-flow is discretized (i.e. integrable lattices or partial difference equations), have been considered on the classical level in a number of papers [5, 6]. Recently, they have become of interest in connection with the construction of integrable *mappings*, i.e. finite-dimensional reductions of these integrable lattice equations, [7, 8]. Their integrability is to be understood in the sense that the discrete time-flow is the iterate of a canonical transformation, preserving a suitable symplectic structure, leading to invariants which are in involution with respect to this symplectic form. A theorem à la Liouville then tells us in analogy with the continuous-time situation that one can linearize the discrete-time flow on a hypertorus which is the intersection of the level sets of the invariants, [9]. Integrable mappings have been considered from a slightly different perspective also in the recent literature, cf. e.g. [10]-[13].

Integrable two-dimensional lattices arise, both on the classical as well as on the quantum level, as the compatibility conditions of a discrete-time ZS

(Zakharov-Shabat) system

$$L'_n(\lambda)\cdot M_n(\lambda) \;=\; M_{n+1}(\lambda)\cdot L_n(\lambda)\;, \tag{1.1}$$

in which λ is a spectral parameter, L_n is the lattice translation operator at site n, and the prime denotes the discrete time-shift corresponding to a translation in the second lattice direction. As L and M, in the quantum case, depend on operators, the question of operator ordering becomes important. Throughout this paper we impose in the quantum case as a normal order the order which is induced by the lattice enumeration, with n increasing from the left to the right. Finite-dimensional mappings are obtained from (1.1) imposing a periodicity condition

$$L_n(\lambda) \;=\; L_{n+P}(\lambda)\quad,\quad M_n(\lambda) \;=\; M_{n+P}(\lambda) \tag{1.2}$$

for some $P \in N$.

In a recent paper [14], we introduced a novel quantum structure that is appropriate for obtaining an integrable quantization of mappings of the so-called KdV-type, i.e. mappings derived from a lattice version of the KdV equation, cf. [7]. In this paper we will review the construction of such integrable quantum mappings and their quantum invariants and we consider also the quantum mappings associated with the lattice version of the MKdV equation, cf. also [15]. A related system with continuous time is the quantum Volterra system, cf. [16]. As was indicated in [8], it turns out that these mappings and their underlying integrable lattices are –on the classical level– symplectic with respect to a so-called *non-ultralocal* Poisson structure, cf. also [17]. In the continuous-time case such (classical) non-ultralocal r-matrix structures have been studied in a number of papers, cf. e.g. [18]-[22]. The discrete version of the non-ultralocal Poisson bracket structure reads, cf. ref. [8],

$$\begin{aligned}\{L_{n,1}, L_{m,2}\} = -\delta_{n,m+1}\, L_{n,1}\, s^{+}_{12}\, L_{m,2} \;+\; \delta_{n+1,m}\, L_{m,2}\, s^{-}_{12}\, L_{n,1} \\ +\, \delta_{n,m}\left[r^{+}_{12}\, L_{n,1}L_{m,2} \;-\; L_{n,1}L_{m,2}\, r^{-}_{12}\right]\;,\end{aligned} \tag{1.3}$$

Throughout this paper we adopt the usual convention that the subscripts $1, 2, \cdots$ in (1.3) denote the factors in a matricial tensor product, i.e. $A_{i_1,i_2,\ldots,i_M} = A_{i_1,i_2,\ldots,i_M}(\lambda_1, \lambda_2, \ldots, \lambda_M)$ denotes a matrix acting nontrivially only on the factors labeled by $i_1, i_2, \ldots, i_M$ of a tensor product $\otimes_\alpha V_\alpha$, of vector spaces V_α and trivially on the other factors, cf. also e.g. [4, 17]. For example, in eq. (1.3), the subscripts $\alpha, \beta = 1, 2, \cdots$ for the operator matrices $L_{n,\alpha}$ denote the corresponding factor on which this L_n acts (acting trivially on the other factors), i.e. $L_{n,1} = L_n(\lambda_1) \otimes 1$, $L_{n,2} = 1 \otimes L_n(\lambda_2)$. We suppress the explicit dependence on the spectral parameter $\lambda = \lambda_1$ respectively $\lambda = \lambda_2$, assuming that each value accompanies its respective factor in the tensor product.

Eq. (1.3) defines a proper Poisson bracket provided that the following relations hold for $s^\pm = s^\pm(\lambda_1, \lambda_2)$ and $r^\pm = r^\pm(\lambda_1, \lambda_2)$:

$$s^{-}_{12}(\lambda_1, \lambda_2) \;=\; s^{+}_{21}(\lambda_2, \lambda_1)\quad,\quad r^{\pm}_{12}(\lambda_1, \lambda_2) \;=\; -r^{\pm}_{21}(\lambda_2, \lambda_1)\;, \tag{1.4}$$

to ensure the skew-symmetry, and

$$\left[r^{\pm}_{12}, r^{\pm}_{13}\right] + \left[r^{\pm}_{12}, r^{\pm}_{23}\right] + \left[r^{\pm}_{13}, r^{\pm}_{23}\right] = 0\,, \tag{1.5a}$$

$$\left[s^{\pm}_{12}, s^{\pm}_{13}\right] + \left[s^{\pm}_{12}, r^{\pm}_{23}\right] + \left[s^{\pm}_{13}, r^{\pm}_{23}\right] = 0, \tag{1.5b}$$

to ensure that the the Jacobi identities hold for the Poisson bracket (1.3). The relation (1.5a) for $r^{\pm}$ is nothing but the usual classical Yang-Baxter equation (CYBE). As a consequence of the ZS system (1.1) we have on the classical level a complete family of invariants of the mapping ,namely by introducing the associated monodromy matrix $T(\lambda)$, obtained by gluing the elementary translation matrices L_j along a line connecting the sites 1 and $P+1$ over one period P, namely

$$T(\lambda) \equiv \overleftarrow{\prod_{n=1}^{P}} L_n(\lambda)\,. \tag{1.6}$$

In order to be able to integrate (1.3) to obtain Poisson brackets for the monodromy matrix we need in addition to these relations the extra relation

$$r^{+}_{12} - s^{+}_{12} = r^{-}_{12} - s^{-}_{12}\,. \tag{1.7}$$

In the classical case the traces of powers of the monodromy matrix are invariant under the mapping as a consequence of

$$T'(\lambda) = M_{P+1}(\lambda) T(\lambda) M_1^{-1}(\lambda) \tag{1.8}$$

and the periodicity condition $M_{P+1} = M_1$, thus leading to a sufficent number of invariants which are obtained by expanding the traces in powers of the spectral parameter λ. The involution property of the classical invariants follows from the Poisson bracket

$$\{trT(\lambda), trT(\lambda')\} = 0 \tag{1.9}$$

which can be derived from (1.3).

For the quantum mappings we will use the structure of [14, 23] which is the quantum analogue of this non-ultralocal Poisson structure. In the continuous-time case such a novel quantization scheme was proposed in ref. [24], in connection with the quantum Toda theory. Similar structures with continuous time flow have been introduced also for the quantum Wess-Zumino-Novikov-Witten (WZNW) theory with discrete spatial variable, cf. [25, 26]. When considering discrete-time flows some interesting new features arise, as was indicated in [14. 23]. In fact, the conventional point of view, that the M-part of the Lax equations does not need to be considered explicitly in order to construct quantum invariants, is no longer true. Therefore, one needs to establish the complete quantum algebra, containing commutation relations between the L-operators as well as between the L- and the M-operators, and between the M-operators themselves. As a consequence we will find, in the quantum mappings under consideration, non-trivial quantum corrections in the quantum invariants of the mappings. From an algebraic point of view, the basic algebraic relations for the monodromy matrices, that are relevant in the context of non-ultralocal models, are similar to the algebras of currents of a quantum group, [3], that have appeared also in different contexts, [27]-[29]. Interestingly enough, the relations between the monodromy matrix and the time-part of the Lax representation are very similar to the relations associated with the description of the cotangent bundle of a quantum group $(T^*G)_q$ [30].

The outline of this paper is as follows. In section 2 we introduce the basic ingredients of the non-ultralocal quantum R, S-structure. In section 3 we investigate the canonical structure of quantum gauge- or similarity transformations, leading to (integrable) quantum mappings. This leads to a 'full' Yang-Baxter structure including the discrete-time part of the Lax representation. In section 4 we present two examples of this structure: quantum mappings associated with the lattice KdV and with the lattice MKdV equation. In order to establish the quantum integrability of these mappings, we then develop in section 5 the 'full' quantum structure for the monodromy matrix, and show how to construct commuting families of exact quantum invariants for these mappings. Finally, in section 6 we give a construction of the generating operator of the quantum mappings, as a canonical transformation on the quantum phase space, in the special cases of the KdV and MKdV systems.

2 Non-ultralocal Yang-Baxter structure

We now define a quantum Yang-Baxter structure that is adequate for the mappings in this paper, i.e. discrete-time systems arising (both on the classical as well as quantum level) from compatibility equations of the form of (1.1).

We introduce the *quantum* L-operator $L_n(\lambda)$ at each site n of a one-dimensional lattice, which is a matrix whose entries are quantum operators (acting on some properly chosen Hilbert space). The operators $L_n(\lambda)$ are supposed to have only non-trivial commutation relations between themselves on the same and nearest-neighbour sites, namely as follows

$$R_{12}^{+}\, L_{n,1} \cdot L_{n,2} = L_{n,2} \cdot L_{n,1}\, R_{12}^{-} \tag{2.1a}$$

$$L_{n+1,1} \cdot S_{12}^{+}\, L_{n,2} = L_{n,2} \cdot L_{n+1,1}\ , \tag{2.1b}$$

$$L_{n,1} \cdot L_{m,2} = L_{m,2} \cdot L_{n,1}\ ,\ | n-m | \geq 2\ . \tag{2.1c}$$

These relations are the quantum analogue of the *non-ultralocal* Poisson bracket (1.3), defining what in [26] is referred to as the Kac-Moody algebra on the lattice. We will show in section 4 that the quantum mappings provide examples of such a non-ultralocal quantum R, S-matrix structure.

The compatibility relations of the equations (2.1a)-(2.1c) lead to the following consistency conditions on $R^{\pm}$ and S

$$R_{12}^{\pm}\, R_{13}^{\pm}\, R_{23}^{\pm} = R_{23}^{\pm}\, R_{13}^{\pm}\, R_{12}^{\pm}\ , \tag{2.2a}$$

$$R_{23}^{\pm}\, S_{12}^{\pm}\, S_{13}^{\pm} = S_{13}^{\pm}\, S_{12}^{\pm}\, R_{23}^{\pm}\ , \tag{2.2b}$$

where $S^+_{12} = S^-_{21}$. Eq. (2.2a) is the quantum Yang-Baxter equation (QYBE's) for $R^\pm$ coupled with an additional equation (2.2b) for $S^\pm$. For a derivation of eqs. (2.2) see appendix A.

In order to establish that the structure given by the commutation relations (2.1) allows for suitable commutation relations for the monodromy matrix, we need to impose in addition to (2.2) that

$$R^\pm_{12}\, S^\pm_{12} \;=\; S^\mp_{12}\, R^\mp_{12}\,. \tag{2.3}$$

Using these relations it is easy to establish that each sign of eqs. (2.2a),(2.2b) can be combined into a single equation as follows

$$R^\pm_{12}\,(R^\pm_{13}S^\pm_{13})\,S^\pm_{12}\,(R^\pm_{23}S^\pm_{23}) \;=\; (R^\pm_{23}S^\pm_{23})\,S^\mp_{12}\,(R^\pm_{13}S^\pm_{13})\,R^\mp_{12}\,. \tag{2.4}$$

At this point it is useful to introduce the following decomposition of the monodromy matrix (1.6)

$$T \;=\; T^+_n \cdot T^-_n\;, \tag{2.5}$$

in which

$$T^+_n(\lambda) \;=\; \prod_{j=n+1}^{\overleftarrow{P}} L_j(\lambda)\;\;,\;\; T^-_n(\lambda) \;=\prod_{j=1}^{\overleftarrow{n}} L_j(\lambda)\;, \tag{2.6}$$

First, one derives for the monodromy matrices T^+_n and T^-_n the following set of relations

$$R^+_{12}\,T^\pm_{n,1}\cdot T^\pm_{n,2} \;=\; T^\pm_{n,2}\cdot T^\pm_{n,1}\,R^-_{12}\;, \tag{2.7a}$$

$$T^+_{n,1}\cdot S^+_{12}\,T^-_{n,2} \;=\; T^-_{n,2}\cdot S^-_{12}\,T^+_{n,1}\;, \tag{2.7b}$$

for $2 \le n \le P-1$.

Next, taking into account the periodic boundary conditions we obtain for the monodromy matrix the commutation relations

$$R^+_{12}\,T_1\cdot S^+_{12}\,T_2 \;=\; T_2\cdot S^-_{12}\,T_1\,R^-_{12}\,. \tag{2.8}$$

Some details of the derivation of eqs. (2.7), (2.8) are presented in appendix B.

Remark: The classical limit of the quantum structure (2.1)-(2.2) is easily obtained by considering the quasi-classical expansion

$$\begin{aligned} S^\pm_{12} &= 1\otimes 1 \;-\; h\, s^\pm_{12} \;+\; \mathcal{O}(h^2)\,,\\ R^\pm_{12} &= 1\otimes 1 \;+\; h\, r^\pm_{12} \;+\; \mathcal{O}(h^2)\,, \end{aligned} \tag{2.9}$$

In this limit the quantum commutation relations (2.1) yield the non-ultralocal Poisson bracket structure given in eqs. (1.3)-(1.5).

3 Quantum Mappings

We are interested in the canonical structure of discrete-time integrable systems, i.e. systems for which the time evolution is given by an iteration of mappings. If the mapping contains quantum operators, the commutation relations with the monodromy or Lax matrices become nontrivial and it is not a priori clear in this case that the Yang-Baxter structure is preserved. Furthermore, the traces of powers of the monodromy matrix are no longer trivially invariant as the cyclic property of the traces is no longer true for operator-valued arguments. To deal with these new features, it is necessary to take the M-part of the Lax or ZS system also into account, and investigate the *full* quantum structure involved in these systems, consisting of commutation relations between the L-part as well as of the M-part of the Lax pair.

In [23] we have introduced such a full Yang-Baxter structure taking account of the spatial as well as the time part of the Lax pair. The structure is obtained by supplying in addition to eqs. (2.1) the following equations.

$$M_{n+1,1} \cdot S_{12}^{+} L_{n,2} = L_{n,2} \cdot M_{n+1,1} \ , \tag{3.1a}$$
$$L'_{n,2} \cdot S_{12}^{-} M_{n,1} = M_{n,1} \cdot L'_{n,2}, \tag{3.1b}$$

and

$$R_{12}^{+} M_{n,1} \cdot M_{n,2} = M_{n,2} \cdot M_{n,1} R_{12}^{-} \ , \tag{3.2a}$$
$$M'_{n,1} \cdot S_{12}^{+} M_{n,2} = M_{n,2} \cdot M'_{n,1} \ . \tag{3.2b}$$

The trivial commutation relations are the following

$$M_{n,1} \cdot L_{m,2} = L_{m,2} \cdot M_{n,1} \quad , \quad |n-m| \geq 2 \ , \tag{3.3a}$$
$$M_{n+1,1} \cdot L'_{m,2} = L'_{m,2} \cdot M_{n+1,1} \quad , \quad |n-m| \geq 2 \ , \tag{3.3b}$$
$$M_{n,1} \cdot M_{m,2} = M_{m,2} \cdot M_{n,1} \quad , \quad |n-m| \geq 2 \ , \tag{3.3c}$$

in combination with

$$M''^{\cdots}_{n,1} \cdot M_{n,2} = M_{n,2} \cdot M''^{\cdots}_{n,1} \ , \tag{3.4a}$$
$$M'_{n+1,1} \cdot M_{n,2} = M_{n,2} \cdot M'_{n+1,1} \ , \tag{3.4b}$$
$$M'_{n+1,1} \cdot L_{n,2} = L_{n,2} \cdot M'_{n+1,1} \ , \tag{3.4c}$$

for multiple applications of the mapping. We shall not specify other commutation relations, as they do not belong to the Yang-Baxter structure. More precisely, one may notice in the explicit examples of section 4 that the commutation relations

$$[L_n \overset{\otimes}{,} M_n] \ , \quad [L_{n+1} \overset{\otimes}{,} M_n] \ , \quad [M_{n+1} \overset{\otimes}{,} L'_n] \ , \quad [M_{n+1} \overset{\otimes}{,} L'_{n-1}] \ , \quad [M_{n+1} \overset{\otimes}{,} M_n] \ ,$$

are nontrivial, and they depend on the details of the system satisfying the Yang-Baxter equations. However, in order for the Yang-Baxter structure to be preserved under the mapping, we do not need information on these latter commutation relations. The main statement now is that the commutation relations between the matrices L_n are invariant under the mapping

$$L_n \to L'_n = M_{n+1}\, L_n\, M_n^{-1}\ , \tag{3.5}$$

see appendix C for some details.

For the quantum mappings under consideration here the operator L_n has a composite structure, i.e.

$$L_n = V_{2n} \cdot V_{2n-1} \tag{3.6}$$

and the commutation relations of the Yang-Baxter structure involving the L_n can be inferred from the commutation relations among the V_n themselves, as well as the commutation relations between the V_n and M_m. In fact, imposing the commutation relations

$$V_{n+1,1} \cdot S^+_{12}(n)\, V_{n,2} = V_{n,2} \cdot V_{n+1,1}\ , \tag{3.7a}$$

$$R^+_{12}\, V_{n,1} \cdot V_{n,2} = V_{n,2} \cdot V_{n,1}\, R^-_{12}\ , \tag{3.7b}$$

$$V_{n,1} \cdot V_{m,2} = V_{m,2} \cdot V_{n,1}\ ,\ |n-m| \geq 2\ , \tag{3.7c}$$

we obtain the relations (2.1) as can be easily verified.

In eq. (3.7a) the $S^+_{12}(2n)$ is independent of n and is equal to the S^+_{12} occuring in eq. (2.1b). For odd values of n, $S^+_{12}(n)$ may be a different solution of eqs. (2.1)-(2.3). The proof of eq. (2.1a) from eq. (3.7b) is essentially the same as the proof in appendix B showing how eq. (2.7a) is obtained from eqs. (2.1).

Next we impose the commutation relations between the operators V_n and M_n. The only nonvanishing commutation relations involving M_n are taken to be the following ones

$$\begin{matrix} V_{2n+1} \\ V_{2n} \\ V_{2n-1} \\ V_{2n-2} \end{matrix} \quad \leftrightarrow M_n \leftrightarrow \quad \begin{matrix} V'_{2n-1} \\ V'_{2n-2} \\ V'_{2n-3} \\ V'_{2n-4} \end{matrix}\ . \tag{3.8}$$

and in addition we impose simple commutation relations between M_n and V_{2n-2} and V'_{2n-1}, respectively

$$M_{n+1,1} \cdot S^+_{12} V_{2n,2} = V_{2n,2} \cdot M_{n+1,1}\ , \tag{3.9a}$$

$$V'_{2n-1,2} S^-_{12} \cdot M_{n,1} = M_{n,1} \cdot V'_{2n-1,2}\ , \tag{3.9b}$$

With the use of eqs. (3.8), (3.9) and (3.6) it is straightforward to derive eqs. (3.1) and (3.3). Eq. (3.4c) can also be shown replacing L_n by $M^{-1}_{n+1}\, L'_n\, M_n$ and taking account of the invariance of commutation relations under the mapping. The relations (3.7)-(3.9) are satisfied by the quantum mappings which will be considered in section 4. In section 5 we construct commuting families of quantum invariants on the basis of the full Yang-Baxter structure given above.

4 The Quantum Lattice KdV and MKdV

Here we consider two examples of integrable quantum mappings coming from the lattice analogues of the KdV and MKdV equations.

a) The first example of a concrete integrable family of quantum mappings that exhibit the structure outlined above, is the mapping of the KdV type (i.e. mappings arising from the periodic initial value problem of lattice versions of the KdV equation, see [7] and section 6 below). These are rational mappings $R^{2P} \to R^{2P} : (\{v_j\}) \mapsto (\{v'_j\})$ of the form

$$v'_{2j-1} = v_{2j} \ , \ v'_{2j} = v_{2j+1} + \frac{\epsilon\delta}{v_{2j}} - \frac{\epsilon\delta}{v_{2j+2}} \qquad (j = 1, \cdots, P), \tag{4.1}$$

imposing the periodicity condition $v_{i+2P} = v_i$. The mapping (4.1) has the Casimirs

$$\sum_{j=1}^{P} v_{2j} = \sum_{j=1}^{P} v_{2j-1} = c \ , \tag{4.2}$$

where c is chosen to be invariant under the mapping, in which case we obtain a $(2P-2)$-dimensional generalization of the McMillan mapping [10].

To obtain the Yang-Baxter structure it is worthwhile to note that eq. (4.1) arises as the compatibility condition of a ZS system (1.1) with

$$L_j = V_{2j} \cdot V_{2j-1} \ , \ M_j = \begin{pmatrix} a_j & 1 \\ \lambda_{2j} & 0 \end{pmatrix} \ , \tag{4.3}$$

$$V_i = \begin{pmatrix} v_i & 1 \\ \lambda_i & 0 \end{pmatrix}$$

in which $\lambda_{2j} = k^2 - q^2$, $\lambda_{2j+1} = k^2 - p^2$ and $\epsilon\delta = p^2 - q^2$. In fact, from the ZS condition (1.1) one obtains

$$a_j = v_{2j-1} - \frac{\epsilon\delta}{v_{2j}} \tag{4.4}$$

as well as the mapping (4.1). The corresponding classical invariants, obtained by expanding the trace of the monodromy matrix (1.6) in powers of k^2, are in involution, cf. eq. (1.9), with respect to the Poisson structure [7]

$$\{v_j \, , v_{j'}\} = \delta_{j+1,j'} - \delta_{j,j'+1} \ , \tag{4.5}$$

which was obtained using a Legendre transformation on an appropriately chosen Lagrangian [7]. This ensures that the mapping (4.1) is symplectic, i.e. the same Poisson brackets hold also for the primes variables v'_j. This property can also be checked easily by direct computation. On the basis of this a canonical transformation to action-angle variables can be found following ideas from [9], thereby showing complete integrability in the sense of Liouville [7, 8]. In the quantum case the variables v_j become hermitean operators on which we impose the following Heisenberg type of commutation relations, as a natural quantization of the Poisson relations (4.5), cf. [14, 23],

$$[v_j \, , v_{j'}] = h \left(\delta_{j,j'+1} - \delta_{j+1,j'}\right) \ , \tag{4.6}$$

(where $h = i\hbar$). It is easy to check that the quantum mapping (4.1) preserves the commutation relations (4.6), and in section 6, we show that the mapping is a canonical transformation with respect to these commutation relations.

The special solution of the quantum relations (2.2), (2.3), which constitutes the R, S-matrix structure for the quantum mapping (4.1), together with the commutation relation (4.6), is given by

$$\begin{aligned} R^+_{12} &= R^-_{12} - S^+_{12} + S^-_{12} \\ R^-_{12} &= 1 \otimes 1 \; + \; h \frac{P_{12}}{\mu_1 \; - \; \mu_2} \\ S^+_{12} &= 1 \otimes 1 \; - \frac{h}{\mu_2} F \otimes E \;\; , \;\; S^-_{12} = S^+_{21} \; , \end{aligned} \tag{4.7}$$

in which $\mu_\alpha = k_\alpha^2 - q^2, \alpha = 1, 2$ and the permutation operator P_{12} and the matrices E and F are given by

$$P_{12} = \begin{pmatrix} 1 & 0 & 0 & 0 \\ 0 & 0 & 1 & 0 \\ 0 & 1 & 0 & 0 \\ 0 & 0 & 0 & 1 \end{pmatrix} ,$$

$$E = \begin{pmatrix} 0 & 1 \\ 0 & 0 \end{pmatrix} \; , \; F = \begin{pmatrix} 0 & 0 \\ 1 & 0 \end{pmatrix} . \tag{4.8}$$

We mention the useful identity

$$R^+_{12} \; = \; \Lambda_1 \Lambda_2 \, R^-_{12} \Lambda_1^{-1} \Lambda_2^{-1} \; , \tag{4.9}$$

where $\Lambda_\alpha = \mu_\alpha F + E$, $(\alpha = 1, 2)$, from which it is evident that it is not strictly necessary to introduce two different R-matrices $R^\pm$.

The complete Yang-Baxter structure can now be derived from the mapping (4.1), the relation (4.4) for a_j and the commutation relation (4.6). In fact, from (4.6) one immediately obtains eq. (3.7a) with

$$S^+_{12}(n) = 1 \otimes 1 - \frac{h}{\lambda_{n,2}} F \otimes E \tag{4.10}$$

with $\lambda_{2j,2} = k_2^2 - q^2$, $\lambda_{2j-1,2} = k_2^2 - p^2$, and also eqs. (3.7b) and (3.7c). These relations are at the basis of the L part, i.e. eqs. (2.1), of the Yang-Baxter structure. To derive the commutation relations (3.8), (3.9) one first checks by explicit calculation that the only nonvanishing commutation relations between the matrix M_n and the matrices V_m, V'_m are indeed those indicated by eq. (3.8). Furthermore, one has the commutation relations

$$[M_{n+1} - V_{2n+1} \overset{\otimes}{,} V_{2n}] = 0 \;\; , \;\; [M_{n+1} - V'_{2n} \overset{\otimes}{,} V'_{2n+1}] = 0 \tag{4.11}$$

which with eq. (3.7a) and its counterpart in terms of the primed operators immediately yield eqs. (3.9). Finally the nontrivial commutation relations (3.2) follow from

$$[M_n \overset{\otimes}{,} M_n] = 0 \;\; , \;\; [M'_n - V'_{2n-1} \overset{\otimes}{,} M_n] = 0 \tag{4.12}$$

together with (3.9). The trivial commutation relations can be checked in a similar way.

Thus, the mapping (4.1) and its ZS system (4.3) with the commutation relation (4.6) satisfy the complete Yang-Baxter structure treated in sections 2 and 3.

Remark: The KdV mappings considered here are the discrete-time analogue of the quantum Volterra system treated in ref. [16]. Such systems are of interest, in connection with discretizations of the Virasoro algebra [31]-[34].

b) We now consider the example of the MKdV mappings, which is associated with the following R, S-matrices. Introducing

$$R_{12}(\lambda_{12}) = \begin{pmatrix} q\lambda_{12}-1 & 0 & 0 & 0 \\ 0 & \lambda_{12}-1 & q-1 & 0 \\ 0 & \lambda_{12}(q-1) & q(\lambda_{12}-1) & 0 \\ 0 & 0 & 0 & q\lambda_{12}-1 \end{pmatrix} , \tag{4.13a}$$

together with

$$S_{12} = \begin{pmatrix} 1 & & & \\ & 1 & & \\ & & 1 & \\ & & & q \end{pmatrix} , \tag{4.13b}$$

it is straightforward to check that the matrices of (4.13) obey for spectral parameter $\lambda_{12} = \lambda_1/\lambda_2$ the following relations

$$R_{12}\,\Lambda_1\,S_{21}\,\Lambda_2 = \Lambda_2\,S_{12}\Lambda_1\,R_{12} \ , \ \ R_{12}\,\Lambda_1^{-1}\,S_{12}\,\Lambda_2^{-1} = \Lambda_2^{-1}\,S_{21}\Lambda_1^{-1}\,R_{12} \ , \tag{4.14}$$

in which $\Lambda_1 = \Lambda(\lambda_1)\otimes 1$, $\Lambda_2 = 1\otimes\Lambda(\lambda_2)$ and

$$\Lambda(\lambda) = \begin{pmatrix} a & b \\ \lambda & d \end{pmatrix} , \tag{4.15}$$

where a, b and d are arbitrary constants. Eq. (4.20) then yields a solution of the Yang-Baxter relations (2.2), (2.3) with

$$\begin{aligned} R_{12}^- &\equiv R_{12}^-(\lambda_1,\lambda_2) = R_{12}(\lambda_1/\lambda_2) \ , \\ R_{12}^+ &= \Lambda_1\Lambda_2 R_{12}(\lambda_1/\lambda_2)(\Lambda_1\Lambda_2)^{-1} \ , \\ S_{12}^+ &= \Lambda_2 S_{12}\Lambda_2^{-1} \ , \ \ S_{12}^- = S_{12}^+ = \Lambda_1 S_{12}\Lambda_1^{-1} \ . \end{aligned} \tag{4.16}$$

As an example of a quantum mapping associated with this solution of the Yang-Baxter equation we consider

$$\begin{aligned} \varphi_{2n-1}' &= \varphi_{2n} \ , \\ e^{\varphi_{2n}'} &= \frac{(p_{2n}-r)+(p_{2n+1}+r)e^{\varphi_{2n+2}}}{(p_{2n+1}-r)+(p_{2n}+r)e^{\varphi_{2n+2}}}\, e^{\varphi_{2n+1}}\, \frac{(p_{2n-1}-r)+(p_{2n}+r)e^{\varphi_{2n}}}{(p_{2n}-r)+(p_{2n-1}+r)e^{\varphi_{2n}}} \ , \\ n &= 1,\cdots 2P \end{aligned} \tag{4.17}$$

which is a quantum version of the mapping associated with the lattice MKdV equation, cf. [7]. The MKdV mapping (4.17) arises as the compatibility condition of a ZS system (1.1) with $L_n = V_{2n} \cdot V_{2n-1}$, cf. (3.6), and

$$V_n = \Lambda_n \cdot \overline{V}_n \, , \, \overline{V}_n = \begin{pmatrix} 1 & 0 \\ 0 & e^{\varphi_n} \end{pmatrix} ,$$

$$\Lambda_n = \begin{pmatrix} p_n - r & 1 \\ \lambda & p_n + r \end{pmatrix} . \tag{4.18}$$

in which $\lambda = k^2 - r^2$, $p_{2n-1} = p$, $p_{2n} = q$ and

$$M_n = \Lambda_{2n} \cdot \begin{pmatrix} 1 & 0 \\ 0 & e^{\gamma_n} \end{pmatrix} , \tag{4.19}$$

In fact, working out (1.1) with (3.6) and (4.18), (4.19) one finds the mapping (4.17) and

$$e^{\gamma_n} = \frac{(p_{2n} - r) + (p_{2n-1} + r)e^{\varphi_{2n}}}{(p_{2n-1} - r) + (p_{2n} + r)e^{\varphi_{2n}}} e^{\varphi_{2n-1}} . \tag{4.20}$$

In the classical case the mapping is completely integrable with P integrals in involution with respect to the (invariant) Poisson bracket

$$\{\varphi_j, \varphi_{j'}\} = \delta_{j',j+1} - \delta_{j',j-1} \tag{4.21}$$

cf. eq. (1.9) and the expansion of $trT(\lambda)$ in powers of k^2.

In the quantum case we have the commutation relation

$$[\varphi_j, \varphi_{j'}] = h\,(\delta_{j,j'+1} - \delta_{j,j'-1}) \, , \tag{4.22}$$

implying in particular that

$$e^{\varphi_n} e^{\varphi_{n+1}} = q e^{\varphi_{n+1}} e^{\varphi_n} \, , \, q = e^{-h} . \tag{4.23}$$

Starting from (4.23) we find the commutation relations (3.7), in which the $R_{12}^{\pm}, S_{12}^{\pm}$ are given by (4.22) with

$$q = e^{-h} \, , \, x = \frac{k_1^2 - r^2}{k_2^2 - r^2} \tag{4.24}$$

and Λ_n given by (4.18) with $\lambda = k^2 - r^2$.

From the commutation relation (4.23), together with the explicit expression (4.20) for e^{γ_n} it is straightforward to derive the remaining relations of the Yang-Baxter structure, i.e. eqs. (3.8), (3.9) and (3.2), completely analogously to the case of the KdV-type of mappings. The trivial commutation relations can also be checked directly. Thus with the MKdV-type of mappings we have another example of the complete Yang-Baxter structure presented in sections 2 and 3, but here the $R_{12}^{\pm}$ and $S_{12}^{\pm}$ correspond to different (trigonometric) solutions of the Yang-Baxter equations.

5 Quantum Invariants

In the classical case the trace of the monodromy matrix yields a sufficient number of invariants which are in involution. In the quantum case the trace is no longer invariant and we have to consider more general families of commuting operators.

Following the treatment of ref. [27], a commuting parameter-family of operators is obtained by taking (for details, cf. appendix D)

$$\tau(\lambda) = tr\,(T(\lambda)K(\lambda)) \,, \tag{5.1}$$

for any family of numerical matrices $K(\lambda)$ obeying the relations

$$K_1\, {}^{t_1}\big(({}^{t_1}S^{-}_{12})^{-1}\big)\, K_2\, R^{+}_{12} = R^{-}_{12}\, K_2\, {}^{t_2}\big(({}^{t_2}S^{+}_{12})^{-1}\big)\, K_1 \,. \tag{5.2}$$

(We assume throughout that $S^{\pm}_{12}$ and $R^{\pm}_{12}$ are invertible). The left superscripts ^{t_1} and ^{t_2} denote the matrix transpositions with respect to the corresponding factors 1 and 2 in the matricial tensor product. Expanding (5.1) in powers of the spectral parameter λ, we obtain a set of commuting observables of the quantum system in terms of which we can find a common basis of eigenvectors in the associated Hilbert space. We note that a matrix $K(\lambda)$ is commonly introduced in connection with quantum boundary conditions other than periodic ones [27], but in relation to the quantum mappings of the present paper it is essential in the periodic case as well.

Furthermore the Yang-Baxter equations of section 3 lead to the following commutation relations between $M \equiv M_{n=1}$ and the monodromy matrix T,

$$T_1 \cdot M_1^{-1} \cdot S^{+}_{12} M_2 = M_2 \cdot S^{-}_{12} T_1 \cdot M_1^{-1} \,. \tag{5.3}$$

Here we use the notation $M_1 = M \otimes 1$, $M_2 = 1 \otimes M$ as usual for the factors 1 and 2 in the matricial tensor product. The derivation of eq. (5.3) is based on the commutation relation

$$M_{n,1} \cdot \left(L_{n+1} \cdot L_n \cdot M_n^{-1}\right)_2 = \left(L_{n+1} \cdot L_n \cdot M_n^{-1}\right)_2 \cdot S^{-}_{12} M_{n,1} \tag{5.4}$$

which is easily checked noting that

$$L_{n+1} \cdot L_n \cdot M_n^{-1} = M_{n+2}^{-1} \cdot L'_{n+1} \cdot L'_n \tag{5.5}$$

and using the commutation relation (3.1b) and the trivial relations (3.3b), (3.3c). Then with the use of (3.1a) it is found that

$$\begin{aligned} M_1 \cdot S^{+}_{12} \left(L_P \cdot L_{P-1} \cdot \cdots \cdot L_1 \cdot M^{-1}\right)_2 &= L_{P,2} \cdot M_1 \cdot \left(L_{P-1} \cdot \cdots \cdot L_1 \cdot M^{-1}\right)_2 \\ &= (L_P \cdot \cdots \cdot L_3)_2 \cdot \left(L_2 \cdot L_1 \cdot M^{-1}\right)_2 \cdot S^{-}_{12} \cdot M_1 \end{aligned} \tag{5.6}$$

which is just eq. (5.3).

The commutation relation (2.8) for the monodromy matrices is invariant. This can be shown noting that eqs. (2.1) are invariant under the mapping and by repeating the derivation of eq. (2.8), but now with the updated variables $L_j{}'$. It follows also directly from eq. (5.7) and the commutation relation (5.3). In fact,

$$\begin{aligned}
R^+_{12}T'_1 \cdot S^+_{12}T'_2 &= R^+_{12}M_1 \cdot T_1 \cdot M_1^{-1} \cdot S^+_{12}M_2 \cdot T_2 \cdot M_2^{-1} \\
&= R^+_{12}M_1 \cdot M_2 \cdot S^-_{12}T_1 \cdot M_1^{-1} \cdot T_2 \cdot M_2^{-1} \\
&= M_2 \cdot M_1 \cdot S^+_{12}R^+_{12}T_1 \cdot M_1^{-1} \cdot T_2 \cdot M_2^{-1} \\
&= M_2 \cdot M_1 \cdot S^+_{12}R^+_{12}T_1 \cdot S^+_{12}T_2 \cdot M_2^{-1} \cdot M_1^{-1}S^{-\,-1}_{12} \\
&= M_2 \cdot M_1 \cdot S^-_{21}T_2 \cdot S^-_{12}T_1 \cdot R^-_{12}M_2^{-1} \cdot M_1^{-1}S^{-\,-1}_{12} \\
&= T'_2 \cdot S^+_{21}M_1 \cdot M_2 \cdot S^-_{12}T_1M_1^{-1} \cdot M_2^{-1}S^{+\,-1}_{12}R^-_{12} \\
&= T'_2 \cdot S^-_{12}T'_1R^-_{12}\,, \qquad (5.7)
\end{aligned}$$

Our aim is now to describe the integrability of the quantum mappings of section 4 which obey the commutation relations (2.8) and (5.3). For this we need to show that one can find a sufficient family of commuting *invariants* of the mapping. Let us thus use eq. (5.3) to calculate commuting families of quantum invariants in the case of the KdV and MKdV mappings of section 4.

In fact, introducing a tensor

$$K_{12} \;=\; P_{12}K_1K_2\,, \qquad (5.8)$$

where P_{12} is the permutation operator satisfying e.g.

$$P_{12}A_1 = A_2P_{12}\,,\; P_{12}A_2 = A_1P_{12}\,,\; P_{12} = P_{21}\,,\; tr_2P_{12} = 1_1 \qquad (5.9)$$

for matrices A not depending on the spectral parameter, and choosing $\lambda_1 = \lambda_2$, we can take the trace over left- and right hand side of (5.3) contracting with K_{12}. This leads to

$$tr_{12}\left(K_{12}(TM^{-1})_1 \cdot S^+_{12}M_2\right) \;=\; tr_2\left(K_2T_2M_2^{-1}\,tr_1(P_{12}K_2S^+_{12})\,M_2\right) \;=\; tr(KT) \qquad (5.10)$$

provided that

$$tr_1(P_{12}K_2S^+_{12}) = 1_2\,. \qquad (5.11)$$

In eq. (5.10) $tr_{12} = tr_1tr_2$ denotes the trace over the factors 1 and 2 in the direct product space of matrices, whereas tr_1 and tr_2 are restricted to only one of these factors. Under the same condition (5.11) we have that

$$\begin{aligned}
&tr_{12}\left(K_{12}M_2 \cdot S^-_{12}(TM^{-1})_1\right) \\
&\quad=\; tr_1\left(K_1M_1\,tr_2(P_{12}K_1S^-_{12})\,(TM^{-1})_1\right) \;=\; tr(KT')\,. \qquad (5.12)
\end{aligned}$$

A solution to eqs. (5.11) is easily found, namely by taking

$$K_2 \;=\; tr_1\left\{P_{12}{}^{t_1}\left(({}^{t_1}S^+_{12})^{-1}\right)\right\}\,. \qquad (5.13)$$

It can be shown that (5.14) will solve eq. (5.2).

Applying (5.14) to the examples of the KdV and MKdV mappings, we find in the KdV case, using (4.10),

$$K(\lambda) = 1 + \frac{h}{\lambda} S_- \ , \quad S_- = \begin{pmatrix} 0 & 0 \\ 0 & 1 \end{pmatrix} , \tag{5.14}$$

and in the case of the MKdV mapping, with the use of the relation

$$S_{12}^- = 1 \otimes 1 + (q-1) \left(\Lambda_1 \cdot S_- \cdot \Lambda_1^{-1} \otimes S_- \right) , \tag{5.15}$$

cf. (4.13) and (4.16), we find the following solution from (4.18),

$$\begin{aligned} K(\lambda) &= 1 + (q^{-1} - 1) S_- \cdot \Lambda \cdot S_- \cdot \Lambda^{-1} \\ \Lambda &= \begin{pmatrix} q-r & 1 \\ \lambda & q+r \end{pmatrix} , \ \lambda = k^2 - r^2 \end{aligned} \tag{5.16}$$

and again the $K(\lambda)$ in combination with the $R_{12}^{\pm}, S_{12}^{\pm}$ of eqs. (4.19), (4.20) satisfies eq. (5.2).

Hence, in the case of the KdV and MKdV mappings we have obtained a commuting family of quantum invariants that can be evaluated expanding $\tau(\lambda) = trK(\lambda)T(\lambda)$ in powers of k^2.

For instance, in the KdV mapping the explicit expression of the invariants can be inferred from

$$\tau(\lambda) = : \left(\prod_{j=1}^{2P} v_j \right) \left[1 + \sum_{\substack{1 \le J_1 < \cdots < J_N \le 2P \\ J_{\nu+1} - J_\nu \ge 2, J_1 - J_N + 2P \ge 2}} \prod_{\nu=1}^{N} \frac{\hat{\lambda}_{J_\nu}}{v_{J_\nu + 1} v_{J_\nu}} \right] : , \tag{5.17}$$

leading to find a full family of commuting invariants. In (5.17) :: denotes the normal ordering of the operators v_j in accordance with their enumeration, and $\hat{\lambda}_J = \lambda_J$ for $J \neq 2P$, $\hat{\lambda}_{2P} = \lambda_{2P} + h$. Thus the quantum effect is only visible in the boundary terms associated with the factor $1/(v_1 v_{2P})$.

As a very simple example we give the quantum invariant of the original McMillan mapping [10], i.e. (4.1) for $P = 2$, namely

$$x' = y \ , \ y' = -x - \frac{2\epsilon\delta y}{\epsilon^2 - y^2} , \tag{5.18}$$

for $x = v_1 - \epsilon$, $y = v_2 - \epsilon$, (choosing $c = 2\epsilon$) and where $[y, x] = h$, having the invariant

$$\mathcal{I} = (\epsilon^2 - y^2)(\epsilon^2 - x^2) - (\epsilon\delta + h)(2yx - \epsilon\delta) . \tag{5.19}$$

The invariant $\mathcal{I}$ can be viewed as a Hamiltonian generating a continuous-time interpolating flow by $\dot{x} = \frac{1}{h}[\mathcal{I}, x]$, $\dot{y} = \frac{1}{h}[\mathcal{I}, y]$, whose solutions can be considered to be parametrized in terms of what we could call a quantum version of the Jacobi elliptic functions. More general two-dimensional quantum mappings have been studied in ref. [35].

Remark: The construction of quantum mappings can be generalized to a larger class of models, namely those associated with the lattice Gel'fand-Dikii hierarchy, [36], as was explicitly shown in [23]. In [37] we elaborated a fusion procedure for obtaining the commuting families of exact higher-order invariants for these mappings.

6 Quantum Action

In the previous sections, we have established the general structure of mappings of KdV and MKdV type. We have shown that, as a consequence of the Yang-Baxter structure the basic commutation relations are preserved under the discrete-time evolution. Furthermore, we have established their integrability by constructing from the Yang-Baxter structure a complete and commuting set of invariants of the mapping. What was not established yet was the existence of a unitary operator that generates the quantum mappings by conjugation. We will establish here such a generating operator for the quantum mappings associated with the KdV and MKdV lattice starting from the classical action for these lattice equations.

We will restrict ourselves in this note to mappings of KdV type, i.e. mappings coming from reductions of the lattice KdV equation. The lattice potential KdV equation is an integrable partial difference equation, which reads, [6,7],

$$(\delta + u_{n,m+1} - u_{n+1,m})(\epsilon + u_{n,m} - u_{n+1,m+1}) = \epsilon\delta, \tag{6.1}$$

in which $\epsilon = p + q$, $\delta = p - q$ are lattice parameters and the dependent variable u depends on two integer variables $n, m \in Z$. Equation (6.1) is the 'integrated' version of the lattice KdV equation

$$\begin{aligned} &u_{n,m+1} + u_{n,m-1} - u_{n-1,m} - u_{n+1,m} \\ &\quad = \frac{\epsilon\delta}{\epsilon + u_{n,m} - u_{n+1,m+1}} - \frac{\epsilon\delta}{\epsilon + u_{n-1,m-1} - u_{n,m}}, \end{aligned} \tag{6.2}$$

which arises as the Euler-Lagrange equation from the following action

$$S = \sum_{n,m \in Z} \left[u_{n,m}(u_{n+1,m} - u_{n,m+1}) + \epsilon\delta \ln(\epsilon + u_{n,m} - u_{n+1,m+1}) \right], \tag{6.3}$$

i.e. $\delta S/\delta u_{n,m} = 0$ yields eq. (6.2). In [7], cf. also [6], we established a linear problem of Zakharov-Shabat type for eqs. (6.1) as well as for (6.2). Furthermore, in [7] we showed that well-chosen periodic initial value problems on the lattice for (6.1)- and consequently for (6.2)- yield reductions to integrable mappings, specifying one of the lattice directions as the direction of the discrete-time update. Thus a mapping reduction is achieved by taking the m-direction as time and labelling the variables along a 'staircase' in the lattice, i.e. taking

$$u_{j,j} =: u_{2j}, \quad u_{j+1,j} =: u_{2j+1} \Rightarrow u_{j,j+1} = u'_{2j}, u_{j+1,j+1} = u_{2j+2} = u'_{2j+1},$$

denoting by the prime $'$ as before the discrete-time shift, i.e.

$$u'_{2j}(t) = u_{2j}(t+1), \quad u'_{2j+1}(t) = u_{2j+1}(t+1) .$$

Imposing periodic boundary conditions $u_{n+2(P+1)}(t) = u_n(t)$, in which $P+1$ is the period, we obtain a finite-dimensional reduction of the lattice equations (6.1) and (6.2). For this reduction we have a reduced action which reads

$$S_{red} = \sum_{t \in Z_+} \sum_{n=1}^{P+1} \left[u_{2n}(t+1)(u_{2n+2}(t) - u_{2n}(t)) + \epsilon\delta \ln(\epsilon + u_{2n}(t) - u_{2n+2}(t)) \right] , \tag{6.4}$$

in terms of the even variables (an action in terms of the odd variables need not be given separately). For convenience we will use in this section instead of v_j the reduced variables $x_n \equiv u_{2n} - u_{2n+2}$. Using the periodic boundary conditions, leading to

$$\sum_{n=1}^{P} x_n = 0 , \tag{6.5}$$

we can write an action entirely in terms of the variables x_n, $n = 1 \cdots P-1$, namely as

$$S_{red} = \sum_{t \in Z_+} L(x(t), x'(t)) , \tag{6.6a}$$

choosing a discrete-time Lagrangian

$$L(x, x') = \sum_{n=1}^{P-1} \frac{1}{2} \left(x'_n - \sum_{j=1}^{n} x_j \right)^2 - V(x) , \tag{6.6b}$$

$$V(x) = \sum_{n=1}^{P-1} \left[\frac{1}{2} x_n^2 + \frac{1}{2} \left(\sum_{j=1}^{n} x_j \right)^2 - \epsilon\delta \ln(\epsilon + x_n) \right]$$
$$- \epsilon\delta \ln \left(\epsilon - \sum_{j=1}^{P-1} x_j \right) , \tag{6.6c}$$

where x is shorthand for $(x_1, ..., x_{P-1})$ and in which the x_n are varied independently. Although in the original action S_{red} of (6.4) one varies with respect to the variables u_{2n}, it is easily verified that, varying with respect to x_n, the Lagrange equations

$$\frac{\partial L}{\partial x'_n} + \left(\frac{\partial L}{\partial x_n} \right)' = 0 \tag{6.7}$$

yield the proper discrete equations of motion for the $x_n = x_n(t)$, $x'_n = x_n(t+1)$. Thus we can work with the x_n as the reduced canonical variables.

We shall now introduce a different Legendre transformation than the one used in [7] to obtain the generating function of the canonical transformation. In fact, because of the special form of the 'kinetic' part of the Lagrangian (6.6b), it is convenient to choose a form of a Legendre transformation specially adapted to the situation at hand and which is of the form

$$H(x,y') = \sum_{n=1}^{P-1} y'_n(x'_n - \sum_{j=1}^{n} x_j) - L(x,x') , \tag{6.8}$$

introducing momentum variables by

$$y'_n \equiv \frac{\partial L}{\partial x'_n} . \tag{6.9}$$

Variation with respect to the x_n and y'_n variables of (6.8), and using (6.9) together with the Lagrange equations (6.7), we obtain

$$y_n - \sum_{j=n}^{P-1} y'_j = \frac{\partial H}{\partial x_n} , \tag{6.10a}$$

$$x'_n - \sum_{j=1}^{n} x_j = \frac{\partial H}{\partial y'_n} , \tag{6.10b}$$

$(n = 1, ..., P-1)$, which can be interpreted as the discrete-time Hamilton equations.

From eqs. (6.10) it is easily established that the variables x_n and y_n are canonical. In fact, it is a straightforward exercise to show that the symplectic form

$$\Omega = \sum_{n=1}^{P-1} dy_n \wedge dx_n \tag{6.11}$$

is invariant under the mapping $x_n \mapsto x'_n, y_n \mapsto y'_n$ described by eqs. (6.10). Eq. (6.11) implies the Poisson brackets

$$\{x_n, y_m\} = \delta_{n,m} \quad , \quad \{x_n, x_m\} = \{y_n, y_m\} = 0 . \tag{6.12}$$

In principle we can use the Legendre transformation (6.4) in more general situations then only the integrable case of (6.6). In that special case, however, we obtain the discrete-time 'hamiltonian' for the KdV mappings, namely

$$H(x,y') = T(y') + V(x) \quad , \quad T(y) = \sum_{n=1}^{P-1} \frac{1}{2} y_n^2 , \tag{6.13}$$

where $V(x)$ is given by eq. (6.6c). What is especially convenient is the fact that H of (6.13) decomposes into a kinetic and a potential part, a feature that was not there in the hamiltonian description of [7]. However, in order to obtain this feature, we had to modify the Legendre transformation, leading to the additional terms on the left-hand side of eqs. (6.10).

The quantization of discrete-time modes with hamiltonian (6.13) and Poisson brackets (6.12) is obtained by the straightforward quantization prescription $\{\cdot,\cdot\} \to \frac{1}{i\hbar}[\cdot,\cdot]$, replacing the canonical coordinates and we replace x_n, y_n by hermitian quantum operators X_n, Y_n acting on a well-defined Hilbert space. Thus, we obtain the usual Heisenberg algebra for X_n, Y_n,

$$[X_n, Y_m] = i\hbar\delta_{n,m} \quad , \quad [X_n, X_m] = [Y_n, Y_m] = 0 \ . \tag{6.14}$$

Eqs. (6.14) correspond precisely to the commutation relations (4.6) that we need to obtain a quantum R-matrix formulation for the commutation relations of the KdV-mappings.

Now, as for the quantum version of the mapping, we first note that as a consequence of the splitting of H into $T+V$ and the X, Y being canonically conjugate, the form of the mapping need not to be modified in the transition from the classical to the quantum case. This is consistent with the R, S-matrix structure of section 4a. Hence eqs. (6.10) are still valid in terms of the quantum variables X_n, Y_n. Secondly, the splitting of H suggests directly the form of the unitary operator that generates the quantum mapping, the only complication arising from the extra terms on the left-hand side of eqs. (6.10). If these were absent, the form of the unitary operator would simply be the product $\exp(i/\hbar V)\exp(i/\hbar T)$. However, in the presence of these extra terms we now are forced to take:

$$U = e^{\frac{i}{\hbar}V} \left(\prod_{i<j=1}^{\overleftarrow{P-1}} e^{\frac{i}{\hbar}X_iY_j} \right) e^{\frac{i}{\hbar}T} \ , \tag{6.15}$$

in which $V = V(X), T = T(Y)$, and the middle factor on the right-hand side of (6.15) is an ordered product of exponential factors, ordered in lexicographic order, i.e.

$$e^{\frac{i}{\hbar}X_{P-2}Y_{P-1}} \cdots e^{\frac{i}{\hbar}X_2Y_{P-1}} \cdots e^{\frac{i}{\hbar}X_2Y_3} e^{\frac{i}{\hbar}X_1Y_{P-1}} \cdots e^{\frac{i}{\hbar}X_1Y_3} e^{\frac{i}{\hbar}X_1Y_2} .$$

Using (6.15) and the relations

$$e^{\frac{i}{\hbar}X_iY_j} X_j e^{-\frac{i}{\hbar}X_iY_j} = X_i + X_j \quad , \quad e^{\frac{i}{\hbar}X_iY_j} Y_i e^{-\frac{i}{\hbar}X_iY_j} = Y_i - Y_j \quad , \quad (i \neq j),$$

it is straightforward to establish that the transformations

$$X_n \mapsto X_n' = UX_nU^\dagger \quad , \quad Y_n \mapsto Y_n' = UY_nU^\dagger \tag{6.16}$$

yield exactly the quantum mapping provided by the quantization of eqs. (6.10). Eq. (6.15), together with (6.16), demonstrates the fact that the quantum KdV mapping provides indeed a canonical transformation in the full quantum sense, namely as a unitary transformation on the quantum phase space.

Remark: We can use the above given construction for obtaining the generating operator of the canonical transformation also directly for the MKdV mappings of section 4b (we omit the details here). In fact, in that case one starts from the action for the classical MKdV lattice that was given in [7]. The construction can be applied without modification to that case also. Furthermore, these considerations can be generalized to the wider class of mappings corresponding to the lattice Gel'fand-Dikii hierarchy, [23,37]. It would be interesting to look for a direct relation between the operator U and the discrete-time part of the linear problem, which generates the time-shift on the level of the Lax representation. Some results in this direction were announced in [38].

Appendix A

In order to derive the compatibility relations for the Yang-Baxter matrices R and S, i.e. eqs. (2.2), from the commutation relations (2.1) for the Lax matrices L, we encounter four different types of combinations of matrices L. Embedding the L matrices in a tensorial product of three copies of the matrix algebra, i.e. L_n, j, $j = 1,2,3$ acting on vector spaces $V \otimes V \otimes V$, and denoting

$$L_{n,1} = L_n \otimes 1 \otimes 1 \ , \quad L_{n,2} = 1 \otimes L_n \otimes 1 \ , \quad L_{n,3} = 1 \otimes 1 \otimes L_n \ ,$$

we can distinguish the following types of combinations of matrices L involving only coïnciding and/or neighbouring sites:

i) $\underline{L_1 \equiv L_{n+2,1},\ L_2 \equiv L_{n+1,2},\ L_3 \equiv L_{n,3}}$

In this case, no conditions on the R- or S matrices will appear, because

$$L_1 \cdot S^+_{12} L_2 \cdot S^+_{23} L_3 \ = \ L_3 \cdot L_2 \cdot L_1 \ , \tag{A.1}$$

independently of the order in which the relation (2.1a) is applied.

ii) $\underline{L_1 \equiv L_{n+1,1},\ L_2 \equiv L_{n+1,2},\ L_3 \equiv L_{n,3}}$

In this case, we have on the one hand

$$R^+_{12} L_1 \cdot L_2 \cdot S^+_{13} S^+_{23} L_3 \ = \ L_2 \cdot L_1 \cdot R^-_{12} S^+_{13} S^+_{23} L_3 \ , \tag{A.2}$$

whereas on the other hand we find

$$\begin{aligned} R^+_{12} L_1 \cdot S^+_{13} L_2 \cdot S^+_{23} L_3 &= R^+_{12} L_3 \cdot L_1 \cdot L_2 \\ &= L_3 \cdot L_2 \cdot L_1 R^-_{12} = L_2 \cdot S^+_{23} L_1 \cdot S^+_{13} L_3 R^-_{12} \ . \end{aligned} \tag{A.3}$$

Comparing relations (A.2) and (A.3), we have

$$R^-_{12} S^+_{13} S^+_{23} \ = \ S^+_{23} S^+_{13} R^-_{12} \ , \tag{A.4}$$

which after relabelling of the vector spaces becomes eq. (2.2b).

iii) $\underline{L_1 \equiv L_{n+1,1},\ L_2 \equiv L_{n,2},\ L_3 \equiv L_{n,3}}$

Take for this case the combination

$$R^+_{23} L_1 \cdot S^+_{12} L_2 \cdot S^+_{13} L_3 \ = \ L_1 \cdot R^+_{23} S^+_{12} S^+_{13} \cdot L_2 \cdot L_3 \ , \tag{A.5}$$

and compare this with

$$\begin{aligned} R^+_{23} L_1 \cdot S^+_{12} L_2 \cdot S^+_{13} L_3 &= R^+_{23} L_2 \cdot L_3 \cdot L_1 \\ &= L_3 \cdot L_2 \cdot L_1 R^-_{23} \\ &= L_1 \cdot S^+_{13} S^+_{12} L_3 \cdot L_2 R^-_{23} \\ &= L_1 \cdot S^+_{13} S^+_{12} R^+_{23} L_2 \cdot L_3 \ , \end{aligned} \tag{A.6}$$

yielding eq. (2.2b) with the + sign.

iv) $L_1 \equiv L_{n,1},\ L_2 \equiv L_{n,2},\ L_3 \equiv L_{n,3}$

In this case we have the standard braiding type of argument to find as a sufficient condition the quantum R-matrix relations (2.2a) for R^+ and R^-.

Appendix B

In this appendix we establish the commutation relations between the monodromy matrices T and T_n^+, T_n^- of eq. (2.6), using the fundamental commutation relations of the matrices L_n.

i) Using eqs. (2.1a), (2.1b) we can establish

$$\begin{aligned} R_{12}^+\, L_{n+1,1} \cdot L_{n,1} \cdot L_{n+1,2} \cdot L_{n,2} &= R_{12}^+\, L_{n+1,1} \cdot L_{n+1,2} \cdot S_{21}^+\, L_{n,1} \cdot L_{n,2} \\ &= L_{n+1,2} \cdot L_{n+1,1} \cdot R_{12}^-\, S_{21}^+\, L_{n,1} \cdot L_{n,2} \end{aligned} \tag{B.1}$$

which, by imposing the relation (2.3), reduces to

$$\begin{aligned} &= L_{n+1,2} \cdot L_{n+1,1} \cdot S_{12}^+\, L_{n,2} \cdot L_{n,1}\, R_{12}^- \,, \\ &= L_{n+1,2} \cdot L_{n,2} \cdot L_{n+1,1} \cdot L_{n,1}\, R_{12}^- \,. \end{aligned} \tag{B.2}$$

By repeated application of eqs. (B.1) and (B.2) together with eq. (2.3) one shows that

$$\begin{aligned} & R_{12}^+ L_{P,1} \cdot \ldots \cdot L_{n+1,1} \cdot L_{P,2} \cdot \ldots \cdot L_{n+1,2} \\ &= R_{12}^+\, L_{P,1} \cdot L_{P-1,1} \cdot L_{P,2} \cdot L_{P-2,1} \cdot L_{P-1,2} \cdot \ldots \cdot L_{n+1,1} \cdot L_{n+2,2} \cdot L_{n+1,2} \\ &= R_{12}^+\, L_{P,1} \cdot L_{P,2} \cdot S_{21}^+ L_{P-1,1} \cdot L_{P-1,2} \cdot S_{21}^+ \cdot L_{P-2,1} \cdot \ldots \\ &\quad \ldots \cdot L_{n+2,2} \cdot S_{21}^+ L_{n+1,1} \cdot L_{n+1,2} \\ &= L_{P,2} \cdot L_{P,1} S_{12}^+ \cdot L_{P-1,2} \cdot L_{P-1,1} S_{12}^+ \cdot L_{P-2,2} \cdot \ldots \\ &\quad \ldots \cdot L_{n+2,1} S_{12}^+ \cdot L_{n+1,2} \cdot L_{n+1,1}\, R_{12}^- \\ &= L_{P,2} \cdot L_{P-1,2} \cdot \ldots \cdot L_{n+1,2} \cdot L_{P,1} \cdot L_{P-1,1} \cdot \ldots \cdot L_{n+1,1}\, R_{12}^- \,, \end{aligned} \tag{B.3}$$

leading to eq. (2.7a) for T_n^+. A similar argument can be applied for T_n^-. Furthermore, eq. (2.7b) is derived from eqs. (2.1b), (2.1c) by noting that

$$\begin{aligned} & L_{P,1} \cdot \ldots \cdot L_{n+1,1} \cdot S_{12}^+ L_{n,2} \cdot \ldots \cdot L_{1,2} \\ &= L_{P,1} \cdot \ldots \cdot L_{n+2,1} \cdot L_{n,2} \cdot L_{n+1,1} \cdot L_{n-1,2} \cdot \ldots \cdot L_{1,2} \\ &= L_{n,2} \cdot \ldots \cdot L_{2,2} \cdot L_{P,1} \cdot L_{1,2} \cdot L_{P-1,1} \cdot \ldots \cdot L_{n+1,1} \\ &= L_{n,2} \cdot \ldots \cdot L_{2,2} \cdot L_{1,2} S_{21}^+ L_{P,1} \cdot \ldots \cdot L_{n+1,1} \,, \end{aligned} \tag{B.4}$$

where in the last step we have used the commutation relation

$$L_{P,1} \cdot L_{1,2} = L_{1,2} \cdot S_{21}^+ L_{P,1} \tag{B.5}$$

taking into account the periodic boundary conditions.

Finally, from (2.7) we immediately obtain

$$\begin{aligned} R_{12}^{+} T_1 \cdot S_{12}^{+} T_2 &= R_{12}^{+} T_{n,1}^{+} \cdot T_{n,2}^{+} \cdot S_{21}^{+} T_{n,1}^{-} \cdot T_{n,2}^{-} \\ &= T_{n,2}^{+} \cdot T_{n,1}^{+} \cdot R_{12}^{-} S_{21}^{+} T_{n,1}^{-} \cdot T_{n,2}^{-} \\ &= T_{n,2}^{+} \cdot T_{n,1}^{+} \cdot S_{12}^{+} T_{n,2}^{-} \cdot T_{n,1}^{-} R_{12}^{-} \\ &= T_2 \cdot S_{21}^{+} T_1 R_{12}^{-} \, , \end{aligned} \tag{B.6}$$

which is eq. (2.8).

Appendix C

i) We prove that eqs. (3.1a),(3.1b), together with (3.2b) are sufficient to ensure that the basic commutation relations between the matrices L_n, eqs. (2.1a), are preserved under the mapping

$$L_n \mapsto L'_n = M_{n+1} \cdot L_n \cdot M_n^{-1}.$$

In fact,

$$\begin{aligned} L'_{n+1,1} \cdot S_{12}^{+} L'_{n,2} &= L'_{n+1,1} \cdot S_{12}^{+} M_{n+1,2} \cdot L_{n,2} \cdot M_{n,2}^{-1} \\ &= M_{n+1,2} \cdot L'_{n+1,1} \cdot L_{n,2} \cdot M_{n,2}^{-1} \\ &= M_{n+1,2} \cdot M_{n+2,1} \cdot L_{n+1,1} \cdot M_{n+1,1}^{-1} \cdot L_{n,2} \cdot M_{n,2}^{-1} \\ &= M_{n+1,2} \cdot M_{n+2,1} \cdot L_{n+1,1} S_{12}^{+} L_{n,2} \cdot M_{n+1,1}^{-1} \cdot M_{n,2}^{-1} \\ &= L'_{n,2} \cdot M_{n,2} \cdot L'_{n+1,1} \cdot M_{n,2}^{-1} \\ &= L'_{n,2} \cdot L'_{n+1,1} \, , \end{aligned} \tag{C.1}$$

and similarly

$$\begin{aligned} R_{12}^{+} L'_{n,1} \cdot L'_{n,2} &= R_{12}^{+} M_{n+1,1} \cdot L_{n,1} \cdot M_{n,1}^{-1} \cdot L'_{n,2} \\ &= R_{12}^{+} M_{n+1,1} \cdot L_{n,1} \cdot L'_{n,2} \cdot M_{n,1}^{-1} (S_{12}^{-})^{-1} \\ &= R_{12}^{+} M_{n+1,1} \cdot M_{n+1,2} \cdot S_{12}^{-} L_{n,1} \cdot L_{n,2} \cdot M_{n,2}^{-1} \cdot M_{n,1}^{-1} (S_{12}^{-})^{-1} \\ &= M_{n+1,2} \cdot M_{n+1,1} \cdot S_{12}^{+} R_{12}^{+} L_{n,1} \cdot L_{n,2} \cdot M_{n,2}^{-1} \cdot M_{n,1}^{-1} (S_{12}^{-})^{-1} \\ &= M_{n+1,2} \cdot M_{n+1,1} \cdot S_{12}^{+} L_{n,2} \cdot L_{n,1} \cdot M_{n,1}^{-1} \cdot M_{n,2}^{-1} (S_{12}^{+})^{-1} R_{12}^{-} \\ &= M_{n+1,2} \cdot L_{n,2} \cdot M_{n+1,1} \cdot L_{n,1} \cdot M_{n,1}^{-1} \cdot M_{n,2}^{-1} (S_{12}^{+})^{-1} R_{12}^{-} \\ &= L'_{n,2} \cdot M_{n,2} \cdot L'_{n,1} \cdot M_{n,2}^{-1} (S_{12}^{+})^{-1} R_{12}^{-} \\ &= L'_{n,2} \cdot L'_{n,1} R_{12}^{-} \, . \end{aligned} \tag{C.2}$$

Finally, we have that

$$\begin{aligned} M'_{n+1,1} \cdot S_{12}^{+} L'_{n,2} &= M'_{n+1,1} \cdot S_{12}^{+} M_{n+1,2} \cdot L_{n,2} \cdot M_{n,2}^{-1} \\ &= M_{n+1,2} \cdot M'_{n+1,1} \cdot L_{n,2} \cdot M_{n,2}^{-1} \\ &= M_{n+1,2} \cdot L_{n,2} \cdot M'_{n+1,1} \cdot M_{n,2}^{-1} \\ &= L'_{n,2} \cdot M'_{n+1,1} \, . \end{aligned} \tag{C.3}$$

It is straightforward to check along similar lines that all trivial commutation relations remain so after applying the mapping.

Appendix D

In order to show that eq. (5.2) provides a commuting family of operators, let us give an argument similar to the one given by Sklyanin in [27]. Denoting by T_i and K_i, $(i = 1, 2)$ the monodromy matrix resp. matrix K for two different values of the spectral parameter λ_1 resp. λ_2 and acting in two different factors of a matricial tensor product, and denoting by τ_1,τ_2 the invariants (5.1) evaluated at these respective values of the spectral parameter, we have on the one hand, assuming that $[K_i \overset{\otimes}{,} T_j] = 0$,

$$\begin{aligned} \tau_1 \tau_2 &= tr_1 (T_1 K_1)\, tr_2 (T_2 K_2) \;=\; tr_{1,2} \left\{ T_1 K_1 \, {}^{t_2}T_2 \, {}^{t_2}K_2 \right\} \\ &= tr_{1,2} \left\{ {}^{t_2}(T_1 \, S^+_{1,2} T_2) \, {}^{t_1}({}^{t_1}K_1 \, {}^{t_1}\left(({}^{t_2}S^+_{1,2})^{-1}\right) \, {}^{t_2}K_2) \right\} \\ &= tr_{1,2} \left\{ R^{+\,-1}_{1,2} \, T_2 \, S^-_{1,2} T_1 \, R^-_{1,2} \, {}^{t_{12}}\left[{}^{t_1}K_1 \, {}^{t_1}\left(({}^{t_2}S^+_{1,2})^{-1}\right) \, {}^{t_2}K_2 \right] \right\} \qquad \text{(D.1)} \\ &= tr_{1,2} \left\{ T_2 \, S^-_{1,2} T_1 \, {}^{t_{12}}\left[({}^{t_{12}}R^+_{1,2})^{-1} \, {}^{t_1}K_1 \, {}^{t_1}\left(({}^{t_2}S^+_{1,2})^{-1}\right) \, {}^{t_2}K_2 \, {}^{t_{12}}R^-_{1,2} \right] \right\} \;, \end{aligned}$$

whereas on the other hand we have

$$\begin{aligned} \tau_2 \tau_1 &= tr_2 (T_2 K_2)\, tr_1 (T_1 K_1) \\ &= tr_{1,2} \left\{ {}^{t_1}(T_2 \, S^-_{1,2} T_1) \, {}^{t_2}\left[{}^{t_2}K_2 \, {}^{t_2}\left(({}^{t_1}S^-_{1,2})^{-1}\right) \, {}^{t_1}K_1 \right] \right\} \qquad \text{(D.2)} \\ &= tr_{1,2} \left\{ T_2 \, S^-_{1,2} T_1 \, {}^{t_{12}}({}^{t_2}K_2 \, {}^{t_2}\left(({}^{t_1}S^-_{1,2})^{-1}\right) \, {}^{t_1}K_1) \right\} \;, \end{aligned}$$

from which it is clear that (D.1) and (D.2) can be identified provided that we have the following condition on the matrices K

$$\left({}^{t_{12}}R^+_{12} \right)^{-1} \, {}^{t_1}K_1 \, {}^{t_1}\left(({}^{t_2}S^+_{12})^{-1}\right) \, {}^{t_2}K_2 \;=\; {}^{t_2}K_2 \, {}^{t_2}\left(({}^{t_1}S^-_{12})^{-1}\right) \, {}^{t_1}K_1 \, \left({}^{t_{12}}R^-_{12} \right)^{-1} . \qquad \text{(D.3)}$$

Eq. (D.3) is a very general condition for operator valued matrices K of which the entries commute with the entries of T, which is sufficient to ensure that the $T(\lambda)$ form a parameter family of commuting operators. For numerical matrices $K(\lambda)$ eq. (D.3) leads to the condition (5.2) given in the main text.

References

[1] V.G. Drinfel'd, *Quantum Groups*, Proc. ICM Berkeley 1986, ed. A.M. Gleason, (AMS, Providence, 1987), p. 798.

[2] M. Jimbo, Lett. Math. Phys. **10** (1985) 63, ibid. **11** (1986) 247, Commun. Math. Phys. **102** (1986) 537.

[3] L.D. Faddeev, N.Yu. Reshetikhin and L.A. Takhtadzhyan, Algebra i Analiz. **1** (1988) 178 [in Russian].

[4] L.D. Faddeev, in *Développements Récents en Théorie des Champs et Mécanique Statistique*, eds. J.-B. Zuber and R. Stora, (North-Holland Publ. Co., 1984), p.561.

[5] M.J. Ablowitz and F.J. Ladik, Stud. Appl. Math. **55** (1976) 213, **57** (1977) 1; R. Hirota, J. Phys. Soc. Japan **43** (1977) 1424, 2074, 2079; **50** (1981) 3785; E. Date, M. Jimbo and T. Miwa, J. Phys. Soc. Japan **51** (1982) 4116, 4125, **52** (1983) 388, 761, 766.

[6] F.W. Nijhoff, G.R.W. Quispel and H.W. Capel, Phys. lett. **97A** (1983) 125; G.R.W. Quispel, F.W. Nijhoff, H.W. Capel and J. van der Linden, Physica **125A** (1984) 344.

[7] V.G. Papageorgiou, F.W. Nijhoff and H.W. Capel, Phys. Lett. **147A** (1990) 106; H.W. Capel, F.W. Nijhoff and V.G. Papageorgiou, Phys. Lett. **155A** (1991) 377.

[8] F.W. Nijhoff, V.G. Papageorgiou and H.W. Capel, *Integrable Time-Discrete Systems: Lattices and Mappings*, in Proc. of the Intl. Workshop on Quantum Groups, The Euler Intl. Math. institute, Leningrad, ed. P.P. Kulish, Springer Lecture Notes Math. **1510** (1992) 312.

[9] M. Bruschi, O. Ragnisco, P.M. Santini and G.-Z. Tu, Physica **49D** (1991) 273.

[10] E.M. McMillan, in *Topics in Physics*, eds. W.E. Brittin and H. Odabasi, (Colorado Associated Univ. Press, Boulder, 1971), p. 219.

[11] G.R.W Quispel, J.A.G. Roberts and C.J. Thompson, Phys. Lett. **A126** (1988) 419, Physica **D34** (1989) 183.

[12] A.P. Veselov, Funct. Anal. Appl. **22** (1988) 83; Theor. Math. Phys. **71** (1987) 446; P.A. Deift and L.C. Li, Commun. Pure Appl. Math. **42** (1989) 963. J. Moser and A.P. Veselov, Preprint ETH (Zürich), 1989.

[13] Yu.B. Suris, Phys. Lett. **145A** (1990) 113; Algebra i Analiz **2** (1990) 141 [in Russian].

[14] F.W. Nijhoff, H.W. Capel and V.G. Papageorgiou, Phys. Rev. **46A** (1992) 2155.

[15] F.W. Nijhoff and H.W. Capel, in Proc. of NATO ARW on Applications of Analytic and Geometric Methods to Nonlinear Differential Equations, Exeter, July 1992 (Kluwer Publ. Co., to be published).

[16] A.Yu Volkov, "The Quantum Volterra Model", Preprint HU-TFT-92-26.

[17] L.D. Faddeev and L.A. Takhtadzhyan, *Hamiltonian Methods in the Theory of Solitons*, (Springer Verlag, Berlin, 1987).

[18] J.M. Maillet, Phys. Lett. **162B** (1985) 137, Nucl. Phys. **B269** (1986) 54.

[19] A.G. Reyman and M.A. Semenov-Tian-Shanskii, Phys. Lett. **130A** (1988) 456.

[20] O. Babelon and C. Viallet, Phys. Lett. **237B** (1990) 411.

[21] J. Avan and M. Talon, Nucl. Phys. **B352** (1991) 215.

[22] L.-C. Li and S. Parmentier, C.R. Acad. Sci. Paris, t. **307** Série I (1988) 279; Commun. Math. Phys. **125** (1989) 545.

[23] F.W. Nijhoff and H.W. Capel, Phys. Lett. **163A** (1992) 49.

[24] O. Babelon and L. Bonora, Phys. Lett. **253B** (1991) 365; O. Babelon, Commun. Math. Phys. **139** (1991) 619. .

[25] A. Alekseev, L.D. Faddeev and M.A. Semenov-Tian-Shanskii, Peprint CERN-TH-5981/91.

[26] A. Alekseev, L.D. Faddeev and M.A. Semenov-Tian-Shanskii, in Proc. of the Intl. Workshop on Quantum Groups, The Euler Intl. Math. institute, Leningrad, ed. P.P. Kulish, Springer Lecture Notes Math. **1510** (1992) 148.

[27] E.K. Sklyanin, J. Phys. **A21** (1988) 2375.

[28] N. Yu. Reshetikhin and M.A. Semenov-Tian-Shanskii, Lett. Math. Phys. **19** (1990) 133.

[29] G.I. Olshanskii, *Twisted Yangians and infinite-dimensional classical Lie algebras*, Preprint CWI (Amsterdam), 1990.

[30] A. Alekseev and L.D. Faddeev, Commun. Math. Phys. **141** (1991) 413.

[31] J.L. Gervais, Phys. Lett. **160B** (1985) 277,279.

[32] L.D. Faddeev and L.A. Takhtadzhyan, Springer Lect. Notes Phys. **246** (1986) 166.

[33] A.Yu. Volkov, Theor. Math. Phys. **74** (1988) 96.

[34] O. Babelon, Phys. Lett. **215B** (1988) 523, ibid. **238B** (1990) 234.

[35] G.R.W. Quispel and F.W. Nijhoff, Phys. Lett. **161A** (1991) 419.

[36] F.W. Nijhoff, H.W. Capel, G.R.W. Quispel and V.G. Papageorgiou, Inverse Probl. **8** (1992) 597.

[37] F.W. Nijhoff and H.W. Capel, *Integrability and Fusion Algebra for Quantum Mappings*, Preprint INS (Clarkson University) # 212/92.

[38] L.D. Faddeev and A.Yu. Volkov, *Quantum Inverse Transform Method with Discrete Space-Time*, LOMI Preprint, St. Petersburg, 1992.

Some Aspects of the q-Deformed Oscillator Algebras as Quantum Groups

Suemi Rodriguez-Romo[1] and Dieter W. Ebner[2]

[1] Facultad de Estudios Superiores Cuautitlán,
Universidad Nacional Autónoma de México,
Apdo 142, Cuautitlán Izcalli, Estado de México, 54700 México.

[2] Physics Department, University of Konstanz,
Box 5560, D-7750 Konstanz, Germany

Abstract. We study the q-deformation of the Clifford algebras that come out in a natural way for fermions and bosons in Fock space. An analysis of three particular cases; the transformation of fermion (bosons) among themselves, the linear combination of fermions (bosons) in order to get bosons (fermions) and a supersymmetric transformation, is carried out.

1 Introduction

Since two decades it is well known that the Yang-Baxter equation (YBE) can be considered as the master equation of integrable models in statistical mechanics and in two dimensional quantum field theory.[1]

Faddeev [2] and coworkers proposed the quantum inverse method as an unification of classical and quantum integrable models. In this case the solutions of YBE give the commutation properties among operators.

This approach has led to the idea of introducing deformation over groups or Lie algebras, so called quantum groups. In the course of constructing trigonometric solutions of YBE the deformation of the universal enveloping algebra of $SL(2)$ was introduced. This is one of the first ocurrences of new algebraic objects, now called "quantum groups".

The purpose of the present work is to propose generalized commutation relations for fermions and bosons in Fock space (q-deformed oscillator algebras) and their subsequent analysis using the Manin's viewpoint [3] for three particular cases. The first of them considers a linear transformation $GL_q(N)$ among bosons (fermions), the second one studies a linear transformation $GL_q(N)$ on fermions (bosons) in order to get bosons (fermions). At last, an analysis of a supersymmetric linear transformation is presented.

The paper is organized as follows : in section II we present a short review of Clifford algebras to introduce the usual oscillator algebras in Fock space. In section III we q-deform the second ones keeping Pauli principle invariant. In part IV, we consider quantum groups as effecting linear transformations on the space of the q-oscillator algebras defined in section III; the conditions for such a mapping to be an endomorphism constitute the quantum group relations, finally we study the three particular cases stated above.

2 Clifford Algebra in Fock Space and Oscillator Algebras

Let K be a commutative field and let Q be a quadratic form on the $K-$ module V. Let

$$T(V) = K \oplus V \oplus V \otimes V \oplus ...$$

be the tensor algebra over V, and let $I(Q)$ be the two sided ideal generated by the elements $X \otimes X - Q(X) \cdot 1$ in $T(V)$. The quotient algebra $T(V)/I(Q)$ is called the Clifford algebra of Q and is denoted by $\mathbf{C}(Q)$ [3].

If $e_1, ..., e_n$ is a basis of V, then

$$1,\ e_i\ , e_{ij}(i < j), ..., e_1 e_2 ... e_n$$

form a basis of $\mathbf{C}(Q)$.

In a general context the definition stated above corresponds to a symmetric Clifford algebra [4]. We define now the symplectic Clifford algebra [5], as the quotient associative algebra $T(V)/I(F)$, where the two sided ideal $I(F)$ is generated by the elements $X \otimes Y - Y \otimes X - F(X,Y)$; $X, Y \in V$, V is a vector space over K and F a symplectic form.

Therefore, the boson algebra is defined as the symplectic Clifford algebra generated by $\{b_1, b_2, ..., b_n, b_1^\dagger, b_2^\dagger, ..., b_n^\dagger\}$ basis set of V and a unit element satisfying the following boson relations

$$[b_i, b_j] = b_i b_j - b_j b_i = 0\ ,\ \left[b_i^\dagger, b_j^\dagger\right] = b_i^\dagger b_j^\dagger - b_j^\dagger b_i^\dagger = 0 \qquad (1)$$
$$\left[b_i, b_j^\dagger\right] = b_i b_j^\dagger - b_j^\dagger b_i = \hbar \delta_{ij}.$$

In a similar way the fermion algebra is defined as the symmetric Clifford algebra generated by $\{a_1, a_2, ..., a_n, a_1^\dagger, a_2^\dagger, ..., a_n^\dagger\}$ basis set of V' and a unit element, satisfying the following fermion relations

$$\{a_i, a_j\} = a_i a_j + a_j a_i = 0,\ \{a_i^\dagger, a_j^\dagger\} = a_i^\dagger a_j^\dagger + a_j^\dagger a_i^\dagger = 0 \qquad (2)$$
$$\{a_i, a_j^\dagger\} = a_i a_j^\dagger + a_j^\dagger a_i = \hbar \delta_{ij}$$

A final comment in this section; we stress the possibility of constructing a superalgebra endowed with a nonsingular bilinear form which is either symplectic (boson case) or symmetric (fermion case).

3 Quantum Superspace

Quantum groups have been extensively studied from different points of view since the papers of Faddeev [2, 6, 7, 8], Drinfeld [9, 10, 11], Jimbo [12, 13, 14], Woronowicz [15], Manin [16, 17] and coworkers appeared.

Manin described a class of quantum groups as natural symmetries of non-commutative algebraic varieties defined by quadratic equations. According with this idea we present in this section a q-deformation of the Clifford (oscillator) algebras above defined, as the non-commutative algebraic varieties whose symmetries define a quantum group.

The quantum superspace A_q studied by Manin [17] is generated by operator valued coordinates $X_1, ..., X_n$ with parity assignment $X_i = \hat{\imath}$ and commutation rules

$$X_i^2 = 0 \text{ for } \hat{\imath} = 1 \tag{3}$$

$$X_i X_j - q_{ij}^{-1}(-1)^{\hat{\imath}\hat{\jmath}} X_j X_i = 0 \text{ for} J \ i < j,$$

and its dual, A_q^* is generated by operator valued coordinates $\xi^1, ..., \xi^n$ with $\xi^k = 1 - \hat{k}$ parity assignment and commutation rules

$$\left(\xi^k\right)^2 = 0 \ \text{ for } \ \hat{k} = 0 \tag{4}$$

$$\xi^k \, \xi^l - q_{kl}(-1)^{(\hat{k}+1)(\hat{l}+1)} \xi^l \, \xi^k = 0 \ \text{ for } \ k < l.$$

Here $q = \{q_{ij}\}$ are non-zero elements of a field K.

The symmetry of the non-commutative algebraic manifolds (3) and (4) is given by the quadratic equations produced by the following constrains

(i) $X' = MX$ and $X'' = M^T X$ verify (3)

(ii) X' verifies (3) and $\xi' = M\xi$ verifies (4) where $X = (X_1, ..., X_n)$, $\xi = (\xi^1, ..., \xi^n)$, $q_{ij}^2 \neq -1$ and M is any quantum matrix, namely matrices whose elements are themselves non commutative.

We define a q-bosonic space generated by operator valued coordinates X_i^α, ($i = 1, ..., n$; $\alpha = (a, b)$, a is the tag that identifies the annihilation operators and b is the tag for the creation ones) such that they fulfill the following relation

$$\left[X_i^\alpha, X_j^\beta\right] = X_i^\alpha X_j^\beta - q^{-1} X_j^\beta X_i^\alpha = \hbar \delta_{ij} \eta^{\alpha\beta} \ \text{ where } \ i < j \tag{5}$$

and

$$\eta^{\alpha\beta} = \begin{bmatrix} 1 & \text{if} & \alpha \text{ corresponds to an annihilation operator,} \\ & & \text{whereas } \beta \text{ is a creation one} \\ 0 & \text{otherwise} & \end{bmatrix}$$

Hereafter q is any non-zero real number.

Besides, we define a q-fermionic space generated by operator valued coordinates ξ_i^α ($i = 1, ..., n; \alpha = (a, b)$ a for annihilation operators and b for creation ones) such that the following relations are fulfilled

$$\{\xi_i^\alpha, \xi_j^\beta\} = \xi_i^\alpha\, \xi_j^\beta \;+\; q\xi_j^\beta\, \xi_i^\alpha = \hbar\delta_{ij}\eta^{\alpha\beta} \;\; \text{where} \;\; i < j \tag{6}$$

and

$$(\xi_i^\alpha)^2 = \left(\xi_i^\beta\right)^2 = 0$$

In our approach the Pauli principle invariance is kept as mandatory structure of nature. In the limit $q \to 1$, (5) and (6) transform to (1) and (2) respectively; on the other hand, if we consider the "classical" limit $\hbar \to 0$ we recover a structure of type (3) and (4).

Summarizing our proposition in a supersymmetric short notation, we have a similar structure to the so called quantum hypervolume graded rings [17]

$$S_i^\alpha\, S_j^\beta + F\, S_j^\beta\, S_i^\alpha = \hbar\eta^{\alpha\beta}\delta_{ij} \tag{7}$$

where $i < j$ and

$$F = \begin{cases} -q^{-1} & \text{if} \quad S_i^\alpha \quad \text{is such that } (S_i^\alpha)^2 \neq 0, \text{being} \\ & \qquad S_i^\alpha = X_i^\alpha \\ q & \text{if} \quad S_i^\alpha \quad \text{is such that } (S_i^\alpha)^2 = 0, \text{being} \\ & \qquad S_i^\alpha = \xi_i^\alpha \end{cases}$$

4 Yang-Baxter Approach

The Yang-Baxter equation is a functional equation for a four indices function ${}_{\alpha\beta}R_{\gamma\delta}$ of at least one parameter u that is called the spectral parameter

$$\begin{aligned} &{}_{\alpha\alpha'}R(u-v)_{\gamma\gamma'\,\gamma\alpha''}R(u)_{\beta\gamma''\,\gamma'\gamma''}R(v)_{\beta'\beta''} = \\ &{}_{\alpha'\alpha''}R(v)_{\gamma'\gamma''\,\alpha\gamma''}R(u)_{\gamma\beta'\,\gamma\gamma'}R(u-v)_{\beta\beta'} \end{aligned} \tag{8}$$

A solution of the Yang-Baxter equation (YBE) we shall call a Yang-Baxter bundle (YBB) [18].

In this section we will study how the algebra given by the relations that define the quantum space of linear endomorphisms of the quantum hypervolume (7) (for the cases $\hbar \to 0$ and $\hbar \neq 0$) can be rewritten in the tensor product form like

$$ {}_{ij}R_{kl}\, M_{km}\, M_{ln} = M_{jl}\, M_{ikkl}\, R_{mn} \tag{9} $$

up to a sign, where ${}_{ij}R_{kl}$ can be obtained from a YBB and M is a quantum matrix and a linear transformation operator over the Fock space simultaneously. We will present some consequences of this property.

Case 1. Let us consider a complex vector space S_i^α of dimension $4n, n \geq 2$, with the following basis

$$ \left\{X_1, X_2, ..., X_n, X_1^\dagger, X_2^\dagger, ..., X_n^\dagger, \xi_1, \xi_2, ..., \xi_n, \xi_1^\dagger, \xi_2^\dagger, ..., \xi_n^\dagger\right\} \tag{10} $$

Let us consider the following quantum matrix

$$ M_1 = \begin{pmatrix} GL_q(2n) & \\ & GL_q(2n) \end{pmatrix} \tag{11} $$

M_1 is the group of quantum matrices that mix bosons X_i^α in order to get new bosons $(X_i^\alpha)'$ and fermions ξ_i^α in order to get new fermions $(\xi_i^\alpha)'$. If they are isometries of the bilinear form (7) for $\hbar = 0$ (classical limit) then they define an associative algebra that can be written like the tensor product (9). In this case $R = [{}_{\alpha\beta}R_{\gamma\delta}]$ is given by $\lim_{u \to 0}$ of

$$ \begin{aligned} R(u) = (q^{-1} - uq) \sum E_{\alpha\alpha} \otimes E_{\alpha\alpha} + (1-u) \sum_{\alpha \neq \beta} E_{\alpha\alpha} \otimes E_{\beta\beta} + \\ +(q^{-1} - q) \Big(\sum_{\alpha<\beta} + u \sum_{\alpha>\beta}\Big) E_{\alpha\beta} \otimes E_{\beta\alpha} \end{aligned} \tag{12} $$

where $E_{\alpha\beta}$ denotes the matrix $(\delta_{i\alpha}\delta_{j\beta})$, u is the spectral parameter and q denotes, a real number. The matrix $R(u)$ in (12) corresponds to the $A_n^{(1)}$ matrices given by Jimbo [13].

In the quantum case ($\hbar \neq 0$) the M_1 transformations are restricted by the following relations

$$ \sum_{j=1}^{(4n-1)} r_{ij} r_{ij+1} = \sum_{j=1}^{(4n-1)} r_{ij+1} r_{ij} = 0 \quad i = 1, ..., (4n-1) \tag{13} $$

$$ \sum_{l=1} (r_{il} r_{jl+1} - q^{-1} r_{jl} r_{il+1}) = 1 \quad \begin{array}{l} i = 1, \cdots, (2n-1), (2n+1), \cdots (4n-1) \\ j = i+1, \cdots, (4n) \end{array} $$

where l is an odd number and r_{ij} is the $ijth$ entry of the M_1 quantum matrix. We should remark that M_1 is the most general group of quantum matrices which mix fermions (bosons) in order to get new fermions (bosons) for the classical

limit and the quantum case if we consider restrictions (13). A mixture fermions-bosons is not allowed in this case.
Besides, if $\hbar \neq 0$ and $q \neq 1$, eq. (7) can be considered as a generalized commutation relation and the corresponding M_1 matrix as its natural symmetry. These relations have been recently studied and, in the quantum-mechanical context, they can contain not only Fermi and Bose statistics but Greenberg's infinite statistics ($q = 0$) as special cases [19].

Case 2. Let us consider a complex vector space S_i^α of dimension $4n, n \geq 2$, with the basis set given by (10).
Let us consider the following quantum matrix

$$M_2 = \begin{pmatrix} 0 & GL_q(2n) \\ GL_q(2n) & 0 \end{pmatrix} \tag{14}$$

M_2 is the group of quantum matrices that mix bosons X_i^α in order to get fermions $(\xi_i^\alpha)'$ and fermions ξ_i^α in order to get bosons $(X_i^\alpha)'$. If they are isometries of the bilinear form (7) for $\hbar = 0$ (classical limit) then they define an associative algebra under multiplication that can be written like the tensor product (9), up to a sign. In this case $R = [_{\alpha\beta}R_{\gamma\delta}]$ is given by $\lim_{u \to -1}$ of the matrix $R(u)$ in (12).
In the quantum case ($\hbar \neq 0$) the M_2 transformation are restricted by the following relations

$$\sum_{j=1}^{(4n-1)} s_{ij}s_{ij+1} = \sum_{j=1}^{(4n-1)} s_{ij+1}s_{ij} = 0 \quad i = 1, ..., (4n-1) \tag{15}$$

$$\sum_{l=1} (s_{il}s_{jl+1} + q s_{jl}s_{il+1}) = 1 \quad \begin{array}{l} i = 1, \cdots, (2n-1), (2n+1), \cdots (4n-1) \\ j = i+1, ..., (4n) \end{array}$$

where l is an odd number and s_{ij} is the $ij-th$ entry of the M_2 quantum matrix. We should remark that M_2 is the most general group of quantum matrices which mix bosons (fermions) in order to get fermions (bosons) for the classical limit and the quantum case if we consider the restrictions given in (15).
On the other hand, the $R(v)$ YBB constructed from $R(u)$ in (12) using the following conformal mapping $u \to \frac{1}{q^{-1}-q}e^v$ (where either $0 < Imv < \pi$ and $Imu > 0$ or $0 < Imv < \pi, Rev < 0$ and $Imu > 0, |u| < 1$) is a regular solution of YBE. It means

$$R(v)/_{v=0} = P \quad , \quad P = [_{\alpha\beta}P_{\gamma\delta}] \tag{16}$$

where P is the permutation operator

$$_{\alpha\beta}P_{\gamma\delta} \; = \; \delta_{\alpha\delta}\, \delta_{\beta\gamma} \tag{17}$$

Furthermore, from a given YBB a set of completely integrable models (those possessing as many commuting and conserved physical magnitudes as degrees of freedom, an infinite number for a field theory or statistical model in the thermodynamical limit) can be constructed [18]. Each regular YBB can be considered as a Lax operator for a one dimensional periodic system of M bound states each one of them has N quantum states in the following initial value problem [18]

$$\frac{\partial}{\partial X_2} T(X_1, X_2; v) = L(X_2 v) T(X_1, X_2; v) \tag{18}$$

for the transition (or monodromy) matrix T, where X_i are coordinates of the quantum chain.

Let us present the one dimensional quantum system associated to $R(v)$. The quantum system in question is a closed ring of bound states [1, 18], thus the space of quantum states is $V_1 \otimes V_2 \otimes ... \otimes V_M$. The L−operator $L(u)$ is considered as a matrix whose elements are operators in V_N.

$$_{\alpha\alpha'} L(u)_{\beta\beta'} =_{\alpha\alpha'} R(u)_{\beta\beta'} \tag{19}$$

the indices α, β being the matrix ones and α', β' being the quantum ones.

The infinite number of commuting and conserved physical magnitudes for this one dimensional field model can be constructed using a generating function and is related with an infinite dimensional symmetry transformation group or gauge transformation.

M. Lüscher [20] has shown that

$$J_n = \frac{d^n}{dv^n} ln[t(0)^{-1} t(v)] /_{v=0} \tag{20}$$

is the generating function for the commuting local quantities. Here $t(v) = trT(1, M; v) = tr\{L(M - \Delta : v)...L(1; v)\}$ and $T(1, M; v)$ is the transition matrix for the chain such that $X_1 = 1$ and $X_2 = M$.

For our case

$$J_n = \frac{1}{(q^{-1} - q)^M} \frac{d^n}{dv^n} ln[(q^{-1} - e^{iv})^M + (1 - e^{iv})^M] /_{v=0} \tag{21}$$

Case 3. Let us consider a complex vector space S_i^α of dimension $4n, n \geq 2$, with the basis set given by (10). Additionally, let us consider the quantum matrices $M \in GL_q(4n)$, namely a supersymmetric transformation mixing fermions and bosons.

We study the set of M matrices, isometries of the bilinear form (7) for $\hbar = 0$ (classical limit) and of

$$\left[\xi_i^\alpha \; X_j^\beta\right] = \xi_i^\alpha X_j^\beta - p X_j^\beta \xi_i^\alpha = 0 \tag{22}$$

where p is a real number. Although, we would like that they define an associative algebra under multiplication that would be written like the tensor product

(9), up to a sign (where $R = [{}_{\alpha\beta}R_{\gamma\delta}]$ might be given by the limit of a linear combination of the R-matrices reported by Jimbo [13]) we are not able to write the $GL(4n)$ symmetry in the (9) tensor product form. On the contrary, only for particular cases of M we can write the symmetry in the proper way.
Let us consider the following example; M is defined like

$$M_3 = m_\mu \sigma^\mu \qquad \alpha = 0, +, -, 3 \qquad \text{where} \tag{23}$$

$$\sigma^0 = \frac{1}{q\sqrt{Q}} \begin{pmatrix} GL_q(2n) & 0 \\ 0 & GL_q(2n) \end{pmatrix} \qquad \sigma^+ = \frac{1}{\sqrt{q}} \begin{pmatrix} 0 & GL_q(2n) \\ 0 & 0 \end{pmatrix} \tag{24}$$

$$\sigma^- = \sqrt{q} \begin{pmatrix} 0 & 0 \\ GL_q(2n) & 0 \end{pmatrix} \qquad \sigma^3 = \frac{1}{\sqrt{Q}} \begin{pmatrix} q^{-1}GL_q(2n) & 0 \\ 0 & -qGL_q(2n) \end{pmatrix},$$

q is any real number, $Q = q + q^{-1}$ and m_μ are real coefficients such that $m_0 \neq -m_3$. If the set of M_3 matrices are isometries of the bilinear form (7) for $\hbar = 0$ (classical limit) then they define an associative algebra under multiplication that can be written like the tensor product (9).
In this case $R = [{}_{\alpha\beta}R_{\gamma\delta}]$ is given by

$$\begin{aligned} R = &\sum_{\alpha\neq\beta} E_{\alpha\alpha} \otimes E_{\beta\beta} + q^{-1} \sum_{\substack{\alpha>\beta \\ \alpha\neq\alpha'}} E_{\alpha\beta} \otimes E_{\beta\alpha} + p^{-1} \sum_{\alpha>\beta} E_{\alpha\beta} \otimes E_{\beta\alpha} - \\ &-p \sum_{\substack{\alpha>\beta \\ \alpha'=\beta}} E_{\alpha\beta} \otimes E_{\beta\alpha} - q \sum_{\substack{\alpha>\beta \\ \beta'=\beta}} E_{\alpha\beta} \otimes E_{\beta\alpha} + q \sum_{\substack{\alpha<\beta \\ \alpha'=\beta}} E_{\alpha\beta} \otimes E_{\beta\alpha} + \\ &+p \sum_{\alpha<\beta} E_{\alpha\beta} \otimes E_{\beta\alpha} - p^{-1} \sum_{\substack{\alpha>\beta \\ \alpha=\alpha'}} E_{\alpha\beta} \otimes E_{\beta\alpha} - q^{-1} \sum_{\substack{\alpha<\beta \\ \beta'=\alpha}} E_{\alpha\beta} \otimes E_{\beta\alpha} \end{aligned} \tag{25}$$

for all (α', β') parameters.
Furthermore, the set of matrices σ^μ constitutes a representation for q-deformed Pauli matrices [21]. On the other hand eq.(25) resembles, provided we choose properly the spectral value and free parameters, the $D^{(2)}_{n+1}$ quantum matrix reported by Jimbo [13]. Anyway, we claim that at least it constitutes a linear combination of the $R(x)$ matrices presented by Jimbo for a particular set of spectral values.

5 Summary and Conclusions

We use the Fermi-Dirac and Bose-Einstein statistics in Fock space as the first step to define a q-mutator algebra for any q real number. A quantum group structure as symmetry of this q-deformed statistics, which keeps the Pauli principle invariant, is found out. In the "classical" limit $\hbar \to 0$ we recover the usual Manin quantum superspaces, in the limit $q = 1$ we have the usual Clifford algebra in n dimensional free Fock space.

Three different cases for the quantum symmetry are studied. The first one corresponds to the well established symmetry transformations among fermions (bosons), namely the mixed states fermions-bosons are not considered.

Like the quantum symmetry for this case (in the classical limit) can be considered as a limit of a regular Yang Baxter bundle, we construct the totally integrable one dimensional field model associated with it. The generating function of the infinite number of conserved and commuting physical magnitudes for this system is given. For the quantum structure ($\hbar \neq 0$), some restrictions to the quantum matrices M_1 are presented.

In the second case, we present the symmetry transformation of bosonic states into fermionic ones and vice versa . Like in case 1 the quantum symmetry up to a sign (in the classical limit) can be considered as a limit of the same regular Yang Baxter bundle obtained for case 1. For the quantum structure ($\hbar \neq 0$), some restrictions to the quantum matrices M_2 are presented.

The last case is related with a linear transformation between bosons and fermions, the so called supersymmetry. We can not relate the general case to the Yang-Baxter tensor product but only particular forms for M_3 are allowed to be used, an example (in the q-deformed Pauli matrices basis) is given.

We remark that all the matrices so obtained might be associated to six-vertex or ice-type models in statistical mechanics. Additionally, locality has been lost either in the sense of space-like commutativity or in the sense that observables are point-like functionals of the fields [22].

6 Acknowledgments

The technical assistance of Miss C. Brichard is greatly appreciated.

References

[1] J.B. Mc Guire, J. Math. Phys. 5, **622**, 1964.
C.N. Yang, Phys. Rev. Lett. 19, **1312**, 1967.
R.J. Baxter, Ann. Phys. 70, **193**, 1972 and 76 **1**, 1973.

[2] L.D. Faddeev, E.K. Sklyanin and L.A. Takhtajan, Theor. Math. Phys. 40,**194**, 1979.

[3] M.F. Atiyah, R. Bott and A. Shapiro, Topology, vol. 3 suppl. 1, **3**, 1964.

[4] C. Chevalley, The algebraic theory of spinors (Columbia University press, 1954).

[5] A.Crumeyrolle, Clifford Algebras and their Applications in Mathematical Physics; J.S.R. Chisholm and A.K. Common editors, proceedings of the NATO and SERC Workshop in Canterbury U.K. (Reidel Dordrech, 1986).

[6] L.D. Faddeev, Soviet Scientific Reviews (Harvard Academic, London) C1, **107**, 1980.

[7] L.D. Faddeev, Integrable models in (1+1) dimensional quantum field theory, Les Houches Session XXXIX (Elsevier, Amsterdam, 1982), **563**.

[8] L.D. Faddeev and N.Y. Reshetilkhin, Ann. Phys. 167, **227**, 1986.

[9] V.G. Drinfeld, Sov. Math. Dokl. 27, **68**, 1983.

[10] V.G. Drinfeld, Sov. Math. Dokl. 32, **254**, 1985.

[11] V.G. Drinfeld, Quantum Groups, ICM Proceedings, Berkley, **798**, 1986.

[12] M. Jimbo, Lett. Math. Phys. 10, **63**, 1985.

[13] M. Jimbo, Commun. Math. Phys., 102, **537**, 1986.

[14] M. Jimbo, Lett. Math. Phys. 11, **247**, 1986.

[15] S.L. Woronòwicz, Commun. Math. Phys. III, **613**, 1987.

[16] Y.I. Manin, Ann. Inst. Fourier (Grenoble) 37, 4, **191**, 1987.

[17] Y.I. Manin, Commun. Math. Phys. 123, **163**, 1989.

[18] Integrable Quantum Field Theories, proceedings. Lecture Notes in Physics, Springer Verlag, 1982.

[19] S. Chaturvedi, A.K. Kapoor, R. Sandhya, V. Srinivasan and R. Simon, Phys. Rev. A, 43, **4555**, 1991

[20] Lüscher M., Nucl. Phys. B117, **475**, 1976.

[21] U. Carow-Watamura, M. Schlieker, M. Scholl and S. Watamura, Int. Jour of Mod. Physics A, Vol. 6 N!17, **3081**, 1991.

[22] K. Fredenhagen, Commun. Math. Phys. 79, **141** , 1981.

List of Participants

G.G.A. Bäuerle	Institute of Theoretical Physics University of Amsterdam
T. van Bemmelen	Department of Applied Mathematics University of Twente
P.J.M. Bongaarts	Lorentz Institute University of Leiden
R.H. Dijkgraaf	Department of Mathematics Universtiy of Amsterdam
J.A. van Gelderen	Department of Mathematics Delft University of Technology
J. de Graaf	Department of Mathematics Eindhoven University of Technology
J. Harnad	Centre de recherches mathématiques Université de Montréal, Canada
G.F. Helminck	Department of Applied Mathematics University of Twente
P.C.J. van den Heuvel	Department of Applied Mathematics University of Twente
N.W. van den Hijligenberg	Department of Applied Mathematics University of Twente
M.V. de Hoop	Schlumberger Cambridge Research, Cambridge, England
J. Huebschmann	U.F.R. de Mathématiques Université des Sciences et Technologie de Lille - Flandres - Artois, France
A. Kasman	Mathematics Department Boston University, U.S.A.
H. Knörrer	Mathematik Eidgenössische Technische Hochschule Zürich, Switzerland

A.P.E. ten Kroode	Shell Research Rijswijk
J.W. van de Leur	Department of Mathematics University of Utrecht
H. Lemei	Department of Mathematics Delft University of Technology
R. Martini	Department of Applied Mathematics University of Twente
F.W. Nijhoff	Department of Mathematics and Computer Science and Institute for Nonlinear Studies Clarkson University, Potsdam, U.S.A.
H.G.J. Pijls	Department of Mathematics University of Amsterdam
G.F. Post	Department of Applied Mathematics University of Twente
E. Previato	Mathematics Department Boston University, U.S.A.
G.R.W. Quispel	Department of Mathematics La Trobe University, Melbourne, Australia
H. Rijnks	Department of Mathematics Delft University of Technology
G.H.M. Roelofs	Department of Applied Mathematics University of Twente
J.W. de Roever	Department of Applied Mathematics University of Twente
S. Rodriguez-Romo	Facultad de Estudios Superiores Cuautitlán Universidad Nacional Autónoma de México, Mexico City, Mexico
D.J. Smit	Shell Research Rijswijk
C.A. Tracy	Department of Mathematics and Institute of Theoretical Dynamics University of California, Davis, U.S.A.
M.A. Wisse	Département de mathématiques et de statistique Université de Montréal, Canada

Springer-Verlag and the Environment

We at Springer-Verlag firmly believe that an international science publisher has a special obligation to the environment, and our corporate policies consistently reflect this conviction.

We also expect our business partners – paper mills, printers, packaging manufacturers, etc. – to commit themselves to using environmentally friendly materials and production processes.

The paper in this book is made from low- or no-chlorine pulp and is acid free, in conformance with international standards for paper permanency.

Lecture Notes in Physics

For information about Vols. 1–384
please contact your bookseller or Springer-Verlag

Vol. 385: A. Treves, G. C. Perola, L. Stella (Eds.), Iron Line Diagnostics in X-Ray Sources. Proceedings, Como, Italy 1990. IX, 312 pages. 1991.

Vol. 386: G. Pétré, A. Sanfeld (Eds.), Capillarity Today. Proceedings, Belgium 1990. XI, 384 pages. 1991.

Vol. 387: Y. Uchida, R. C. Canfield, T. Watanabe, E. Hiei (Eds.), Flare Physics in Solar Activity Maximum 22. Proceedings, 1990. X, 360 pages. 1991.

Vol. 388: D. Gough, J. Toomre (Eds.), Challenges to Theories of the Structure of Moderate-Mass Stars. Proceedings, 1990. VII, 414 pages. 1991.

Vol. 389: J. C. Miller, R. F. Haglund (Eds.), Laser Ablation-Mechanisms and Applications. Proceedings. IX, 362 pages, 1991.

Vol. 390: J. Heidmann, M. J. Klein (Eds.), Bioastronomy - The Search for Extraterrestrial Life. Proceedings, 1990. XVII, 413 pages. 1991.

Vol. 391: A. Zdziarski, M. Sikora (Eds.), Ralativistic Hadrons in Cosmic Compact Objects. Proceedings, 1990. XII, 182 pages. 1991.

Vol. 392: J.-D. Fournier, P.-L. Sulem (Eds.), Large-Scale Structures in Nonlinear Physics. Proceedings. VIII, 353 pages. 1991.

Vol. 393: M. Remoissenet, M.Peyrard (Eds.), Nonlinear Coherent Structures in Physics and Biology. Proceedings. XII, 398 pages. 1991.

Vol. 394: M. R. J. Hoch, R. H. Lemmer (Eds.), Low Temperature Physics. Proceedings. X,374 pages. 1991.

Vol. 395: H. E. Trease, M. J. Fritts, W. P. Crowley (Eds.), Advances in the Free-Lagrange Method. Proceedings, 1990. XI, 327 pages. 1991.

Vol. 396: H. Mitter, H. Gausterer (Eds.), Recent Aspects of Quantum Fields. Proceedings. XIII, 332 pages. 1991.

Vol. 398: T. M. M. Verheggen (Ed.), Numerical Methods for the Simulation of Multi-Phase and Complex Flow. Proceedings, 1990. VI, 153 pages. 1992.

Vol. 399: Z. Švestka, B. V. Jackson, M. E. Machedo (Eds.), Eruptive Solar Flares. Proceedings, 1991. XIV, 409 pages. 1992.

Vol. 400: M. Dienes, M. Month, S. Turner (Eds.), Frontiers of Particle Beams: Intensity Limitations. Proceedings, 1990. IX, 610 pages. 1992.

Vol. 401: U. Heber, C. S. Jeffery (Eds.), The Atmospheres of Early-Type Stars. Proceedings, 1991. XIX, 450 pages. 1992.

Vol. 402: L. Boi, D. Flament, J.-M. Salanskis (Eds.), 1830-1930: A Century of Geometry. VIII, 304 pages. 1992.

Vol. 403: E. Balslev (Ed.), Schrödinger Operators. Proceedings, 1991. VIII, 264 pages. 1992.

Vol. 404: R. Schmidt, H. O. Lutz, R. Dreizler (Eds.), Nuclear Physics Concepts in the Study of Atomic Cluster Physics. Proceedings, 1991. XVIII, 363 pages. 1992.

Vol. 405: W. Hollik, R. Rückl, J. Wess (Eds.), Phenomenological Aspects of Supersymmetry. VII, 329 pages. 1992.

Vol. 406: R. Kayser, T. Schramm, L. Nieser (Eds.), Gravitational Lenses. Proceedings, 1991. XXII, 399 pages. 1992.

Vol. 407: P. L. Smith, W. L. Wiese (Eds.), Atomic and Molecular Data for Space Astronomy. VII, 158 pages. 1992.

Vol. 408: V. J. Martínez, M. Portilla, D. Sàez (Eds.), New Insights into the Universe. Proceedings, 1991. XI, 298 pages. 1992.

Vol. 409: H. Gausterer, C. B. Lang (Eds.), Computational Methods in Field Theory. Proceedings, 1992. XII, 274 pages. 1992.

Vol. 410: J. Ehlers, G. Schäfer (Eds.), Relativistic Gravity Research. Proceedings, VIII, 409 pages. 1992.

Vol. 411: W. Dieter Heiss (Ed.), Chaos and Quantum Chaos. Proceedings, XIV, 330 pages. 1992.

Vol. 412: A. W. Clegg, G. E. Nedoluha (Eds.), Astrophysical Masers. Proceedings, 1992. XX, 480 pages. 1993.

Vol. 413: Aa. Sandqvist, T. P. Ray (Eds.); Central Activity in Galaxies. From Observational Data to Astrophysical Diagnostics. XIII, 235 pages. 1993.

Vol. 414: M. Napolitano, F. Sabetta (Eds.), Thirteenth International Conference on Numerical Methods in Fluid Dynamics. Proceedings, 1992. XIV, 541 pages. 1993.

Vol. 415: L. Garrido (Ed.), Complex Fluids. Proceedings, 1992. XIII, 413 pages. 1993.

Vol. 416: B. Baschek, G. Klare, J. Lequeux (Eds.), New Aspects of Magellanic Cloud Research. Proceedings, 1992. XIII, 494 pages. 1993.

Vol. 417: K. Goeke P. Kroll, H.-R. Petry (Eds.), Quark Cluster Dynamics. Proceedings, 1992. XI, 297 pages. 1993.

Vol. 418: J. van Paradijs, H. M. Maitzen (Eds.), Galactic High-Energy Astrophysics. XIII, 293 pages. 1993.

Vol. 419: K. H. Ploog, L. Tapfer (Eds.), Physics and Technology of Semiconductor Quantum Devices. Proceedings, 1992. VIII, 212 pages. 1993.

Vol. 420: F. Ehlotzky (Ed.), Fundamentals of Quantum Optics III. Proceedings, 1993. XII, 346 pages. 1993.

Vol. 421: H.-J. Röser, K. Meisenheimer (Eds.), Jets in Extragalactic Radio Sources. XX, 301 pages. 1993.

Vol. 422: L. Päivärinta, E. Somersalo (Eds.), Inverse Problems in Mathematical Physics. Proceedings, 1992. XVIII, 256 pages. 1993.

Vol. 423: F. J. Chinea, L. M. González-Romero (Eds.), Rotating Objects and Relativistic Physics. Proceedings, 1992. XII, 304 pages. 1993.

Vol. 424: G. F. Helminck (Ed.), Geometric and Quantum Aspects of Integrable Systems. Proceedings, 1992. IX, 224 pages. 1993.

New Series m: Monographs

Vol. m 1: H. Hora, Plasmas at High Temperature and Density. VIII, 442 pages. 1991.

Vol. m 2: P. Busch, P. J. Lahti, P. Mittelstaedt, The Quantum Theory of Measurement. XIII, 165 pages. 1991.

Vol. m 3: A. Heck, J. M. Perdang (Eds.), Applying Fractals in Astronomy. IX, 210 pages. 1991.

Vol. m 4: R. K. Zeytounian, Mécanique des fluides fondamentale. XV, 615 pages, 1991.

Vol. m 5: R. K. Zeytounian, Meteorological Fluid Dynamics. XI, 346 pages. 1991.

Vol. m 6: N. M. J. Woodhouse, Special Relativity. VIII, 86 pages. 1992.

Vol. m 7: G. Morandi, The Role of Topology in Classical and Quantum Physics. XIII, 239 pages. 1992.

Vol. m 8: D. Funaro, Polynomial Approximation of Differential Equations. X, 305 pages. 1992.

Vol. m 9: M. Namiki, Stochastic Quantization. X, 217 pages. 1992.

Vol. m 10: J. Hoppe, Lectures on Integrable Systems. VII, 111 pages. 1992.

Vol. m 11: A. D. Yaghjian, Relativistic Dynamics of a Charged Sphere. XII, 115 pages. 1992.

Vol. m 12: G. Esposito, Quantum Gravity, Quantum Cosmology and Lorentzian Geometries. XVI, 326 pages. 1992.

Vol. m 13: M. Klein, A. Knauf, Classical Planar Scattering by Coulombic Potentials. V, 142 pages. 1992.

Vol. m 14: A. Lerda, Anyons. XI, 138 pages. 1992.

Vol. m 15: N. Peters, B. Rogg (Eds.), Reduced Kinetic Mechanisms for Applications in Combustion Systems. X, 360 pages. 1993.

Vol. m 16: P. Christe, M. Henkel, Introduction to Conformal Invariance and Its Applications to Critical Phenomena. XV, 260 pages. 1993.

Vol. m 17: M. Schoen, Computer Simulation of Condensed Phases in Complex Geometries. X, 136 pages. 1993.